FLEXIBLE KALINA CYCLE SYSTEMS

FLEXIBLE KALINA CYCLE SYSTEMS

Tangellapalli Srinivas, PhD
N. Shankar Ganesh, PhD
R. Shankar

APPLE
ACADEMIC
PRESS

Apple Academic Press Inc.
3333 Mistwell Crescent
Oakville, ON L6L 0A2
Canada

Apple Academic Press Inc.
1265 Goldenrod Circle NE
Palm Bay, Florida 32905
USA

© 2019 by Apple Academic Press, Inc.

First issued in paperback 2021

Exclusive worldwide distribution by CRC Press, a member of Taylor & Francis Group

No claim to original U.S. Government works

ISBN 13: 978-1-77463-408-0 (pbk)
ISBN 13: 978-1-77188-713-7 (hbk)

Library and Archives Canada Cataloguing in Publication

Title: Flexible kalina cycle systems / Tangellapalli Srinivas, PhD, N. Shankar Ganesh, PhD, R. Shankar.

Names: Srinivas, Tangellapalli, author. | Ganesh, N. Shankar, author. | Shankar, R., 1987- author.

Description: Includes bibliographical references and index.

Identifiers: Canadiana (print) 2018906787X | Canadiana (ebook) 20189067888 | ISBN 9781771887137 (hardcover) | ISBN 9780429487774 (PDF)

Subjects: LCSH: Fluids—Thermal properties.

Classification: LCC QC145.4.T5 S75 2019 | DDC 530.4/2—dc23

CIP data on file with US Library of Congress

Apple Academic Press also publishes its books in a variety of electronic formats. Some content that appears in print may not be available in electronic format. For information about Apple Academic Press products, visit our website at **www.appleacademicpress.com** and the CRC Press website at **www.crcpress.com**

ABOUT THE AUTHORS

Tangellapalli Srinivas

Tangellapalli Srinivas, PhD, is an Associate Professor of the Department of Mechanical Engineeirng at Dr. B. R. Ambedkar National Institute of Technology, Jalandhar, Punjab, India. He previously worked as a Professor at the CO2 Research and Green Technologies Centre of the School of Mechanical Engineering at the Vellore Institute of Technology, Vellore, India, and in the Faculty of Engineering and Applied Sciences, University of Ontario Institute of Technology, Canada, as a postdoctoral fellow.

Dr. Srinivas has published 190 research papers in various international and national journals as well as conference proceedings. He has contributed six book chapters in the area of energy efficiency and integrated systems and has published two full books, titled *Flexible Kalina Cycle Systems* and *Desalination and Cooling Integration.* He has also published a patent on a combined power and cooling system and has filed two patents on trigeneration and single axis self-tracker. He has completed two funded projects on vapor power and trigeneration systems, respectively with SERB and CSIR, India. His research area includes power-free solar trackers, hybrid solar-biomass systems, polygeneration, conducting exergy scrutinizing, heat recovery options in cement factory, desalination, solar cooling, advanced combined cycle systems, and ORC and KCS plants.

Dr. Srinivas has 17 years of teaching experience and has guided five PhD scholars. He worked in the machine tools industry in the field of production and quality control for four years. Dr. Srinivas received his MTech (Thermal Engineering) and his PhD in advanced combined power cycles from JNTUH College of Engineering Hyderabad in 2002 and 2008 respectively. He secured a university first rank in the MTech program and was the overall branch topper during the diploma programs. He received an outstanding scientist award (VIFRA) in 2015 from Venus International Foundation, India.

N. Shankar Ganesh, PhD

N. Shankar Ganesh, PhD, is a Professor in the Department of Mechanical Engineering at Kingston Engineering College, Vellore, Tamil Nadu, India. He has published 30 research papers in various international and national journals as well as in conference proceedings. He has contributed two book chapters. He has also published a patent on a thermodynamic cycle-based power generation system. His research areas include processes simulations for advanced combined cycle power plants, modeling of heat recovery steam generators, and solar thermal power technology. He received his MTech degree in Energy Systems Engineering and PhD in Power Generation Cycles from Vellore Institute of Technology in 2002 and 2014, respectively.

R. Shankar

R. Shankar has worked as a senior research fellow for CSIR and SERB funded projects and recently he submitted his doctoral thesis.

CONTENTS

ABBREVIATIONS

AP	approach point
BFP	boiler feed pump
BPT	bubble-point temperature
CFP	condensate feed pump
COP	coefficient of performance
CSF	cooling sharing factor
CW	cooling water
DF	dryness fraction
DNI	direct normal irradiance
DPT	dew-point temperature
ECO	economizer
ENUF	energy utilization factor
EVA	evaporator
GEN	generator
HE	heat exchanger
HRVG	heat recovery vapor generator
HT	high temperature
HTKCS	high-temperature Kalina cycle system
HTR	high-temperature regenerator
KCCC	Kalina cooling cogeneration cycle
KCS	Kalina cycle system
LT	low temperature
LTKCS	low-temperature Kalina cycle system
LTR	low-temperature regenerator
MT	medium temperature
MTKCS	medium-temperature Kalina cycle system
MXR	mixer
MXT	mixture turbine
ORC	organic Rankine cycle
PP	pinch point
PSF	power sharing factor
PTTO	power to total output

SH   superheater
SOFC/GT-KC   hybrid solid oxide fuel cell/gas turbine-Kalina cycle
TTD   terminal temperature difference
VAR   vapor absorption refrigeration

NOMENCLATURE

c	specific heat (kJ/kg K)
e	specific exergy (kJ/kg)
h	specific enthalpy (kJ/kg)
m	mass (kg)
n	number
P	pressure (bar)
Q	heat (kJ)
q	specific heat (kJ/kg)
r	ratio of energies
R	solar radiation (W/m^2)
s	specific entropy (kJ/kg K)
T	temperature (K)
W	work (kJ)
w	specific work output (kJ/kg)
x	mass fraction of ammonia (kg/kg mixture)
y	ammonia concentration in vapor phase
ε	effectiveness
η	efficiency
Φ	rim angle

Subscripts

a	ambient
b	beam
bp	bubble point
bs	base
c	collector
cogen	cogeneration
cw	cooling water
f	fluid
fi	fluid inlet
h	hot
hf	hot fluid
m	mechanical
net	net output
p	pump

ps   parallel segments
rs   reflective sheet
ss   strong solution
t turbine
tot   total
w water
v vapor
0   reference state

PREFACE

A binary fluid system has an efficient heat recovery compared to a single-fluid system due to a better temperature match between hot and cold fluids. This book provides a good understanding of binary fluid systems, highlighting new dimensions to the existing Kalina cycle system (KCS), a thermodynamic process for converting thermal energy into usable mechanical power. The volume illustrates that providing new flexibility leads to new research outcomes and possible new projects in this field. The information provided in the book simplifies the application of KCS with an easy-to-understand and thorough explanation of properties development, processes solution, subsystem work, and total system work. There are currently no other books available in the area of binary fluid system in the field of KCS with added flexibility in the operation and process design. Currently, decentralized power systems are gaining more attention due to shortages in power. Also, the cooling demands are crossing other electrical loads. This book fills this information gap, providing insight into a new dimension for designers, practicing engineers, and academicians in this field.

KCS is a vapor absorption power cycle developed from the improvement of the Rankine cycle. In this book, KCS configurations are studied to suit low-temperature (LT), intermediate-temperature (IT), and high-temperature (HT) heat recoveries. The LT heat source is restricted from 100°C to 200°C to suit the KCS process conditions. The IT heat recovery is limited up to 300°C. Above this temperature, the heat recovery is termed as HT level. Thermodynamic processes in binary mixture plants are formulated and evaluated for thermodynamic model after properties generation. A detailed methodology is presented to solve binary vapor processes and its vapor cycles. The key operational parameters have been identified, and its influence has been analyzed on energy performance to recommend the efficient running conditions for the configurations considered. The results are used to select the operation conditions in boiler, separator, and turbine to maximize the power output and efficiency of the plant.

KCS at low-temperature heat recovery (LTHR) is exhaustively investigated by identifying the individual component's role, namely, superheater, LT regenerator, high-temperature regenerator (HTR), and dephlegmator. Two possible boiler connections are studied to relax the boiler load. The

HTR, economizer, and evaporator are serially connected in one option of KCS at LTHR. But it results in low performance due to low heat recovery in HTR. It is solved by sharing of heat load between HTR and boiler (economizer + partial evaporator) proportional to source heat with parallel connections serial flow of fluids. The results show that parallel operation of boiler and HTR results high efficiency compared to series heaters. A new KCS configuration has been developed to suit the IT heat recovery with power augmentation. Nearly 30% of extra working fluid is noticed in turbine compared to other configurations. Similarly, to suit HT heat recovery, KCS is configured and studied from 400°C to 600°C of heat source.

Cooling needs are increasing rapidly in hot climatic countries with increased global warming. Some commercial units and industries need more cooling than power needed for such things as cold storage, shopping complexes, etc. The existing vapor compression refrigeration system demands electricity for its operation, which is more expensive. The available combined power and cooling cycle (Goswami cycle) operates with an ammonia–water mixture as the working fluid, having low cooling with saturated vapor at the inlet of evaporator. It also demands a high ammonia concentration in the turbine for cooling.

In this book, a new cooling cogeneration cycle has been proposed and solved to generate more cooling with adequate power generation from a single source of heat. Two cogenerations named as split cycle, and once through are compared with each other. The operational process conditions for the proposed cooling cogeneration plants are different compared to the power-only and cooling-only (vapor absorption refrigeration) options. A suitable range for source temperature has been developed, and the optimum range has been recommended to maximize the total output. Finally, the cogeneration with once throughflow is recommended for higher generation than the existing separate plants.

ACKNOWLEDGMENTS

My first journey with thermodynamics started with an academic institute, Gudlavalleru Engineering College, Gudlavalleru, Andhra Pradesh, India, and my warm and sincere thanks go to the institute and its members for providing a strong platform to equip myself, develop the concepts, and the opportunity to teach thermodynamics. I feel proud to be associated with the Vellore Institute of Technology (VIT), Vellore, Tamil Nadu, a place to learn and have a chance to grow. I convey my sincere thanks to VIT, its administration, faculty, staff, and students for supporting me in sharpening the concepts and ideas into reality. I am glad to be affiliated with the National Institute of Technology (NIT), Jalandhar, Punjab, as a faculty member to enlighten, refine, and sharpen the new ideas into reality and introduce to society.

My sincere gratitude goes to my research advisor, Dr. P. K. Nag, Professor, IIT Kharagpur, for showing the road map, giving inspiration, and providing awareness on the Kalina cycle system. I am extremely happy to express my deepest thanks to my PhD guide, Dr. A. V. S. S. K. S. Gupta, Professor, JNT University, Hyderabad, India, for molding me by refining my thoughts and work. It is my good fortune to have had a dynamic and energetic guide like Dr. Bale Viswanadha Reddy, Professor at University of Ontario Institute of Technology (UOIT), Canada, for my postdoctoral work. I wish to express my gratitude to Dr. B. V. Reddy for his continuous support and encouragement in my profession by helping with goal settings.

The Kalina cycle system has wide applications in power generation, cogeneration, trigeneration, polygeneration, and so on. I greatly acknowledge the Council of Scientific and Industrial Research (CSIR), New Delhi, India (22(0627)/13/EMR-II) for financial support in the development of the trigeneration plant at VIT, Vellore, where vapor absorption refrigeration is one of the areas of research. I would like to extend my sincere acknowledgment of the project grant from the Science and Engineering Research Board (SERB), New Delhi, India (SB/S3/MMER/008/2014) for demonstration of the Kalina cycle system at VIT, Vellore. The financial support greatly helped our research team to shape our thoughts into product.

The first step of the Kalina cycle work was guided by a doctoral scholar, Dr. Shankar Ganesh, awarded during 2014. I am happy to convey my thanks to all my research scholars, faculty, staff, and students as part of my research work. My special thanks to all the staff from the CO2 Research and Green Technologies Centre, VIT, for assisting in plant development, erection, and testing processes.

I would like to express my deep sense of gratitude and respect to my parents, who shaped my hard work and strong determination in the field of engineering from the beginning. My heart-felt gratitude to my wife, Kavitha Devi, elder son Rahul, and younger son Jignesh, for without their regular support and encouragement, I could not have done this much.

Finally, thanks to all who involved in this work directly or indirectly who helped shape this book in into a fruitful form.

—Dr. Tangellapalli Srinivas

CHAPTER 1

INTRODUCTION

ABSTRACT

Kalina cycle system (KCS) is a binary fluid power generation cycle suitable to operate at low-temperature source or heat recovery. Similarly, vapor absorption refrigeration (VAR) is also a binary fluid refrigeration cycle operating on low-temperature thermal source instead of mechanical energy supply. VAR seems to be a reverse KCS in operation. Therefore, cooling cogeneration or combined power and cooling cycles are developed and studied from the conceptual operation of KCS and VAR. This chapter gives the introduction of KCS, VAR, and its derived cooling cogenerations. The need of diluting the binary fluid at the entry of absorber for condensation is illustrated for easy understanding. The conventional power plant and KCS are differentiated with the air of property charts drawn to scale.

1.1 INTRODUCTION

Once upon a time, the telephonic talk was a costly process. Nowadays, it is very cheap and everyone is using mobile phones as a common practice. Energy is more expensive and its cost is continuously increasing from a long period. For example, the cost of electricity, petrol, diesel, cooking, natural gas, etc. is rising, so they are expensive. Therefore, two possible solutions are discussed to address these problems. To solve these issues, innovative ideas and updating old technologies are required such as 4G (fourth generation) and/or increasing the nonconventional energy sharing. Development of new solutions and systems operated by waste heat recovery is aimed in this chapter. These needs can be fulfilled by taping natural resources. To develop the energy conversion devices, initially thermodynamic evaluation is a preliminary step. According to Sadi Carnot, father of thermodynamic science, man controls the entire universe even though he is weak compared to animals because of electricity. He suggested doing thermodynamic analysis before starting the experimentation on thermal systems. The best thermo-dynamic cycle is a circle. Minimum four basic thermodynamic processes are required to complete a thermodynamic cycle. Carnot also recommended high-source and low-sink temperature to achieve the high thermal efficiency. Vapor absorption cycles can be categorized into power cycles and cooling cycles. Unlike other cycles, vapor absorption cycles have special features of merging these two cycles into a single plant, which is called combined power and cooling. Apart from the power, the vapor absorption system can be designed for multioutput generation.

1.2   CURRENT ENERGY SCENARIO

We need energy for cooling, lighting, air-conditioning, transportation, farming, industry, entertainment, etc. We cannot imagine our life without the use of energy. Otherwise, it looks like living in a forest or remote area. As reported in International Energy Outlook 2013 (IEO, 2013), the world's energy demand is increasing drastically by 56% in 2040. It is, respectively, 524, 630, and 820 quadrillion British Thermal Units in 2010, 2020, and 2040. To support and improve the status of the world's energy scenario, a committee is formed by the developed countries, which is called Organization for Economic Cooperation and Development (OECD). The countries other than the OECD countries are called non-OECD countries. The per-capita energy consumption and growth in OECD countries are much higher than the non-OECD countries. The energy growth in non-OECD and OECD countries are, respectively, 17% and 90%. The power generation in the world is 20.2 trillion units and it is expected to reach to 40 trillion units by 2040. Now also, many villages in non-OECD countries are suffering from lack of electricity. The strategy and supply of power in OECD countries are well established.

Presently, majority of the electricity generation is coming from the fissile fuels. More than 50% of power is generated from coal at national and international level. On the other side, the renewable energy sharing in power generation also increases due to increased awareness on global warming. In addition to fossil fuels in power plants, its role in transportation is also one of the reasons for global warming. The sharing of renewable and nuclear energy in power production is increasing by 2.5% per year. The use of natural gas is increasing tremendously (1.7% per year) compared to others. In future, the use of coal will increase compared to petroleum and liquid fuels. In 1990–2010, the Indian economy rate increased with an average of 6.4% per year. In future, coal occupies a second rank in the energy sharing. Its use is increasing 1.3% per year. But the use of coal involves lot of issues such as thermal pollution, ash, dust, carbon dioxide, etc. Therefore, there is a need to suppress the coal use in the energy field. The alternative solution is the waste heat recovery and renewable energy development. In some countries such as China, the use of coal is discouraged by developing the gas resources. Many countries also framed the policies and regulations to minimize the use of fossil fuels and encourage using the natural resources.

The power section involves generation, transmission, and distribution. Few state administrates are called electricity boards and others are called corporations. New trends with a focus on energy efficiency and environmental protection in the lines of generation, transmission, and distribution will change the shape of this sector. As mentioned in the earlier section, many villages are not connected with electricity. Farmers will stay at villages. To secure our farmers, focus must be given on rural electrifications. Decentralized power and smart-gird systems are fabulous solutions for rural electrification. Apart from the focus on power production, awareness on energy conservation avoids the wastage of electricity. Citizens also should support the government by eliminating the power robbers. New technologies and encouragement on research and development in energy field should be developed meeting the current challenges. The cogeneration, trigeneration, and polygeneration make a nation independent and practice the sustainability. In summer, the power failures and interruptions are more compared to winter. There is more solar energy intensity in summer. So, people may not feel the seasonal scarcity by involving the sun in energy conversion.

Energy is a part of human life, which involves in the daily activities such as lighting, transportation, electricity, etc. Energy meets the requirements of all the functions and is the main part of industries, domestic, commercial forms, agricultural, etc. Most of the energy needs are fulfilled by fossil fuels; its reservoirs are diminishing. The fossil fuels are exhaustive in nature and so alternative energy sources are natural resources (Sustainable Urban Energy, 2012). The irresponsible tapping of these fossil fuels causes harmful gases and damages the wealthy surroundings. The urban area in the world is less than 3% and it causes 67% of emission by consuming 75% of resources.

The renewable energy sharing is increasing every year with the fear of environmental damage. During 2011–2013, the nonconventional energy sources' sharing is 3.8% in India. Currently, its sharing is nearly 15% and in future will shoot up as per the expectations. Burning fossil fuels causes the generation of more carbon dioxide, but the burning of biomass neutralizes the carbon dioxide production by generation and by absorbing through trees. The increase in thermal energy conversion efficiency suppresses the fuel use and so indirectly is held in the emission control. New regulations are framing to encourage the renewable energy conversion, reducing the carbon dioxide emissions. A single-energy output plant has lower conversion efficiency compared to the multioutput plant. The efficiency of a thermal power plant is considerably low compared to integration with process heat. Therefore, polygeneration (production of double or more than double) systems are gaining more attention recently to meet the multiple needs from the single resource or hybridized sources.

Figure 1.1 differentiates the energy conversion for single output and multibenefits. In the multigeneration plant such as cogeneration, trigeneration, polygeneration, etc., the efficiency is increasing drastically by minimizing the energy losses. Nearly 9% of energy demand is met by cogeneration in 2011. Out of this cogeneration, 26% of plants are operating by industrial waste heat. Fuel saving, compact, high-energy conversion, etc. are the driving elements to promote the cogeneration plants.

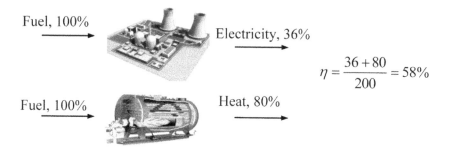

Fuel, 100%   Electricity, 36%

Fuel, 100%   Heat, 80%

$$\eta = \frac{36 + 80}{200} = 58\%$$

(a) Thermal efficiency of non-integrated plant

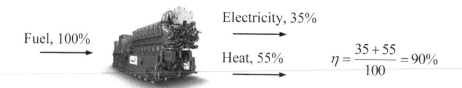

Fuel, 100%   Electricity, 35%

                    Heat, 55% $$\eta = \frac{35 + 55}{100} = 90\%$$

FIGURE 1.1 Comparison of noncogeneration and cogeneration systems: (a) thermal efficiency of nonintegrated plant and (b) thermal efficiency of integrated plant.

The majority of power in Denmark and Latvia is supplied by cogeneration plants. Similarly, Japan is also encouraging significantly to increase the cogeneration plants to meet the power and process heat demands. India has a great potential for cogeneration as it is an agricultural-based country. The population is increasing drastically to 1.3 billion and now it occupies 17% of the global population. It ranks seventh in the world and is now trying to develop enough energy through the cogeneration mode.

The coal reserves are depleting very fast. It may be available only for a few decades. The coal is used not only in thermal power plants but also in steel plants, cement factories, and other industries. On the other hand, the majority of the liquid fuels are used in transport sector and little is used in power sector. So, there is a serious need to address the scarcity of energy and the emission. The gaseous fuels are used in cooling, industries, power, and transport sector due to clear in nature. Still, the natural gas sources are also exhaustive and need alternative solutions such as hydrogen generation, producer gas, etc.

The sharing of renewable energy sources is increasing with government encouragement and support with the subsidies. High cost, nonavailability of biomass, challenges in hydrogen storage and transportation, dilute nature of solar energy, and many challenges are involved in the collection, storage, and utilization of natural resources. The conventional energy sources are rich, even though they are exhaustive in nature. The mix of conventional and nonconventional energies solves the issues in the renewable energies to some extent. The mixing of different renewable energies also solves the above-listed challenges. For example, the solar energy is nonuniform and dilutes in supply. The nonavailability of biomass suppresses its commercialization. The hybrid use of solar and biomass subsidizes these setbacks.

To solve the climatic problems caused by emissions from engines, power plants, industries, etc., a lot of innovations and research is required. Carbon dioxide may be used as a raw material to create a new fuel. For every problem, there might be many solutions, but they are hidden. The researchers have to contribute to find these unknown or hidden solutions. Earlier, our forefathers followed many environmental friendly practices such as green buildings, natural lightings, hydro-energy in agriculture, etc. The new generation also needs to relook into these passive practices to avoid the use of electricity for every need. The major thermal power plants in India are operating at lower efficiency as they have thermal losses. By properly utilizing this thermal waste such as in cogeneration, the overall energy conversion rate will increase. It also saves the environment by increasing the fuel conversion efficiency and minimizing the thermal pollution. Due to the lack of technology development for the Indian climatic conditions, installation of cogeneration installation capacity is limited in India. From the report of global cold-storage capacity 2012, the United States stands first in the cold-storage industries, India occupies the second position, and China the third position. In the year 2008–2012, the capacity of the cold storage increased globally, especially India showed the highest increase in cold-storage capacity of 25% from the last 14 years (1998–2012).

According to the International Institute of Refrigeration (IIR) in Paris, 15% of electricity produced in the world is used only for refrigeration and air-conditioning. The increase of the cold-storage area again increases the energy demand because most of the refrigeration and air-conditioning works by using electricity. According to the report of IIR, the global temperature will increase to 3.6°C in 2017 due to more emission of CO_2 if we fail to limit the present temperature which is 2°C. To limit global warming, the halogenated refrigerants are banned all over the world. To avoid the electricity use in the compressor, the vapor absorption refrigeration (VAR) systems are developed which is run by the use of heat. There must be a concept to meet the energy demand and electricity use for the refrigeration to run a successful day-to-day human life.

1.3   THERMAL POWER PLANTS

In thermal power plant, heat is used as a source of energy and this heat (low-grade energy) is converted into electricity (high-grade energy). Thermal power can be operated using renewable energy sources or conventional energy sources. Since many challenges are involved in the tapping of renewable energy sources, conventional energy sources can be complemented with nonconventional sources such as cofiring of coal with biomass, solar water heating before coal firing, etc. In conventional thermal power plants, coal, natural gas, diesel, etc. are used to operate the heat engine such as steam turbine, gas turbine, diesel engine, and similar. Most of the thermal power plants work on the basic principle of fuel energy (chemical energy) conversion into thermal, thermal into mechanical, and finally mechanical into electrical energy conversions. In a solar thermal power plant, sunbeams are focused on a line or point to convert the water into steam and further the steam expands in turbine to generate electricity. In a solar Stirling engine, which is an external combustion engine, sunbeams are focused, the working fluid expands and rejects heat at the condenser where it contracts. In some power plants, the thermal energy is directly converted into electricity without any mechanical root. They would not obey the heat engine principles. These direct energy conversion plants are solar photovoltaic (PV) plants, thermo-ionic generator, magneto-hydrodynamic power plants, etc. But in fuel cells, the chemical energy of fuel is directly converted into electricity without any thermal or mechanical shades.

Thermal power plant obeys the law of thermodynamics and the Carnot principles and works with the Carnot limitation. In the thermal power plant, the source and sink are the mandatory items. Without sink, it violates the second law of thermodynamics and becomes the perpetual motion machine which is impossible. The thermal efficiency of a heat engine can be increased by increasing the source temperature and/or decreasing the sink temperature. It has been found from the thermodynamic evaluation that the decrease in sink temperature has more influence on efficiency than the rise in source temperature. A thermodynamic engineer (thermodynamist) always tries to improve the current status of the performance with modifications and finally gets increased specific output and thermal efficiency. For this, a benchmark engine is required for the reference which is called Carnot engine. The practice engines can try to reach nearer but below the efficiency of Carnot engine to get the perfection. Heat engines work on thermodynamic cycles. For example, steam power plant works on Rankine cycle, diesel engine on diesel cycle, petrol engine on Otto cycle, gas power plant on Brayton cycle, etc. So, the study and optimization of a thermodynamic cycle is a crucial work in the development of a thermal power plant. Thermodynamic power cycles can be grouped into vapor power cycles and gas power cycles. The turbine inlet temperature of working fluid in steam power plant is from 200°C to 600°C. The lower range of source temperature is used in the thermal power plant operating with waste-heat recovery mode. The higher range of temperature at the turbine inlet is used in the thermal power plants with the fuel-firing furnace such as coal burning.

Figure 1.2 shows a simple layout of Rankine cycle for power production. Solar thermal or waste heat can be used to operate this cycle in place of coal firing. It consists of four components and processes, and they are pump, boiler, turbine, and condenser. The four processes are pumping, vapor generation, expansion, and condensation. In case of solar thermal-based steam power plant, the steam can be generated either directly or indirectly heated boiler. In the direct solar boiler, water is fed to the solar receiver and converted into steam by absorbing the focused sunbeams. In the indirect method, thermic fluid is used in the solar receiver, and it carries the heat to the steam boiler. Thermic fluid is known as heat transfer fluid. The majority of solar power plants are erected with the solar PV panels. Due to heavy maintenance in the solar thermal power plants, its commercialization is now questionable. On other hand, solar thermal power plants have more flexibility in operation and integration with multi-outcome design. In future, its significance and innovations increase the chance to meet the current and future demands.

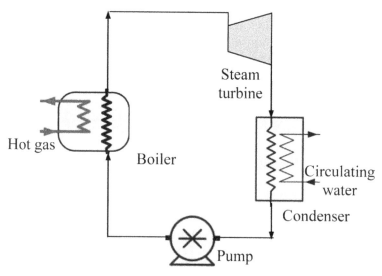

FIGURE 1.2   Thermal power plant operating on Rankine cycle.

Steam Rankine cycle can also be operated by incineration of biomass. The biomass power plants are suitable for low-capacity plants due to lower calorific value and nonavailability. The calorific value of coal is much more than the biomass and its quantity in mines is enormous and thus suitable for high-capacity plants. Steam power plant can be maintained with the hybridization of biomass and solar thermal sources. It avoids the use of solar thermal storage which is more expensive.

At low temperature of working fluids, organic Rankine cycle (ORC) and Kalina cycle system (KCS) are the suitable power plants. The efficiency of a steam power plant can be improved by increasing the boiler pressure, turbine inlet temperature through the degree of superheat, reheating, regeneration, and reheat–regeneration. The increase in thermal efficiency with the reheating may give the positive result or negative results depending on the reheat pressure. It has been found that the steam power plant efficiency increases with the pressure ratio of reheater and boiler in the range of 0.2–0.25 for ideal Rankine cycle. In the regenerator, feedwater heaters are used. The two categories of feedwater heaters are open feedwater heaters and closed feedwater heaters. In a typical thermal power plant, both open and closed feedwater heaters are used. With the exit of steam turbine, the working fluid temperature is not much high to add the heat recovery before condensation. But in ORC and KCS, the working fluid temperature is enough high to tap the heat before the condenser and so it can be used to decrease the capacity of condenser.

Compared to gas power cycles, vapor power cycles works at relatively low-temperature sources. Sensible heat transfer occurs in the gas power cycles, and in vapor power cycles, latent heat transfer takes place. To cover the high-temperature and low-temperature range, gas power cycles and vapor power cycles can be coupled. Gas turbine–steam turbine (GT–ST)-combined cycle power plant is such a system that combines the merits of gas power and vapor power cycles. The gas power cycles have the advantage of high-temperature source but heat rejection also at high temperature which is far above the atmospheric temperature. The vapor power cycles have the drawback of low-temperature heat source but heat rejection at a lower temperature. In GT–ST-combined cycle power plant, these two negatives are neutralized. The power cycles are subclassified into single fluid system and multifluid system. Binary fluid system is one of the examples of multifluid system. In the binary fluid system, two working fluids are used to perform the task. The selected fluids should not harm the environment, is noncorrosive, nontoxic, and is of low cost. The first binary fluid power plant invented is mercury–steam power plant. It has good merits of high thermal properties of mercury. But the cost of mercury is more and the total investment is expensive compared to the existing. It is not popular due to these problems even though there are considerable potentials. Ammonia–water solves these challenges, and so now, a more popularized binary fluid is used in power plants and refrigeration plants. The binary fluid system has an advantage of temperature glide which is not possible with the single fluid system. The temperature glide during the heat transfer matches with the heat source fluid temperature drop and also matches with the cooling-medium temperature rise in the sink. Binary fluid power plants can also be compared with the ORC plants as both are suitable for low-temperature heat sources. In ORC, the working fluid is organic, high molecular weight with the low boiling point suitable to the source. As per the source temperature, the organic fluid can be selected.

1.3.1 KALINA POWER CYCLE

To overcome the more need of power for both domestic and industrialist and to increase the efficiency of power plant, Kalina developed a low-grade heat power plant to recover the waste heat. The Kalina cycle is working on low-heat temperature using vapor absorption refrigerant pair which is having a condenser above the atmospheric pressure. The simple Kalina cycle is shown in Figure 1.3 running by aqua–ammonia as working fluid. The strong solution of aqua–ammonia (1) is pumped to high pressure (HP) of generator pressure (4) to the boiler/generator via solution heat exchanger. Due to its high boiling point temperature difference, it is separated into ammonia vapor (5) and weak solution (8). The ammonia vapor with little percentage of water is again heated in the superheater and the superheated ammonia vapor (6) is used to run the turbine from HP to sink pressure.

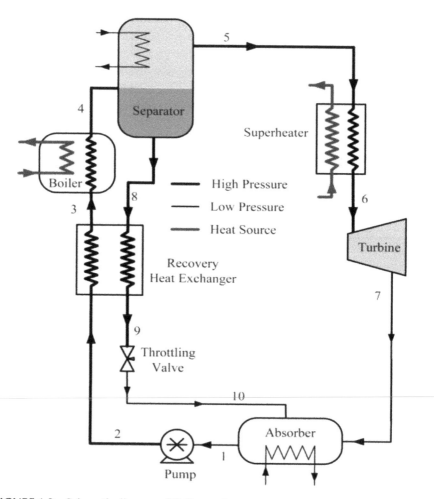

FIGURE 1.3   Schematic diagram of Kalina cycle system.

The weak solution moves to the absorber by sharing its heat to the strong solution in solution heat exchanger (recovery heat exchanger). The weak solution after throttling (10) and vapor from the turbine mixed in the absorber to form strong solution concentration. The heat of mixing is rejected to circulating water in absorber, and therefore, the exhausted vapor absorbs into the liquid solution. The dilution of turbine vapor with weak solution allows the vapor to condense at a reasonable sink temperature; otherwise, too low temperature is required to condense the vapor directly. Thus, the cycle repeats for continuous operation. Due to its low boiling temperature and condensation above the atmospheric pressure, Kalina cycle is able to work at different heat source like geothermal, solar thermal, waste heat recovery, etc.

KCS is competitive with the ORC power plant as both are suitable at low-temperature heat recoveries. The temperature at the exit of vapor turbine is high enough in both KCS and ORC plants and it permits the internal heat recovery which is not possible at the exit of steam turbine in steam Rankine cycle. In KCS boiler, the phase change starts at bubble-point temperature and ends at the dew-point temperature. Similarly, condensation starts at dew-point temperature and ends with bubble-point temperature. The close match between hot fluid and the cold fluid decreases the entropy generation in the heat transfer process but it increases the heat exchanger's size.

Figure 1.4a and b shows the temperature entropy plot for steam Rankine and Kalina cycle, respectively. In Figure 1.4a, the heat addition and rejection happens at constant temperature during phase change for Rankine cycle. In Kalina cycle, the heat addition in boiler and heat rejection in condenser happens at variable temperature as shown in Figure 1.4b. State "6" is the bubble-point temperature and state "7" is the dew-point temperature. In condenser, state "3" is the start of phase change and state "4" is the end of the phase change. It is possible to maintain low pinch point in the Kalina system, which is difficult in steam Rankine cycle. Multiple pressure systems in Rankine cycle could recover more energy, resulting in complex design and cost. The condenser pressure can be much higher which eliminates vacuum maintenance at condenser. Expansion in turbine, results in nearly saturated vapor for a two-component fluid cycle compared to wet steam in Rankine cycle, requiring protection of blades in the last few stages. Conventional equipment such as steam turbines and heat recovery steam generators can be used in Kalina cycle plant with proper modifications.

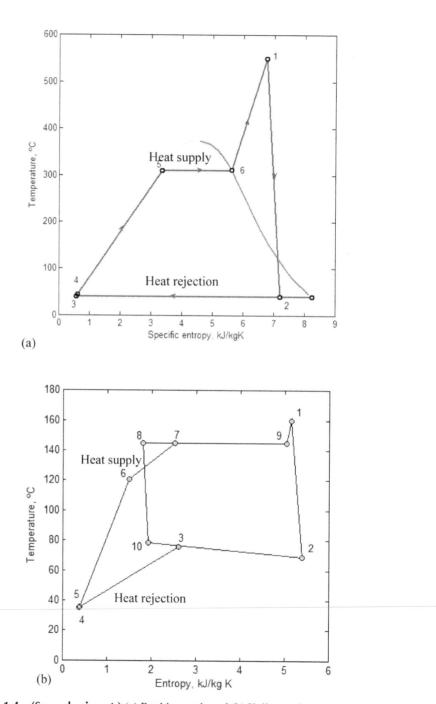

FIGURE 1.4 (See color insert.) (a) Rankine cycle and (b) Kalina cycle.

Figure 1.5 shows the temperature concentration diagram of ammonia–water mixture. For example, vapor at 4 bar and 90% concentration to be condensed directly without dilution, it demands nearly 0°C of sink temperature. On the other hand, the same vapor is diluted by properly mixing with a weak solution supplied from the bottom of vapor generator or a vapor separation drum, the mixture demands 50°C at a typical diluted concentration of 0.4. Condensing above the room temperature is easy and inexpensive compared to the condensation with the chilled water. Therefore, in addition to the Rankine cycle components, Kalina cycle involves separator and dilution of vapor process. In ORC and steam Rankine cycle, there is no separator. One working fluid is continuously moving all the components from starting to ending. In KCS, the expansion in turbine with a high ammonia concentration needs very low condensing temperature. But at this condition, condenser requires low cooling water temperature. It is not possible to condense ammonia at atmospheric conditions. Accomplishing is very difficult, hence requires alternate solution.

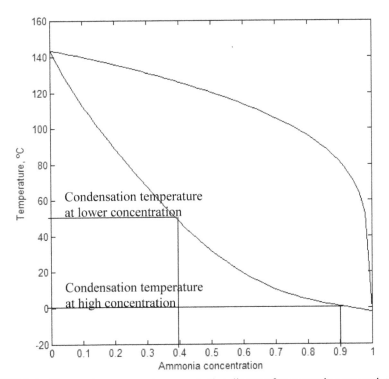

FIGURE 1.5 Temperature–ammonia concentration diagram for ammonia–water mixture at 4 bar pressure.

The difficulty in condensation is overcome by the incorporation of separator as shown in Figure 1.6. The ammonia–water mixture is separated into enriched (high concentration) ammonia mixture and lean (low concentration) ammonia mixture. The lean ammonia mixture is mixed with the high concentration of ammonia mixture from turbine exhaust, thus reduces the concentrations at condenser inlet. The condensation temperature is increased with the decrease in ammonia mixture concentration. Now, the condensation happens with ordinary cooling water in the condenser. The complexity arises because of the separator.

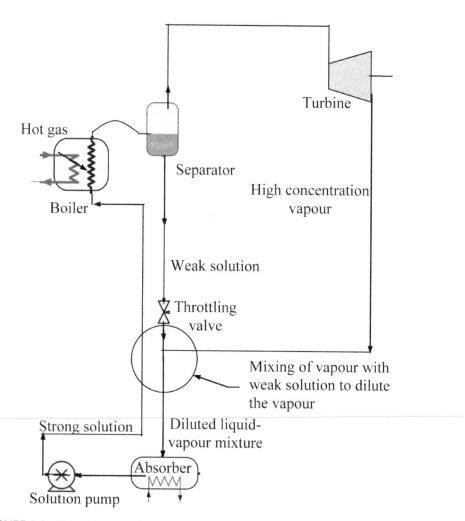

FIGURE 1.6 Use of separator in KCS.

The advantage of KCS is the use of recuperation or regenerator with heat recovery from the turbine exhaust. The temperature at exit of turbine is enough high at 60–70°C, which can be used in regenerator and so the boiler supply can be saved.

In Rankine cycle, the steam is expanded at low temperature. There is no possibility of adding the regenerator after turbine in steam Rankine cycle. The addition of regenerator or heat recovery is possible in Kalina cycle plant to increase the efficiency (Fig. 1.7).

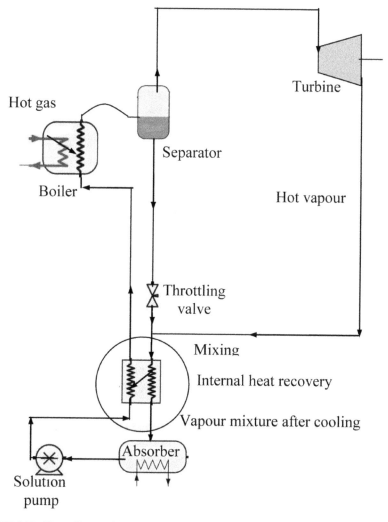

FIGURE 1.7 Use of internal heat recovery in KCS.

1.3.2 EXISTING KALINA PLANTS

KCS plants are suitable to low- and medium-capacity power plants. The first KCS plant is erected at Canoga Park, USA, in 1991 with gas turbine exhaust gas as well as waste heat recovery. It is operated between 1992 and 1997. The second plant is at Fukuoka, Japan, erected in 1998. It is operated by incineration of municipal solid waste. The capacity of Fukuoka plant is 4.5 MW. The next power plant in Japan is Kashima plant in 1999 with a capacity of 3.5 MW operating with waste heat recovery of a steel industry.

After the Japan, Iceland started a KCS plant in 2000 with a small capacity of 2 MW. The heat is supplied from the geothermal power plant. Japan also started one more KCS in 2006 with 4 MW capacity at Fuji using the waste heat recovery of an oil refinery. Germany had two plants with capacities of 3.4 and 0.6 MW at Unterhaching and Bruchsal, respectively, during 2009. It also operated using the geothermal energy. China started a first solar thermal KCS power plant in 2010 with 50 MW capacity. It is one of the biggest plants. One more KCS with the same capacity has been erected in Taiwan using geothermal in 2010. Recently, Pakistan also started a KCS plant with 8.6 MW capacity. India is planning to start KCS plants using the cement factories heat recoveries.

The main operating problem in KCS is the sealing of vapor turbine. In most of the plants, labyrinth seals are used to arrest the leakage. After some period, due to corrosion, the seal may not work and allows the leakage. Similarly, pumps also create some leakage issues. Packing rings are used in the pumps to arrest the fluid leakage.

1.4   VAR CYCLE

Vapor absorption cycles are suitable for power plants as well as the refrigeration cycles. Reversing the power plant cycle's direction, refrigeration cycle can be derived. In a novel way, the features of vapor power and vapor refrigeration cycles can be merged to evolve a new cooling cogeneration cycle. The electricity used for the vapor compression refrigeration system (VCR) is very high and this process cannot be used directly for the heat recovery system and renewable energy sources. Hence, the VAR system is introduced and the compressor in VCR is replaced by the boiler/generator (3–4) and absorber as shown in Figure 1.8.

The high-pressure pump is used for pumping the mixture of refrigerant and absorbent from the absorber (2) to the generator through solution heat exchanger. The pump consumes less energy compared to VCR due to low specific volume of the liquid. The refrigerant is distilled by the generator due to its high boiling point temperature difference between the absorbent and refrigerant.

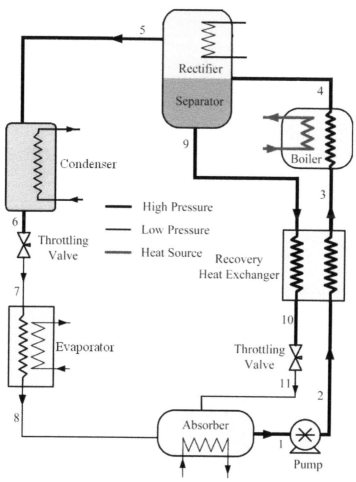

FIGURE 1.8   Schematic diagram of vapor absorption refrigeration cycle.

To achieve the pure ammonia vapor at the inlet of condenser, dephleg-mator is located at the exit of the separator. Dephlegmator/Rectifier removes the water content mixed with ammonia vapor by rejecting heat. Enriching the working fluid at dephlegmator produces more cooling effect at the evaporator.

The strong refrigerant vapor is moved to the condenser for the condensation process. There is a drop in temperature when the high-pressure condensed liquid refrigerant (6) is throttled to low pressure (LP)/sink pressure. It is used as cooling in the evaporator (7–8) by absorbing the heat from the atmosphere. The weak in refrigerant (10) solution from the generator (after throttling) absorbs vapor from the evaporator (8) and the process makes the cycle.

NH_3/H_2O and $H_2O/LiBr$ is the most widely used refrigerant/absorbent pair in VAR system. NH_3/H_2O VAR system has low coefficient of performance (COP) compared to $H_2O/LiBr$ pair due to more heat load consumption in the generator for the separation process as well as the ammonia having less latent heat. Even though $H_2O/LiBr$ VAR system having high COP, it is unable to be used for the cooling temperature less than 4°C because of ice formation in the evaporator. NH_3/H_2O and $H_2O/LiBr$ systems are best suitable for low-temperature heat recovery.

1.5 COOLING COGENERATION CYCLE

Kalina cycle is a reverse VAR cycle to generate the power. Goswami cycle shown in Figure 1.9 has the qualities of both vapor power cycle and VAR cycle due to clubbing of these two into a single cycle to produce simultaneous generation of power and cooling with ammonia–water mixture as working fluid.

In separator, the strong solution is separated into ammonia vapor and weak solution concentration liquid. The excess water vapor content on the ammonia vapor is removed in the dephlegmator; hence, the pure ammonia vapor (5) moves to the superheater. The superheated vapor (6) runs the turbine by expanding from HP to absorber pressure and results more power. Due to the expansion of rich ammonia vapor from HP (5) to sink pressure (1), it produces cooling effect at the exit of the turbine. But it demands low sink temperature to gain cooling after expansion. With higher sink temperature, it is not possible to generate cooling in refrigerator. In this chapter, a new cooling cogeneration model has been developed by addressing this problem.

A new configuration has been developed by addressing the problems associated in the existing combined power and cooling (cooling cogeneration) cycle. At the exit of turbine, the fluid state is vapor and we cannot expect more cooling in the subsequent refrigeration evaporator. In this case, sensible heat transfer occurs in the heat exchanger which is low compared to the latent heat transfer. The degree of cooling is low even through it favor the power by expanding from HP to LP. If the vapor at the exit of turbine condenses into liquid, a subsequent throttling allows more refrigeration in the heat exchanger.

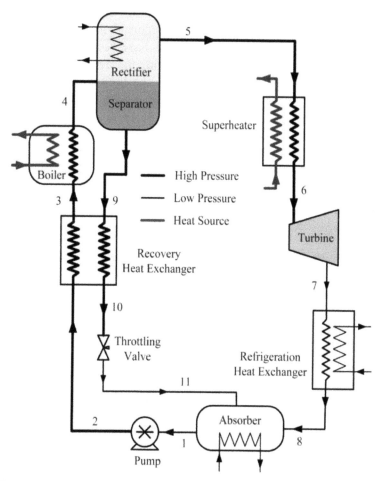

FIGURE 1.9   Schematic diagram of combined power and cooling cycle (Goswami cycle).

1.6   OBJECTIVES

The readers will gain the knowledge about the KCS configurations and be able to generate the thermodynamic properties, modeling, simulation, and analysis. The information provides the flat form to identify the key operational parameters for parametric variations and parametric analysis. The designer can develop the specifications of the plant at the customized conditions. The main objective of this book is to study and investigate KCSs at various heat recovery levels with the possibility of improved performance. The heaters arrangements in source are investigated. Kalina cycle has been focused and evaluated to generate power at hot sink temperature. Parametric analysis has been made on the Kalina cycles considered. The KCS models suitable for three different temperature ranges have been assessed.

The study on KCS is extended by providing the flexibility in operation. The existing cooling cogeneration cycle has the limitation of operation at low sink temperature (10–12°C) and limited amount of cooling production. A new cooling cogeneration cycle, able to operate with high sink temperature (up to 40°C), has been introduced and tested its feasibility with thermodynamic analysis. The work is focused on development of detailed thermodynamic models for ammonia–water cooling cogeneration cycle. The focused results are power generation, refrigeration, power-cycle efficiency, cycle energy utilization factor (EUF), and plant EUF.

1.7   SCOPE OF THE BOOK

Projections for future electricity demand are very uncertain because of the anticipated persistence of India's dynamic development. Gross domestic product development, industry structure, population growth, and income levels are important drivers for energy and electricity demand. Growing energy demand and developing nonpolluting energy conversion systems are the probabilities to meet the challenges of energy crisis. A power cycle is an energy conversion system in which one form of energy is converted into another usable form. Power generation cycle is a set of repetitious processes that has the same starting and ending point in terms of state properties. It is difficult to erect the power plant without the thermodynamic design and the analysis.

Kalina cycle configurations for low, intermediate, and high temperatures heat for power generation have been addressed in this book. In assessing a power cycle, the properties of the working fluid must be known. The properties of binary ammonia–water mixture have been developed for various pressures, temperatures, and concentrations. In determining the properties of the working fluid, computer codes in MATLAB have been prepared. These property codes works as a function program in the main program of the

Kalina systems. At given values of temperature, pressure, and concentration, the property data for enthalpy, entropy, volume, and exergy have been developed. Using mass, energy, and exergy balances, the performance of the KCS has been evaluated at Indian atmospheric conditions.

The binary working fluid method is complicated compared to pure substance as it involves the function of pressure, temperature, and concentration. The current chapter simplifies this complex nature of method and analyzes the basic processes for power generation solutions.

The methodology for KCS at low-temperature heat recovery has been simplified and developed at efficient operational conditions. The regular design of serially connected high-temperature regenerator and boiler in a KCS has been replaced with parallel heaters and compared with each other. The influence of strong solution concentration, vapor concentration, source temperature, and effectiveness of dephlegmator is studied on performance of the plant. The key parameters are optimized to get a high performance.

Thermodynamic development and assessment of a KCS at medium-temperature heat recovery has been examined to augment the power from a medium-temperature heat recovery. There is no throttling device in this plant as the separator is located at low-pressure side. The key parameters have been addressed for performance results.

Thermodynamic development and assessment for a KCS at high temperature heat recovery suitable up to 600°C with an improved heat recovery has been focused. The cycle has three pressure levels, that is, HP, intermediate pressure (IP), and LP. The superheated vapor expands from HP to LP, and the separator is located at IP. The work develops a simple method for thermodynamic evaluation. Strong solution concentration, turbine concentration, hot-gas inlet temperature, and separator temperature are identified as key parameters for the plant evaluation.

The literature survey shows that not much effort has been carried out on development of single cycle for both power and cooling. The cooling demand is increasing more compared to power demand because of increasing in global warming, load in cold storage, air-conditioning needs, etc. High vapor concentration and low sink temperatures/pressure are required to operate a Goswami cycle plant. These are the identified problems in the existing combined power and cooling cycle. These problems are addressed and solved in the proposed cooling cogeneration cycle with feasible solutions. Thorough investigation on flexible KCS has been done to identify and justify the merit of cooling cogeneration option.

1.8   LAYOUT OF THE BOOK

Chapter 1: Current energy scenario, the basic working principles of power generation, VAR, and combined power and cooling generation plants are briefed. The concept of existing thermal power plant is outlined. The difference between the existed single fluid system and current double fluid system is differentiated. The significance of fluid dilution in KCS is critically analyzed. The current cooling cogeneration cycle is outlined with the material flow diagram.

Chapter 2: The literature work has been reviewed. The research developments in the generation of thermodynamic properties of ammonia–water mixture, Kalina power generation, and Kalina cooling cogeneration are summarized and arranged in the order of periodical improvements. VAR is discussed to understand the combined power and cooling plant. A literature has been reviewed in the area of aqua–ammonia properties, KCS, exergy analysis, innovation in power cycles, and combined power and cooling systems. The developments in KCS at low temperature heat are reported. Diverse innovations in Kalina cycle have been reviewed. A review on exergy analysis is carried out to highlight the importance of second law of thermodynamic evolution to Kalina cycle. Based on the reported literature on KCS plants, the research gaps are identified in KCS. The research component in the work is highlighted.

Chapter 3: Vapor absorption cycles involve many complex thermodynamic processes such as heating, cooling, separation, mixing, dephlegmation (distillation), pumping, and heat recovery. All these thermodynamic problems are derived to understand the methodology and cycle. Before developing the formulation suitable to KCS and its derivatives, the fundamental rules and its basic concepts are outlined in this chapter. It will clear the reader to understand the methodology clearly and understand smoothly. The thermodynamic properties, processes, and thermodynamic cycles are formulated in the subsequent chapters and the main underground work is the applied thermodynamics.

Chapter 4: In this chapter, the properties of ammonia–water mixture are developed. The correlation equations are listed with the coefficients. The property diagrams are plotted and described. The thermodynamic formulae to find the bubble-point temperature, dew-point temperature, specific volume, specific enthalpy, and specific entropy at liquid and vapor phase for ammonia water mixture are presented. With concentration change and pressure change, charts are developed for specific volume, specific enthalpy, specific entropy, and specific exergy.

Chapter 5: In this chapter, the methodology involved to solve processes is outlined. The formulations are developed with the derivations. This section is useful to understand the further description and methodology of thermodynamic cycle. The formulae involved in separator, mixing, turbine, pump, and heat exchanger have been detailed. Detailed solutions for binary mixture-based thermodynamic processes are developed and discussed to understand the methodology of proposed power and cooling cogeneration cycle solutions. A methodology to design a new cooling cogeneration cycle is developed for aqua–ammonia-based systems. A step-by-step mathematical modeling for solving the proposed design is discussed along with h–x and T–s diagram.

Chapter 6: In this chapter, the KCS at low-temperature heat recovery is developed. The chapter outlined the description of the cycle, methodology, analytical results, and conclusions. The role of each and every component in KCS is highlighted with the parametric analysis. The mass balance and energy balance equations are derived and framed. The performances of the cycle and plant are studied. Heat source arrangements are compared.

Chapter 7: In this chapter, KCS configuration suitable to intermediate temperature heat recovery is studied. The methodology is outlined with the simplification of mass balance equations and energy balance formulations. The performance levels of KCS with the identified operation conditions are analyzed and discussed. Specifications are developed at the optimized operational conditions.

Chapter 8: This chapter is focused on KCS configuration operating with high-temperature heat recovery. The cycle processes and components are discussed with a schematic flow diagram. The methodology is elaborated to understand the formulations developed for the performance evaluation. The performance variations are plotted with the identified key operation conditions. Conclusions are drawn to find the higher power generation and efficiency from the heat recovery. Specifications are presented at the identified operation conditions.

Chapter 9: This chapter focused on outcomes of two cooling cogeneration cycle with justifications and validations. The flexibility in KCS at low temperature heat recovery is identified and generated four possible configurations, namely, KCS, VAR, Kalina cooling cogeneration cycle (KCCC) with once through configuration and KCCC with split option. Performance variations are developed to analyze the cycle EUF, plant EUF, and output of combined power and cooling.

Chapter 10: It is the appendix and aimed on property tables and diagrams. Tables are developed with variations in pressure, temperature, and concentration. The saturated fluid properties and wet fluid (liquid vapor mixture) are focused on tables and charts. The steps involved to found the fluid region are narrated.

KEYWORDS

- **thermodynamic evaluation**
- **vapor absorption cycles**
- **combined power**
- **cooling**

CHAPTER 2

HISTORY OF DEVELOPMENTS

ABSTRACT

The first step in thermodynamic simulation is the development of fluid properties. The development of coefficients and use of equations for binary fluid (ammonia–water mixture) are reviewed and arranged. The developed properties are presented in tables and charts in appendix section. After the properties are developed, the review is focused on Kalina cycle system (KCS). The performance and capacity of KCS are collected at different source temperatures. The KCS finding with exergy approach and its innovations are outlined in the chronological order. Similarly, the developments in vapor absorption refrigeration and cooling cogeneration are presented.

2.1 INTRODUCTION

The primary innovation of vapor absorption principle in two fluids is seen through separation of liquids by heating and blending through cooling. This essential standard is driving the specialists to imagine numerous developments such as refrigeration, power plants and cogeneration, trigeneration, etc. In this segment, the research commitments in vapor absorption cycles are seen. In understanding a power and/or cooling cycle, the decision of working liquid is basic. The thermodynamic properties of the working liquid considered must be built up for flow model simulation, investigation, and enhancement. The developments in the systems using the binary fluids are identified and summarized. The possible future findings are narrated based on the existing research. The research gaps are recognized after the exhaustive audit of detailed works in the area of binary vapor systems.

The writing study has been combined in the accompanying regions.

1. Thermodynamic properties of ammonia–water (NH_3–H_2O) mixture
2. Kalina cycle system (KCS)

3. Vapor absorption refrigeration (VAR)
4. Cooling cogeneration cycle

2.2 THERMODYNAMIC PROPERTIES OF AMMONIA–WATER MIXTURE

In rating of vapor absorption plant components, thermodynamic properties play a vital role. Using empirical correlations or inbuilt properties such as ASPEN PLUS tool, the properties of the mixture can be developed. Temperature, pressure, single liquid mole fraction, and single vapor mole fraction (Reid et al., 1987) are the variables considered in analyzing vapor–liquid equilibrium for binary mixtures. The simplified thermodynamic model for excess enthalpy of mixtures is represented by Ruiter (1990). At infinite pressure and zero pressure, the combination of activity coefficient models with equations of state can be applied (Orbey and Sandler, 1995). Abovsky (1996) correlated the properties for ammonia–water to a pressure of 230 bar and a temperature of 200–640 K. The thermodynamic properties of ammonia–water mixture in the superheated region are estimated utilizing two constant equation of state (Najjar, 1997). Kiselev and Rainwater (1997) evaluated the ammonia–water mixture properties using crossover equation. For polar and nonpolar fluids, Peng–Robinson equation of state is surveyed (Tsai and Chen, 1998).

Using virial cross coefficients, the excess enthalpy of ammonia–water mixture is resulted in gaseous phase. The values are evaluated up to 493.15 K (Wormald and Wurzberger, 2001). Inoue et al. (2002) conducted an experimental study on ammonia–water mixture and addressed the difficulty in predicting the heat transfer coefficients. In the near and supercritical regions, the pressure, volume, and temperature properties for ammonia–water mixture are calculated up to 280 bar and 634 K. Farrokh Niae et al. (2008) suggested a new accurate cubic three-parameter equation of state for calculating pressure, volume, temperature, and vapor–liquid equilibrium equations for high polar fluids. The properties of ammonia–water are evaluated at critical point using simple polynomial correlation (Akasaka, 2009). Polikhronidi et al. (2009) measured the pressure, volume, and temperature properties of ammonia–water mixture in the supercritical regions up to a temperature of 361°C and 280 bar.

The thermodynamic properties essential to investigate the energy and exergy performances are enthalpy, entropy, internal energy, exergy, fugacity, etc. The boiling point of ammonia is low; hence, it boils at low temperature.

The nonazeotropic mixture, ammonia–water mixture, has the nature of boiling and condensation in a range of temperatures which permits a closer match between the heat source and working fluid. The vapor absorption plants are evaluated using the correlations developed by Ziegler and Trepp (1984) and Xu and Goswami (1999) in finding the thermodynamic properties of ammonia–water mixture. The exergy values of binary mixture are developed in evaluating second law.

The development of binary mixture properties and its values with detailed methodology are provided in subsequent chapters and the Appendix. The improvement in ammonia–water mixture properties and the contributions are detailed in Table 2.1.

2.3   KALINA CYCLE SYSTEMS

The Kalina cycle is a power generation cycle using binary mixture as working fluid in the conversion of heat energy to mechanical power. The cycle uses two different components with almost similar molecular weights as working fluid. The ratio of the ammonia–water mixture is changed in each component of the system favoring reduction in the thermodynamic irreversibility of the system. Hence, the overall thermal efficiency is incremented. The Kalina cycle was invented by a Russian Engineer, Dr. Alexander Kalina in 1984 (Kalina, 1984). In recovering waste heat from power and process plants, solar collecting systems, etc., Kalina cycle is considered as a suitable technology. The Kalina cycle system (KCS) is operated at Sumitomo metal steel works and Fuji oil refinery in Tokyo Bay using waste heat recovery. In Husavik, Iceland, and Unterhaching, Germany, KCS runs with geothermal source. Global Geothermal Ltd, the parent of Recurrent Engineering Inc., owns the trademark and patents of KCS.

The performance ranges of KCS with the source details and contributors are summarized in Table 2.2. The series heater arrangement of Kalina plant located at Husavik, Iceland, is thermodynamically evaluated (Mirolli et al., 2002). Mlcak et al. (2002) provided the demonstration report of a 2-MW Kalina cycle at Iceland. For decentralized power generation, Kalina cycle and organic Rakine cycle (ORC) are considered as the suitable power generation technologies. Every system has its own merits and demerits. Presently, Kalina cycle is focused by the researchers attributable to flexibility in operation and environmentally friendly nature. Energy from low-grade waste heat is recovered from Kalina cycle. The design features of Kalina plant and start-up experience were investigated (Mlcak, 2001).

TABLE 2.1 Thermodynamic Properties of Aqua Ammonia.

Sr. no.	Contributors	Maximum pressure (bar)	Maximum temperature (°C)	Correlations
1.	Ziegler and Trepp (1984)	50	227	Gibbs free energy of phase
2.	Renon et al. (1986)	70	230	Redlich–Kwong cubic equation of state
3.	Patek and Klomfár (1995)	20	180	Helmholtz free energy
4.	Abovsky (1996)	230	367	Perturbation theory
5.	Najjar (1997)	20	150	SRK and the PR equations
6.	Nowarski and Friend (1998)	380	225	Haar–Gallagher and Prub–Wagner equations
7.	Thorin et al. (1998)	200	337	Correlations from Stecco and Desideri, and Ibrahim and Klein were used
8.	Tillner-Roth and Friend (1998)	100	327	Helmholtz free energy
9.	Xu and Goswami (1999)	130	150	Gibbs free energy method
10.	Sharma et al. (1999)	20	200	Neural equation of state
11.	Edison and Sengers (1999)	110	227	Helmholtz free energy
12.	Weber (1999)	100	316	Virial coefficients using corresponding states model
13.	Holcomb and Outcalt (1999)	77	106	Helmholtz free energy
14.	Lemmon and Tillner-Roth (1999)	50	140	Equation of state explicit in reduced Helmholtz energy
15.	Thorin (2001)	100	500	Fundamental equation of state for the Helmholtz free energy and Gibbs free energy
16.	Barhoumi et al. (2004)	100	227	Gibbs free energy function, a three-constant Margules model
17.	Conde-Petit (2006)	20	180	Helmholtz free energy
18.	Mejbri and Bellagi (2006)	80	227	Gibbs free enthalpy model, the Patel–Teja cubic equation of state
19.	Alamdari (2007)	100	140	Least-square method for curve fitting

SRK, Soave–Redlich–Kwong; PR, Peng–Robinson.

TABLE 2.2 Performance Details of Kalina Cycle with Source Temperature.

Sr. no.	Authors	Heat source	Temperature (°C)	Power generation (kW)	Thermal efficiency (%)	Remarks
1.	Marston and Hyre (1995)	Gas turbine exhaust	900	531,862	46.75	Bottoming cycle
2.	Dejfors et al. (1998)	Direct fired cogeneration	540	23,500	45.00	NH_3–H_2O cycle with reheat
3.	Rashidi and Yoo (2017)	Waste heat recovery	510	1550	18.30	Combined power cooling cogeneration system
4.	Marston (1990)	Waste heat recovery	500	870	31.00	Stand-alone cycle
5.	Rogdakis and Antonopolos (1991)	Solar	500	1450	44.00	NH_3–H_2O absorption unit
6.	Cao et al. (2014)	Biomass fired system	500	546	26.60	High-temperature KC
7.	Modi and Haglind (2014)	Solar	500	20,847	30.88	KC for a central receiver concentrating solar power plant
8.	Modi et al. (2015)	Solar	500	20,840	30.90	High-temperature KC
9.	Modi and Haglind (2015)	Solar	500	20,000	31.47	Four different KCs
10.	Zhu et al. (2016)	Waste heat recovery	380	1700	27.02	Dual-pressure vaporization KC
11.	Bombarda et al. (2010)	Diesel engine exhaust gas	346	1615	19.70	NH_3–H_2O mixture/experimentation
12.	Wang et al. (2009)	Exhaust gases from cement plant	330	10,500	24.10	NH_3–H_2O mixture/theoretical
13.	Nguyen et al. (2014)	Waste heat recovery applications	330	1910	25.70	Conventional KC and Kalina split cycle
14.	Larsen et al. (2014)	Waste heat recovery on large marine diesel engines	330	1953	23.20	Kalina split cycle, reheat
15.	Zhao et al. (2015)	Waste heat recovery	320	25,401	23.78	Integrated CAES system and KC

TABLE 2.2 *(Continued)*

Sr. no.	Authors	Heat source	Temperature (°C)	Power generation (kW)	Thermal efficiency (%)	Remarks
16.	Guo et al. (2015)	Waste heat recovery	310	946	27.90	Dual-pressure vaporization KC
17.	Galanis et al. (2009)	Geothermal	300	400	22.00	NH_3–H_2O mixture/theoretical
18.	Hua et al. (2015)	Waste heat recovery	300	739	21.69	NH_3–H_2O mixture/theoretical
19.	Zhang et al. (2015)	Waste heat	300	473	20.85	Integrated system of ammonia–water Kalina–Rankine cycle
20.	Chen et al. (2015)	Waste heat	280	495	20.90	Integrated system of ammonia–water Kalina–Rankine cycle
21.	Chew et al. (2014)	Waste heat recovery	262	1309	20.29	Combined BTX–KC
22.	Victor et al. (2013)	Waste heat recovery	250	150	15.01	Low-temperature KC
23.	Yue et al. (2015)	Engine exhaust heat recovery	221	216	18.80	Kalina as a bottoming cycle
24.	Desideri and Bidini (1997)	Geothermal sources	210	80	18.50	NH_3–H_2O mixture/theoretical
25.	Sadeghi et al. (2015)	Waste heat recovery	200	5203	26.32	Double-turbine KC
26.	Mahmoudi et al. (2016)	Waste heat recovery	190	32,300	21.46	Combined augmented KC/gas turbine-modular helium reactor
27.	Bliem (1998)	Geo fluid	182	8	11.80	NH_3–H_2O mixture/theoretical
28.	Spinks (1991)	Geothermal	179	13,000	21.10	NH_3–H_2O mixture/experimentation
29.	Zare and Mahmoudi (2015)	Waste heat recovery	160	27,330	16.60	GT-MHR plant
30.	Mittelman and Epstein (2010)	Geothermal	160	50,000	19.00	Bottoming cycle
31.	Fu et al. (2013a)	Geothermal	145	2200	15.00	KC subsystem

TABLE 2.2 *(Continued)*

Sr. no.	Authors	Heat source	Temperature (°C)	Power generation (kW)	Thermal efficiency (%)	Remarks
32.	Cao et al. (2017)	Low-grade heat source	140	80	16.40	Combined cooling and power cycle
33.	Panea et al. (2010)	Low-temperature geothermal source	140	13	13.00	NH_3–H_2O mixture/theoretical
34.	Tamm (2003)	Geothermal	137	10	23.50	Experimental work
35.	Singh and Kaushik (2013)	Waste heat recovery	134	605	12.95	KC coupled with a coal fired steam power plant
36.	Singh and Kaushik (2012)	Low-temperature heat of exhaust gases	132	605	12.95	Bottoming cycle of a coal fired steam power plant
37.	Arslan (2010)	Geothermal	130	41,200	14.80	NH_3–H_2O mixture/theoretical
38.	Ogriseck (2009)	Waste heat	130	2195	12.30	Combined heat and power plant
39.	Lolos and Rogdakis (2009a)	Solar concentrating collectors	130	190	8.30	KCS
40.	Li and Dai (2014)	Geothermal waste	127	200	10.50	Low-temperature KC at China
41.	He et al. (2014)	Waste heat recovery	127	6	13.00	KC without throttle valve
42.	DiPippo (2004)	Geothermal	124	1696	10.60	NH_3–H_2O mixture/experimentation
43.	Lolos and Rogdakis (2009b)	External heat source	120	250	16.70	Countercurrent absorber
44.	Shokati et al. (2015)	Geothermal	118	4231	14.60	KC
45.	Dorj (2005)	Geothermal	117	1860	14.00	NH_3–H_2O mixture/experimentation
46.	Li et al. (2016)	Geothermal source	116	2593	14.96	Low-temperature geothermal source
47.	Fallah et al. (2016)	Geothermal	116	1672	12.66	KC applied for low-temperature-enhanced geothermal system

TABLE 2.2 *(Continued)*

Sr. no.	Authors	Heat source	Temperature (°C)	Power generation (kW)	Thermal efficiency (%)	Remarks
48.	Zare and Mahmoudi (2015)	Waste heat recovery from gas turbine	116	2186	10.30	Combined GT-MHR and KC
49.	Saffari et al. (2016)	Geothermal	113	1962	20.36	Geothermal KC system using Artificial Bee Colony algorithm
50.	Gholamian and Zare (2016)	Solid oxide cell waste heat	110	2186	12.68	Hybrid SOFC/GT-KC
51.	Li et al. (2013)	Waste heat recovery	110	13	8.36	KC with ejector
52.	Guzovic et al. (2010)	Geothermal	110	3949	10.60	NH_3–H_2O mixture/theoretical
53.	Yari et al. (2015)	Low-grade heat source	109	1108	9.50	Comparison of trilateral Rankine cycle, ORC, and KC
54.	Wang et al. (2013)	Solar	106	11	7.55	Solar-driven KC
55.	Roy et al. (2010)	Industrial waste heat	100	140	18.00	Rankine cycle with ammonia–water mixture
56.	Hettiarachchi et al. (2007)	Geothermal	90	80	12.00	NH_3–H_2O mixture/theoretical
57.	Bai et al. (2004)	Hot spring water	60	4	9.00	Spring thermal energy conversion plant

BTX, benzene, toluene, p-xylene; CAES, compressed air energy storage; GT-KC, gas turbine-Kalina cycle; GT-MHR, gas turbine-modular helium reactor; KC, Kalina cycle; SOFC, solid oxide fuel cell.

The working principles and arrangement of Kalina cycle are reported (Henry, 1996). Jonsson (2003) has made an analysis over the power generation cycles and concluded that the performance of ammonia–water cycle is higher than the stream power plants. In thermal power installations, ammonia with water as a mixture can be used without creating environment difficulties. Brodyanskii (2006) reported that ammonia is easily available and cheap, having no corrosive effect on iron and its alloys. It is easily soluble with water. In binary cycles, to obtain higher isentropic efficiency, radial inflow turbines are superlative (Marcuccilli and Zouaghi, 2007). Comparison over Kalina and conventional ocean thermal energy conversion cycles working with ammonia–water mixture is reported (Asou et al., 2007). Hatem (2007) experimented vapor ammonia in ammonia–water solution. By increasing the heat generation, the efficiency of turbine increases with increased system efficiency (James Hartley, 2001). Prisyazhniuk (2008) reported the options of reducing the fuel consumption from a thermal power plant and reduction of discharge into the environment.

The applications of solar energy are increased worldwide as the solar energy is considered as the primary energy in future. The solar energy applications are restricted in spite of its success, due to the cost. Goswami et al. (2004) suggested that the present situation will be changed with the upcoming developments and innovations in the solar field. Solar power generation system results higher efficiency without environmental hazards (Hu et al., 2010). The benefits of integrating solar energy with fossil-fueled power plants are highlighted. Mills et al. (2004) suggested that the solar thermal power plant for low-temperature applications using linear Fresnel reflector system is cost-effective. The correlations for determining the performance of a solar parabolic trough collector system was performed and the fabrication of a solar parabolic trough collector was performed (Valan Arasu and Sornakumar, 2007). The operation of a Kalina demonstration plant has been initiated in 1992 in the US Department of Energy Technology Engineering Center, California. Leibowitz and Mirolli (1997) reported the Kalina power plant using waste heat works with a maximum pressure and temperature of 110 bar and 516°C, which generates a power of 3 MWe.

The performance of Kalina cycle and the Rankine cycle was assessed (El-Sayed and Tribus, 1985). The design developed by them had more heat exchangers with two streams making the system complicated. The optimization of the power cycle was performed with the development of correlations (Rogdakis, 1996). The higher net power output with lower total exergy loss has been resulted for vapor absorption cycle in comparison to the Rankine cycle (Dejfors and Svedberg, 2010). In a Rankine–Kalina combined cycle, the first and second law analysis was performed (Murugan and Subbarao, 2008). On integrating Rankine cycle with low-temperature Kalina cycle, an improvement of 1.4% of efficiency higher than the condensing Rankine cycle has been resulted.

The exergetic efficiency is not directly related to output. The heat exchanger's surface is directly related to output. The heat recovery vapor generator with water and ammonia–water mixture as working fluids has been compared (Stecco and Desideri, 1992). The heat recovered in the boiler utilizing mixture is 25% higher than the single component. Rankine cycle using geothermal source in New Zealand produces 10 MW net output (Spinks, 1991). The ammonia–water mixture Rankine cycle with the absorber–generator claims an efficiency of 14% at a maximum temperature of 100°C (Styliaras, 1996). Combined ammonia–water mixture Rankine cycle and liquefied natural gas power generation cycle have been investigated (Shi and Che, 2009). At the maximum pressure of 30 bar and temperature of 150°C, the combined system resulted in the energy and exergy efficiency of 33.28% and 48.87%, respectively.

The cost benefit of Kalina and organic Rankine cycles are compared and concluded that the best power and best cost points are different (Valdimarsson and Eliasson, 2003). Wang et al. (2010) reported that organic Rankine cycle, supercritical Rankine cycle, and Kalina power cycle are the competitors for low-grade heat recovery. Carbon dioxide transcritical cycles suitable to low-temperature heat recovery have been developed (Cayer et al., 2009). Baik et al. (2011) made a comparison between carbon dioxide and R125 transcritical cycles. Both cycles utilize low-grade heat source. The results concluded that R125 transcritical cycle exhibits 14% more power than the carbon dioxide transcritical cycle. ORC suitable for medium temperature applications utilizing geothermal heat source is developed (Franco and Villani, 2009). The sizing of the heat exchangers in the organic Rankine cycle for the dual source is reduced compared with the single source (Aneke et al., 2011).

The ORC is developed for combined bottoming cycles (Chacartegui et al., 2009). With the power generators like solar thermal facilities, the developments on ORC are extended. The ORC systems can also be integrated with the combined cycle power plants as bottoming cycles. Srinivas et al. (2008) described the benefits of the Kalina system as bottoming cycle over the steam bottoming cycle. The cycle uses waste heat from the gas turbine exhaust. Mirolli (2001) concluded that in waste heat recovery power plant, the distillation condensation subsystem is the essential component in Kalina cycle favoring higher efficiency. Kalina as bottoming cycle resulted in better performance in comparison to steam bottoming cycle (Heppenstall, 1998).

The performance of power and cooling system has been evaluated using simplified methodology (Srinivas and Vignesh, 2012; Srinivas et al., 2011). The advanced thermal conversion methods for cogeneration and combined power cycles are reported (Korobitsyn, 1998).

The Kalina power system works with wide range of source temperature. Based on the source temperature, KCS can be grouped into low-temperature KCS, intermediate-temperature KCS, and high-temperature KCS. Literature reports that the majority of the work consolidated to low-temperature application and high-temperature applications. There is no much progress in intermediate-temperature applications with the temperature range of 200–300°C as per the report. The Kalina power plant working temperature range for low-temperature application is 150°C; for high-temperature application, it is in the range of 250–600°C.

2.3.1 EXERGY WORKS ON KALINA CYCLE SYSTEMS

The maximum useful work produced upon interaction of a system with the equilibrium state is the exergy. The exergy is the sum of chemical and physical components of exergy. In combustion, chemical exergy plays an important role. Other than combustion, physical exergy deal is enough for the evaluation and analysis on second law point of view of thermodynamics. The flow to cause changes in a system or the measure of the potential of the system is also the exergy. A power generation system cannot be justified by energy analysis. It is necessary to provide exergy efficiency for the systems considered. The losses in the total components of the power system can be estimated and compared with the exergy analysis. Table 2.3 details the exergy efficiency with the source temperature and heat sources of the cycle. In modeling the binary systems, suggestions that lead to the importance of second law efficiency in the evaluation of performance are reported (Kohler and Saadat, 2004). The Kalina power cycle system applicable for intermediate-temperature heat source (190–225°C) arrangements has been designed and examined to provide the advantages compared to the other available configurations. Srinophakun et al. (2001) reported that the ratio of exergy loss with the net generated power is low compared to the Rankine cycle.

In generating the desired products from the inputs, exergy losses represent the true losses of the potential. In determining the irreversibility of a combustion system, exergy analysis is assessed (Som and Dutta, 2008). Keeping the exergy destruction within a reasonable limit in a combustion

TABLE 2.3 Exergy Analysis of Kalina Cycle Systems.

Sr. no.	Authors	Heat source	Temperature range (°C)	Maximum pressure (bar)	Exergy efficiency (%)	Remarks
1.	Modi and Haglind (2014)	Solar	500	160	29.88	KC for a central receiver concentrating solar power plant
2.	Nag and Gupta (1998)	Waste heat recovery	500	100	60.00	KC as bottoming cycle
3.	Galanis et al. (2009)	Industrial waste heat	350	140	70.00	KC
4.	Nguyen et al. (2014)	Waste heat recovery applications	330	102	65.00	Conventional KC and Kalina split cycle
5.	Wang et al. (2009)	Waste heat recovery from a cement industry	330	86	43.00	Cogeneration plant
6.	Zhao et al. (2015)	Waste heat recovery	320	100	46.42	Integrated CAES system and KC
7.	Wagar et al. (2010)	Renewable-driven heat engines	300	200	64.00	KC
8.	Chen et al. (2015)	Waste heat	280	60	44.00	Integrated system of ammonia–water Kalina–Rankine cycle
9.	Victor et al. (2013)	Waste heat recovery	250	39	16.26	Low-temperature KC
10.	Yue et al. (2015)	Engine exhaust heat recovery	221	53	38.20	Kalina as a bottoming cycle
11.	Mahmoudi et al. (2016)	Waste heat recovery	190	53	51.30	Combined augmented KC/gas turbine-modular helium reactor
12.	Franco and Villani (2009)	Geothermal	160	15	45.00	KC
13.	Nemati et al. (2017)	Waste heat recovery	142	50	52.58	Cogeneration system
14.	Panea et al. (2010)	Waste heat from geothermal sources	140	31	56.00	KC

TABLE 2.3 *(Continued)*

Sr. no.	Authors	Heat source	Temperature range (°C)	Maximum pressure (bar)	Exergy efficiency (%)	Remarks
15.	Padilla et al. (2010)	Waste heat from a conventional power cycle	137	32	92.00	NH_3–H_2O mixture cycle
16.	Arslan (2010)	Geothermal	130	30	36.20	NH_3–H_2O mixture/theoretical
17.	He et al. (2014)	Waste heat recovery	127	30	50.61	KC without throttle valve
18.	Li and Dai (2014)	Geothermal	127	39	48.80	KC
19.	Fu et al. (2013b)	Geothermal	122	38	80.00	Cascade utilization system including KC subsystem
20.	DiPippo (2004)	Geothermal	122	5	41.40	Low-temperature Kalina system
21.	Murugan and Subbarao (2008)	Waste heat recovery	120	42	27.22	Combined Rankine–KC
22.	Usvika et al. (2009)	Geothermal	120	8	69.00	Exergetic investigation of KC
23.	Dorj (2005)	Geothermal	117	30	58.90	NH_3–H_2O mixture/Experimentation
24.	Cao et al. (2017)	Low-grade heat source	116	32	25.76	Kalina-based combined cooling and power cycle
25.	Fallah et al. (2016)	Geothermal	116	32	37.87	KC applied for low-temperature-enhanced geothermal system
26.	Eller et al. (2017)	Waste heat recovery	115	32	30.40	KC
27.	Saffari et al. (2016)	Geothermal	113	40	48.20	Geothermal KC system using Artificial Bee Colony algorithm
28.	Gholamian and Zare (2016)	Solid oxide cell waste heat	110	43	59.53	Hybrid SOFC/GT-KC

TABLE 2.3 *(Continued)*

Sr. no.	Authors	Heat source	Temperature range (°C)	Maximum pressure (bar)	Exergy efficiency (%)	Remarks
29.	Yari et al. (2015)	Low-grade heat source	109	43	20.98	Comparison of trilateral Rankine cycle, ORC, and KC
30.	Wang and Yu (2016)	Geothermal	107	23	36.50	KC
31.	Ashouri et al. (2015)	Solar	106	18	67.71	KC
32.	Rodriguez et al. (2012)	Geothermal	100	25	36.50	ORC and KCs
33.	Roy et al. (2010)	Low-temperature heat recovery	100	30	73.20	KC

CAES, compressed air energy storage; GT-KC, gas turbine-Kalina cycle; KC, Kalina cycle; SOFC, solid oxide fuel cell.

process, the primary way is to reduce the irreversibility in heat conduction. In Kalina as bottoming cycle, the energy in the combustion gases is recovered resulting with additional power (Borgert and Velasquez, 2004). Rosen and Dincer (2001) proposed the direction for assessing and examining the environmental problems of varying complexity utilizing the exergy concept. The thermodynamic investigation in an aqua–ammonia vapor-absorption system is carried out to reduce the overall production cost (Mishra et al., 2006). Identifying the effects of design variables on cost and providing the values of design variables, he concluded the overall system as cost-effective. The difficulties associated with energy methods have been avoided by the exergy analysis, thereby allowing understanding and measuring efficiency (Rosen, 2007). The environmental implications of exergy are described additionally. Koroneos and Rovas (2007) checked the possible use of KCS's heat rejection to operate the flash desalination system and showed the improvement in the desalination yield.

The exergy losses in industrial processes are analyzed by a graphic method named as energy utilization diagram (EUD) (Wall et al., 1989). The Kalina cycle is optimized utilizing EUD. The internal phenomenon is provided by representing the exergy losses for every energy transformation. Thermo-economic analysis for a combined cycle/cogeneration plant is carried out with the summarization of electricity cost (Borelli and de Oliveira Junior, 2008). The exergy analysis for lignite-fired thermal plant is carried out (Ganapathy et al., 2009). Due to the irreversibility existing in the combustion process, heat loss, exhaust loss, and incomplete combustion, the exergy loss in the boiler results to 57%. The thermodynamic analysis of a coal-based thermal power plant and gas-based cogeneration power plant are conducted (Reddy et al., 2010).

An exergy analysis for a 120-MW thermal power plant in Malaysia is conducted (Hussein et al., 2001). The highest exergy destruction is reported in the boiler with 54 MW. The high-pressure turbine and intermediate-pressure turbine are resulting larger exergy destruction than the low-pressure turbine in a power plant. In Egypt, the energy and exergy analysis of a steam power plant is assessed (Rashad and El Maihy, 2009). The component-wise performance of the plant is estimated, and a detailed breakup energy and exergy losses at various loads are reported. Exergy analysis with 10 different types of Turkish lignite is made in a thermal power plant. With the conservation of mass and energy, the exergy destruction of each component has been investigated. Boiler has been enlisted as the major source in the total exergy destruction of the entire plant with 299.10 MW and 83.29%. An exergy analysis for a KCS 34 is conducted (Usvika et al., 2009). In

Grassman diagram, the exergy flow of Kalina cycle is briefed. The pressure drop and heat transfer are the factors reasoning for the losses in heat exchanging equipment's from the cycle. Due to mechanical and isentropic efficiencies, turbine favors highest losses. The sizes of vaporizer, condenser, high-temperature recuperator, and low-temperature recuperator are 455.14, 940.18, 50.45, and 144.65 m^2, respectively, for the capacity of 0.92 MW (Rodriguez et al., 2012). Mishra et al. (2006) reported that with the thermo-economic evaluation, the overall product cost is minimized.

2.3.2   INNOVATIONS IN KALINA CYCLE SYSTEMS

Table 2.4 results the innovations and the performance data for KCS. Kalina (1995b) has invented a design for converting heat from geothermal liquid and geothermal steam to electric power. Using the energy potential of both geothermal steam and geothermal liquid, an integrated system has been recommended. For two or more combustion zones, power generation cycle from externally fired power system has been developed (Kalina and Mirolli, 1996). With numerous distillation operations, a thermodynamic cycle transforms energy into usable form (Kalina, 1996). An improved efficiency is achieved by the apparatus with the regeneration subsystem (Kalina and Pelletier, 1997). Transferring heat into useful energy with separate closed loops has been implemented (Kalina and Rhodes, 1998). In the conversion of heat into energy, an improved configuration over the existing one has been proposed (Kalina et al., 1999). A new KCS has been developed utilizing waste heat from cement industry (Mirolli, 2005). He has reported an improvement of about 20–40% in performance against the conventional waste heat. KCS generates electricity using waste heat from the cement plant without the necessity of additional fuel (Mirolli, 2006).

Refurbishing conventional power plants have been implemented (Peletz and Tanca, 2000). The heaters suitable for KCS have been replaced with the Rankine heaters in this technique. Converting heat into power, a new Kalina cycle has been proposed (Peletz, 2001). The fuel gas is heated to the gas turbine using a new method in a combined cycle power plant (Ranasinghe et al., 2001). With the incorporation of Kalina cycle, overall efficiency of the power plant is improved. A new thermodynamic cycle has been proposed for district water heating with increased efficiency in comparison with Rankine cycles (Ranasinghe et al., 2002). A method for power generation system has been reported (Hansen et al., 2000). Hansen et al. (2001a, 2001b) invented a multicomponent working fluid vapor generation system.

TABLE 2.4 Innovations in Kalina Cycle Systems.

Sr. no.	Authors	Heat source	Temperature (°C)	Power generation (kW)	Thermal efficiency (%)	Remarks
1.	Kalina (1986)	Geothermal energy	532	2595	31.78	Closer matching of the working fluid and the heat source enthalpy–temperature characteristics in the boiler has been achieved without reducing the mass flow rate of the cycle
2.	Kalina (1988a)	Thermal source	565	0.0541	47.79	The working fluid has been heated to high temperatures by the returning streams resulting in high heat acquisition temperature
3.	Kalina (1988b)	Thermal source	565	0.175	43.20	The system provides improvement in the efficiency of the cycle by decreasing the irreversibility's by preheating of the expanded stream
4.	Kalina (1991a)	Geothermal	186	80,785	18.04	With heating the multicomponent working stream by the heat released from an expanded spent stream, the improvement in efficiency has been achieved
5.	Kalina (1991b)	Low-temperature heat	103	5502	10.34	This system provides an apparatus for converting low-temperature heat to electric power
6.	Kalina (1992)	Thermal energy	500	96,254	39.72	With generation of two multicomponent liquid working streams, the thermal efficiency of the system has been improved
7.	Kalina (1995a)	Waste heat recovery	280	90,617	39.99	Kalina cycle
8.	Kalina (1995b)	Geothermal	196	79,045	17.86	The conversion of thermal energy into electric power from a geofluid has been implemented

TABLE 2.4 *(Continued)*

Sr. no.	Authors	Heat source	Temperature (°C)	Power generation (kW)	Thermal efficiency (%)	Remarks
9.	Kalina and Mirolli (1996)	Waste heat recovery	550	252,732	36.69	The heat utilization efficiency of a power generation cycle has been improved with the incorporation of distillation condensation subsystem
10.	Kalina and Pelletier (1997)	Combusting fuel	315	12,790	33.74	The improved efficiency has been achieved with the aid of regeneration subsystem
11.	Kalina and Rhodes (1998)	Waste heat recovery	315	4595	28.99	Conversion of thermal energy from the combustion of corrosive fuels has been achieved by two closed loops
12.	Kalina et al. (1999)	Waste heat recovery	100	2820	9.76	The efficiency of the cycle has been increased with the parallel heaters system
13.	Ranasinghe et al. (2002)	Waste heat recovery	120	367,556	50.07	Direct combination of Kalina cycle with district water heating and cooling systems has been implemented results in reduction in costs
14.	Kalina and Hillsborough (2004a)	Geothermal	100	0.0286	13.25	With the multistaged heating process, the improvement in the performance of the power generation cycle has been assessed
15.	Kalina and Hillsborough (2004b)	Low-temperature waste heat	150	0.0242	14.59	The conversion of thermal energy from moderately low-temperature source to more useful forms of energy has been achieved
16.	Kalina and Hillsborough (2005a)	Waste heat recovery	100	0.02867	13.25	With the two-cycle thermodynamic method, the improvement in the overall efficiency of the energy conversion system has been achieved

TABLE 2.4 (Continued)

Sr. no.	Authors	Heat source	Temperature (°C)	Power generation (kW)	Thermal efficiency (%)	Remarks
17.	Kalina and Hillsborough (2005b)	Waste heat recovery	150	0.02422	14.59	For the conversion of greater amount of thermal energy from an external source, two pressure circuits have been utilized
18.	Kalina and Hillsborough (2005c)	Low-temperature waste heat	310	2947	23.93	With the multistaged heating process, the conversion of moderate-to-low-temperature heat sources has been efficiently achieved
19.	Kalina and Hillsborough (2006c)	Waste heat recovery	180	5252	17.24	The apparatus to convert the thermal energy in superheated stream of a multicomponent fluid in high efficient way has been implemented
20.	Kalina and Hillsborough (2006d)	Waste heat recovery	150	12,574	15.02	A method for extracting thermal energy from a fully vaporized boiling stream has been implemented

Different thermodynamic configurations in the conversion of energy from a low-temperature geothermal stream into useable energy have been proposed (Kalina and Hillsborough, 2004a, 2004b, 2004c, 2004d). In the conversion of low-temperature stream energy into a usable form, a new thermodynamic cycle is reported (Kalina and Hillsborough, 2005a, 2005b, 2005c). The reported cycle comprises two different boiling components with a low and high-pressure circuit. An efficient system extracting usable energy from high-temperature waste stream has been invented (Kalina and Hillsborough, 2005d). The proposed multipressure thermodynamic cycle includes a two-stage turbine energy extraction subsystem, distillation–condensation subsystem, and a boiler subsystem.

A single flow cascade system with first-law and second-law efficiencies of 37% and 66% has been proposed (Kalina and Hillsborough, 2006a).

New designs which boil and vaporize a multicomponent fluid have been proposed (Kalina and Hillsborough, 2006b). The nucleate boiling for the entire apparatus length is maintained with the addition of vapor shell, liquid shell, and connecting pipes. The vessel provides high film heat transfer coefficient, thereby protecting the tubes from burn out. Additionally, the device enhances complete vaporization of multicomponent fluids. Kalina and Hillsborough (2008) have invented a better combustion method.

2.4   VAPOR ABSORPTION REFRIGERATION

During the change of state from liquid to vapor in a binary absorbent/refrigerant mixture, the volatile refrigerant vaporizes first. This nature of vaporization provides a better match with hot fluid temperature profile. The first VAR cycle was made in early 1700s (Gosney, 1982). The ice is produced due to the absorption of water by the sulfuric acid kept in vacuum (Herold and Radermacher, 1989). The first aqua–ammonia VAR system was invented by Ferdinand Carrie in 1859. In VAR system, LiBr–water was introduced in 1950s as working fluid for cooling. High boiling point difference, chemically stable, nontoxic, nonflammable, favorable transport properties, and high refrigerant concentration within absorbent are the desirable properties (Holmberg and Berntsson, 1990). Grossman and Gommed (1987) developed subroutine computer codes to solve the components of VAR with seven working fluid pairs. In aqua–ammonia, more than 40 refrigerants and 200 absorbent are available. Among them for absorption cooling, aqua–ammonia and LiBr–water absorbent and refrigerants are the most suitable pairs (Marcriss et al., 1988).

The use of a solution heat exchanger increases the coefficient of performance (COP) up to 60% as per the experimental studies (Aphornratana, 1995). To increase the COP of the system, in 1911, the double and multieffect system was introduced (Chen, 1988; Aphornratana and Eames, 1998). Chung et al. (1984) used DMETEG/R22 and DMETEG/R21 as working fluids in an ejector VAR system to achieve high performance and achieved increase in COP. To avoid the crystallization, an ejector is placed between the generator and condenser (Aphornratana and Eames, 1998). As per the experimentation results in LiBr–water pair, higher generator temperature of 190–210°C is required for obtaining high COP of 0.86–1.04. Improvements are resulted in LiBr–H_2O cooling cycle with LiBr as an absorbent and water/steam as the refrigerant (Talbia and Agnew, 2000; Misra et al., 2003; Aphornratana and Sriveerakul, 2007).

For VAR cycle, the alternate absorbent–refrigerant pairs reported are $LiNO_3/NH_3$ (Best, 1991), LiBr + $ZnBr_2/CH_3OH$ (Idema, 1984), $LiNO_3$ + KNO_3 + $NaNO_3$/water (Grossman and Gommed, 1987), LiCl/water (Grover et al., 1988), and glycerol/water (Bennani et al., 1989). The analyses of water as refrigerant pair with four binary absorbent mixtures, five trinary absorbent mixtures, and seven quaternary absorbent mixtures are investigated (Saravanan and Maiya, 1998). The H_2O–LiCl results higher performance in the circulation point of view and H_2O–LiBr + LiCl + $ZnCl_2$ exhibits high COP. For low-grade heat source VAR system, a new mixture of R134a-DMAC as working fluid is proposed (Muthu et al., 2008). At separator and sink temperatures of 80 and 30°C, respectively, the maximum COP is reported as 0.45 from the experimental findings. The overall performance of the VAR system increases with the addition of subcooler/liquid–vapor heat exchanger, and it is high for R134a-DMAC compared to NH_3/H_2O.

A 10-kW aqua–ammonia gas-fired VAR system is experimented and reported that the system performs well upon lowering the cooling water inlet temperature of the condenser than the absorber (Horuza and Callander, 2004). A monomethylamine–water mixture for solar VAR system, operating with low temperature of −5–10°C, is reported (Pilatowsky et al., 2003). It is preferred for the cold storage of milk products. The VAR system works using the sources of low-grade heat recovery process, solar PV, and thermal energy system (Antonopoulos and Rogdakis, 1992; Grossman, 1991; Kaushik and Kumar, 1987; Kandlikar, 1982).

The solar powered VAR results low COP and also demands more heat supply. More losses in the separator and absorber of VAR system are resulted in exergy analysis. The reason for the losses is due to the heat of mixing and separation which is not available in the mono fluid refrigerant (Talbia and

Agnew, 2000). Saidur et al. (2011) reported that using the nanoparticles, in the refrigerant HFC134a, TiO_2 with mineral oil increases the thermal properties of the refrigerant resulting in the energy conservation to 26% with a mass fraction of 0.1. The heat transfer area of generator increments the driving force upon expansion (Maria et al., 2014). This is indirectly proportional to the driving force of evaporator.

2.5   COOLING COGENERATION CYCLE

For future power generation systems, small-scale distributed cogeneration units (≤ 1 MWe) gain attention more. In combustion of fuel and heat recovery integration, many conventional plants are developed (Deng et al., 2011). In a single effect absorption chiller, waste heat from the microturbine exhaust can be used (Garland, 2003). Austin energy is partnered with Burns and McDonnell in constructing the innovative large-scale cogeneration pilot system in the domain plant of Austin (DeVault, 2005). A prime mover is a microturbine generates 60 kW of electricity at 310°C with an efficiency of 26.9% in University of Maryland. Sun (2013) developed a cooling cogeneration cycle using mid or low-temperature heat sources. A 15-kW cooling system when coupled externally to other cycle minimizes 17.1% with same output compared to individual power and cooling cycle as resulted experimentally. Integrating power and cooling cycle relies an improvement in efficiency as per the theoretical and experimental analysis. The KCS generating power and cooling is modified (Xu et al., 2000). The modified system claims 0.999 ammonia concentration to produce the cooling effect. Saturated ammonia vapor at low temperature is resulted at the exit of the turbine in this cycle, which is used for the purpose of refrigeration.

Ammonia–water mixture hybrid power generating and refrigerating cycles are reported (Amano et al., 2000). To expand the availability of ammonia–absorption cycle, ammonia–water mixture turbine cycle (AWMT) and a single-effect ammonia absorption refrigeration cycle (AAR) are connected. This connected system is used for low-temperature applications. Two Kalina cycles have been considered in AWMT cycle, whereas AAR is a single-stage absorption cycle. The hybrid cycle performance is compared with the individual performance of AWMT and AAR cycles. The AAR shares the working fluid due to which the heat of rectifications has been decreased. Ammonia–water mixture as working fluid in a combined power and cooling cycle is proposed (Xu et al., 2000). In the combined cycle, a parametric analysis has been carried. With the rise in pressure, the cooling capacity increases, and decrease in flow rate, the pressure descends. The parameters influencing the performance are identified as temperature of boiler, condenser, superheater, and absorber. Alternative to fossil fuel technologies are power and cooling cycle (Tamm, 2003). Tamm et al. (2004), Tamm and Goswami (2003), and Hasan and Goswami (2003) examined the performance of a combined power and refrigeration cycle. A new power and refrigeration cycle is recommended (Liu and Zhang, 2007). Comparing to an individual power generation system, the cogeneration system favors 18.2% decrease in energy consumption.

Theoretical analysis for a combined power and refrigeration cycle has been proposed (Lu and Goswami, 2002). The key elements favoring for an increased cycle and plant efficiencies are proposed (Franco and Casarosa, 2002). A combined power and refrigeration cycle using low-temperature heat sources, such as geothermal energy, solar energy with flat plate collectors to generate power, and refrigeration in the same cycle has been proposed (Lu and Goswami, 2003). The adaptability of the vapor generation and absorption condensation processes are experimented (Tamm et al., 2004). A combined power and cooling cycle is examined parametrically (Vijayaraghavan and Goswami, 2006). The cycle configuration has been optimized for maximum resource utilization efficiency. Thermal distillation scheme has been implemented identical to a Kalina cycle. A combined power and refrigeration cycle has been analyzed by the exergy method (Vidal et al., 2006). The properties of the ammonia–water mixture in the combined cycle have been simulated using ASPEN plus. Thermodynamic analysis is carried out in a novel absorption power/cooling combined cycle (APC) using p–T, log p–h, and T–s diagrams (Zheng et al., 2006). The optimum working conditions have been observed in a combined power and cooling thermodynamic cycle (Sadrameli and Goswami, 2007). By expanding the ammonia-rich vapor in an expander to subambient temperatures and heating the exhaust, the output has been resulted. Ammonia–water mixture as working fluid in a power and cooling cogeneration system has been investigated (Padilla et al., 2010). With the decrease in the absorber temperature, the turbine exit quality reduces as per the report. The thermodynamic performance of a combined power and cooling cycle is assessed and optimized (Pouraghaie et al., 2010). The Rankine cycle and the AAR with binary ammonia–water mixture as the working fluid have been reported in a combined power and refrigeration cycle (Wang et al., 2008).

By varying the turbine inlet pressure, a combined power and ejector cooling cycle with R245fa as working fluid have been assessed thermodynamically (Kumar, 2015). Combined Kalina cycle and ammonia–water absorption refrigeration systems are analyzed (Wang et al., 2016). They reported that the high exergy destruction occurs in the heat exchanger.

The performance of a combined cycle utilizing the solar energy as heat source has been investigated and proposed that the second law efficiency of more than 60% is achieved at 77°C. Moreover, the analysis also results that increasing in the source temperature above the specified 150°C will not involve in increase of second law efficiency rather it will help to improve the first law efficiency (Hasan et al., 2002). In a combined power and cooling cycle using ammonia–water mixture as working fluid, an experimental investigation has been carried out (Tamm and Goswami, 2003). The linear increment in refrigeration and thermal efficiency with drop in power output has been noticed by increasing the turbine inlet pressure. The cooling effectiveness in the combined cycle has been assessed (Martin and Goswami, 2006). In achieving the effective COP of 1.1, they have reported that the equal amount of work is needed per unit of cooling production.

A new combined power and cooling cycle with separate absorber and reboiler for both the cycles has been introduced with a maximum exergy efficiency of 37.3% (Zheng et al., 2006). The irreversibility in the system is more, and with huge number of components in the controlling of separate power and cooling cycle, the cost increases. A novel APC and cooling cycle with 24% of cycle efficiency at a maximum temperature of 350°C has been developed (Zheng et al., 2006). The nature of the proposed cycle is complex. The Rankine cycle is combined with the AAR and designed (Wang et al., 2008). At a source temperature of 300°C, the cycle efficiency of 20.5% has been reported. A new combined Rankine power generation cycle and ejector AAR have been proposed resulting in power output and refrigeration output simultaneously. With a source temperature of 300°C, the thermal efficiency reported is 21%. With the decrement in absorber temperature, the power and cooling output increase.

In the performance assessment of a Goswami cycle, the variation in the key parameters has been assessed using a multiobjective technique (Pouraghaie, 2010). A new trigeneration system recovering the heat of the LiBr refrigeration cooling water from heat water boiler resulting in high COP has been developed (Zhang et al., 2011). The Goswami cycle with gas turbine-combined cycle heat recovery system has been investigated (Zare et al., 2012). The unit rate of production cost-integrated power and cooling cycle is reduced by 18.6% and 25.9%, respectively, as reported from the thermodynamic analysis with decrease in the first and second law efficiencies.

By coupling the KCS and VAR system without change in the base cycles, a cooling cogeneration cycle has been developed (Srinivas and

Reddy, 2014). As per the demand, the working fluid common to both power and cooling cycles has been divided. The cooling output is maximized by the high turbine inlet concentration and the rectification process becomes mandatory when the cooling requirement is high. Fontalvo et al. (2013) assessed the rectification with internal and external cooling and reported that the decrement in irreversibility and increment in first and second law efficiencies results because of internal cooling. An aqua–ammonia-based integrated system reports 225 kW of cooling and 80 kW power (Jawahar et al., 2013). At optimum conditions, a maximum combined thermal efficiency of 35–45% and COP of about 0.35 have been attained. Simulation for the split cycle for three working pairs of NH_3/H_2O, $NH_3/LiNO_3$, and $NH_3/NaSCN$ with the option of split ratio concept has been performed (Lopez-Villada et al., 2014). The first law efficiency obtained is less than the Goswami cycle as per the comparison.

In the subtropical climate condition, a sun-driven combined power and ejector refrigeration system has been investigated (Ahmadzadeh et al., 2017). In generating power and cooling, a combined organic Rankine cycle with ejector refrigeration cycle has been investigated (Sabeti and Boyaghchi, 2016). With reduction in the cost rate of products, the exergetic efficiencies have been increased. A combined power and cooling system with ammonia–water mixture as working fluid has been investigated and agreed with good economic benefit (Chen et al., 2017). A combined power and cooling system with reduced CO_2 emissions existing in Helen's fleet has been examined (Tsupari et al., 2017). Naveen and Manavalan (2015) made a comparison between CO_2 transcritical refrigeration cycle and CO_2 transcritical combined power generation system. The combined power and refrigeration cycle have resulted in high COP. In a combined power and cooling cycle operating at a range of 25–55 bar of pressure, the exergy efficiency is higher than energy efficiency (Karaalli, 2017). The critical review is made in a combined power and cooling systems (Liu et al., 2014). In a combined cycle power plant, an exergetic analysis is carried out and presented (Calise et al., 2016). A surprising change in the exergetic execution of the systems is reported. The use of warmth released from the Al-Hamra gas turbine plant has been examined (Mohan et al., 2014). A trigeneration and combined cycle has been compared in exergetic analysis. A $LiBr–H_2O$ mixture-integrated power and cooling plant has been investigated (Shankar and Srinivas, 2014). The expansion in heat source temperature results in improving power with reduction in cooling as per the outcomes demonstrated. The performance assessment of a microgas turbine, source, and final temperatures of the exhaust on the combined power and cooling system has been reported (Ebrahimi and Soleimanpour, 2017). A combined cooling, heating, and power system has been analyzed mathematically (Li et al., 2013). By sensitivity analysis, the changes in the performances are investigated. A new system using fluid nitrogen (LN_2) delivered from sustainable power sources, or abundant power at lower demand times, to provide cooling and energy for the residential houses has been presented (Ahmad et al., 2016). The gas expansion station with an absorption chiller has been coupled, considering the temperature drop for cooling production (Arabkoohsar and Andresen, 2018).

2.6 SUMMARY

The precision of thermodynamic evaluation depends on the use of most accurate thermodynamic properties. Developing the properties is difficult for binary mixture as it has five different regions. To solve the thermodynamic model, the property values in all the regions are necessary. For evaluation, the correlations are developed for a range of temperature and pressure. The details about the correlations developed for binary mixtures as working component and about Kalina cycle are presented in the literature survey. The development in the design of new Kalina cycle is made earlier. For the low and high-temperature heat recoveries, most of the design is developed. For medium-temperature heat recoveries, there are no sufficient works and findings. Based on the literature gap, scope for developing a new system suitable for temperature ranging from 170°C to 250°C has been proposed.

To meet the increased power and cooling loads, it is difficult to erect conventional power plants. The alternative routes to solve the problem are decentralized power system, waste heat recovery and cogeneration system, etc. Considerable save in power consumption in a common cogeneration plant is achieved due to generation of without electrical input. A steam power plant cannot be operated at low-temperature heat source. Steam can be generated in a low-temperature and low-pressure-operated LiBr–H_2O-based VAR system. The COP of this system is higher compared to aqua–ammonia mixture. Generating power and cooling with low-temperature heat source is made possible, if the power and cooling configurations are clubbed together. The overall system performance is increased with the integration of power and cooling systems. The NH_3/H_2O and LiBr–H_2O VAR system are externally coupled in the cogeneration and trigeneration plants. For various cogeneration and trigeneration systems, the theoretical and experimental analyses are made for the external heat recovery system. The ammonia–water mixture as working fluid has high boiling point differences and different condensing pressure. This aids the Kalina cycle power plant to run at low-grade heat sources.

Aqua–ammonia working fluid is used in the development of most of the integrated plant concepts. In combined power and cooling systems, LiBr–water has not been applied as working fluid. This is because of the working of cooling system completely under vacuum. The pressure of the ammonia–water mixture is maintained at 4 bar (at 200°C) instead of vacuum; the difficulty in the LiBr–water system has been eliminated. With these reasons, the benefits of high heat content of steam and high COP are gained.

KEYWORDS

- vapor absorption principle
- innovation
- cooling
- heating
- ammonia–water mixture

CHAPTER 3

BASIC THERMODYNAMICS

ABSTRACT

The study and evaluation of any thermal system need basic thermodynamics principles. They provide the primary and essential steps of system planning, namely, thermodynamic modeling, simulation, analysis and optimization. The mass balance, energy balance, and exergy balance follow the laws of thermodynamics. The properties of working fluid such as pressure, temperature, volume, entropy, internal energy, enthalpy, exergy, etc. can be developed from the thermodynamic relations. The discussed contents in this chapter are concepts, laws of thermodynamics, thermodynamic processes, and fundamental relations.

3.1 INTRODUCTION

Thermodynamics is a branch of science that focuses on heat, work, its conversion, and evaluation. It plays a great role in the thermal system development. The main functions of thermodynamics in engineering are all managerial functions such as planning, coordination, etc. Surrounding us, the nature has tremendous amount of energy. The energy always tries to convert one form to another form. Thermodynamics deals about these energy interactions in a systematic way with reference to certain rules called thermodynamic laws. Sadi Carnot developed many theories in thermodynamics and he is called the father of thermodynamics. Thermodynamics deals with energy, entropy, exergy, and equilibrium. Thermo-economics deals with economics in addition to these four "e"s. All the thermodynamics laws, namely, zero, first, second, and third are designed based on logics and without mathematical proof.

In engineering, the solutions can be categorized into two groups: one is rating (finding the capacity) and the second is sizing. The quantification of energy interactions and capacity finding are the part of rating of an engineering system. Finding the dimensions of a system through a system of methodology is called sizing or design. Thermodynamics deals only with the rating and it can support the sizing which is the limitation of thermodynamics. But the data or information required to do the size can be obtained through the thermodynamic evaluation. The use of thermodynamics is vast and used in many applications such as aerodynamics, automobile engineering, power plants, refrigeration and air conditioning, cogeneration, steam engines, internal combustion engines, steam and gas turbines, fuel cells, thermo-electric and thermo-ionic generators, chemical reactions, phase equilibrium, etc. Thermodynamics is similar to the engineering mechanics which is the main origin of engineering design subjects. Similar to engineering mechanics, thermodynamics is the mother of thermal engineering subjects. In the engineering mechanics, force balance and moment equations solve the structure under the equilibrium condition. In thermodynamics, the mass balance and energy balance solves the performance of thermal system.

Thermodynamics deals the steady state problems, that is, without dealing the time. It cannot deal the unsteady or transient systems. Thermodynamics with added rate equations results many other sciences such as heat transfer, chemical kinetics, hydrodynamics, etc. Most of the thermodynamics is the classical thermodynamics which assumes the matter as continuum and applies the macroscopic properties. The study on microscopic level is a statistical thermodynamics used where the number of particles in a system is less such as space applications which is the limitation of classical thermodynamics. Thermodynamics uses mathematics, physics, and chemical concepts to solve the energy interaction in a system.

Energy is taken from the Greek words *en* (interior) and *ergon* (measurement of work). Energy was introduced by Lord Kelvin and Joule in middle of the 19th century. Afterward, Carnot, Benjamin, Mayer, Thomson, and others worked on finding the relation between heat and work. Energy is the motion of molecules above 0 K, as at this temperature there is no movement. Energy exists as potential energy, kinetic energy, pressure energy or external energy, internal energy, electrical energy, chemical energy nuclear energy, etc.

3.2 CONTINUUM

The concept of continuum is used in macroscopic or classical thermodynamics where the matter is distributed through the space and not as particle-wise or local-wise. The opposite meaning of the continuum is concept of vacuum.

The density in general may change from location to location in a fluid and may also vary with respect to time as in a punctured automobile tire. The continuum idealization is not useful for volumes which are too small. For example, in the space study, the continuum approach is not applicable. If the special variations are too great, a limit may not exist and the continuum method would be invalid.

$$\rho = \frac{\lim}{\delta V \to \delta V'} \frac{\delta m}{\delta V} \tag{3.1}$$

The nature of thermodynamics study can be categorized into macroscopic and microscopic study. Classical thermodynamics or macroscopic thermodynamics uses the measurable properties such as pressure, volume, temperature, etc. The statistical thermodynamics or microscopic thermodynamics studies the individual molecules. The macroscopic study is focused on few coordinates which are readily available to study the system. The microscopic study describes complete coordinates of velocity (u, v, and w), momentum and position (x, y, and z). The probability is the key study is statistical thermodynamics. It deals the system structures and time average behavior.

3.3   THERMODYNAMIC SYSTEM

In thermodynamics, the subject to be focused is called thermodynamic system. Thermodynamic system is a prescribed region with finite matter upon which attention is given for study. To understand the concept of system, Figure 3.1 shows the schematic material flow diagram of a typical thermal power plant operated by coal feed. It consists of four thermodynamic systems, namely, pump, boiler, turbine, and condenser. The feedwater is heated with hot flue gas and converted into superheated steam. In the condenser, the expanded steam is condensed by circulating water in the heat exchanger. The systems are handling various fluids such as steam, circulating water, feedwater, and hot gas. Similarly, the systems also involve heat and work transfers. Therefore, thermodynamic system can be described with mass and energy transactions.

To understand the nature of system, it is required to define the terminology used in system and they are surroundings, boundary, control surface, control volume, etc. Surroundings are everything external to the system also called environment. Boundary is the enclosure that separates the system from the surroundings. The boundary is be existed or imagined for the convenience. It may be fixed or moving. Combined total system and surroundings is called the universe. This concept is used to study the changes on both system and surroundings. Thermodynamic systems can be categorized into open system, closed system, and isolated system.

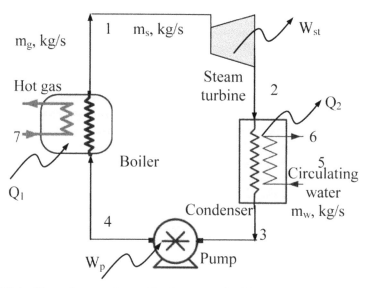

FIGURE 3.1   Thermal power plants with subsystems of turbine, condenser, pump, and boiler.

In open system, the mass and energy cross the boundary. In the open system, the fixed region in space generally called control volume and surface of control volume is called control surface. In open system, the mass and energy crosses the control volume through the control surface. For example in a steam boiler, feedwater enters into the system and leaves superheated steam (mass transfer) by absorbing heat (energy interaction), for example, compressors, turbines, nozzles, diffusers, steam engines, boiler, etc.

In case of a system has mass flow without energy transfer also can be treated as open system. For example, steam is flowing in an insulated pipe. In this case, even though there are no energies crossing the boundary, because of insulated pipe, the fluid carries energy along with the flow which is called kinetic energy. In addition to this kinetic energy, it also possesses flow work. The frictional resistance in the insulated pipe drops the fluid velocity. So within the control volume, the energy exchange occurs without crossing the control surface or boundary. Therefore, in a system with mass flow but without work, heat flow can be treated as open system.

A system is called closed system as it would not allow the matter to enter or leave the energy (heat and work) transfer across its boundary, for example, gas enclosed in a cylinder, water stored in a container, television, etc.

In the closed refrigerator, there are no mass interactions. It receives energy in the form of electricity to operate the refrigerator's compressor. An electric bulb receives energy as electrical energy and releases energy in the form of light and heat. There is mass inflow or outflow from the boundary of bulb. A loudspeaker takes electrical energy and releases sound energy without mass interactions.

In isolated system, neither mass nor energy transfers across its boundary. Examples of isolated system are thermo-flasks and universe. The system and surroundings together is also called universe or isolated system.

3.4 PHASE

The physical structure and chemical composition in a phase is uniform. Water, petrol, steam, oil, etc. are the examples of phase. A liquid mixture of water and ammonia is called aqua–ammonia is in liquid phase which is a zeotropic mixture. A mixture of R134a and R22 is an azeotropic mixture. The basic species can be separated or mixed respectively in heating and cooling. But the azeotropic mixtures will not be separated with heating or cooling. Oxygen and nitrogen are completely mixed in air and both are in gaseous phase. Mixture of alcohol and water is in liquid phase. In this definition, physical uniformity implies that the material is all gas or all liquid or all solid, whereas uniformity of chemical composition never means single chemical species only but it means that chemical composition does not vary from place to place.

On the basis of phases, the system may be classified as homogeneous system and heterogeneous system.

A system consisting of a single phase is called homogeneous system. The examples are ice, water, dry steam, mixture of air and water vapor, water plus nitric acid, solution of ammonia in water, etc.

A system consisting of more than one phase is called a heterogeneous system, for example, mixture of ice and water, vapors in contact with its liquid such as wet steam, solution of immiscible liquids such as water and mercury, etc.

The homogeneous or heterogeneous with single chemical composition is called a pure substance. The pure substance may exist in solid, liquid, or vapor/gas phase or its combination.

The power plant consists of four open systems; the mass (working fluid) is flowing steadily. In the turbine, the superheated steam enters, expands, and generates power. After the expansion, the steam either in wet form or dry form moves into the condenser. In the condenser, heat rejects into the environment hence steam condenses into saturated water. Therefore in the condenser, there is a phase change from vapor to liquid. The pump is a machine type of open system, where the feedwater pressure increases from condenser pressure to boiler pressure by taking power to operate the pump. In the boiler, the energy is transferred in the form of head addition by firing the fuel. The fluid phase changes from liquid to vapor form. The superheated steam enters the turbine and the cycle repeats.

3.5 THERMODYNAMIC EQUILIBRIUM

Thermodynamic equilibrium is related to the internal state of system that there are no net macroscopic flows of mater or energy in the system. At thermodynamic equilibrium, the property is the same at all the points of system. Thermodynamic equilibrium includes mechanical equilibrium, chemical equilibrium, and thermal equilibrium.

A system is said to be in a state of mechanical equilibrium if there exists no imbalanced forces either in the interior of the system or between the system and surroundings. A system is said to be in a state of chemical equilibrium if there exists not chemical reaction or transfer of matter from one part of the system to another. A system is said to be in thermal equilibrium if there exists same temperature throughout the system or between the system and surroundings.

If you leave a cup of coffee in atmosphere, after some time, coffee will reach to the surroundings temperature and gains the equilibrium condition. In a compressor or turbine, the fluid expands from one pressure to other pressure to get the mechanical equilibrium. A chemical reaction occurs in a combustion chamber due to chemically imbalance. Heat will flow from high-temperature body to low-temperature body due to thermally nonequilibrium condition. In hypothetical concept, the process will occurs under thermodynamic equilibrium condition. Really, the process occurs under nonequilibrium conditions. For example, heat will transfer from the hot body to cold body due to temperature difference which is nonequilibrium condition. Theoretically, if it is stated that the process carried out under equilibrium condition, there is no heat transfer between hot fluid and cold fluid and it is infinitesimal small temperature difference. The actual heat transfer process occurs with finite temperature difference. Similarly after compression or expansion, it is reasonable to say that the system reached to mechanical equilibrium condition. After completion or at the beginning of reaction, the system is under chemical equilibrium. Therefore, finally, it can be concluded that the process happens through the change of state via nonequilibrium conditions.

3.6 THERMODYNAMIC PROPERTY

Thermodynamic property of a system is its measurable characteristics describing the system nature. It is independent of the nature of the process and depends only on the state or condition. Therefore, property is a point function. They are classified into two types and they are intensive properties and extensive properties.

If the value of the property does not depend (independent) upon the mass of the system, it is called and intensive property. It is more qualitative in nature, for example, pressure, temperature, density, velocity, height, viscosity, and specific property.

The pressure exerted by a system is defined as the force exerted normal to unit area of the boundary. From the continuum point of view, the pressure at a point is the force per unit area in the limit, where area tends to be very small, that is, approaches to zero.

Pressure as a result of depth of fluid = ρgh

$$P = \frac{w}{A} = \frac{\rho V g}{A} = \rho gh \tag{3.2}$$

Atmospheric pressure is the pressure exerted by the atmospheric air.

P_{atm} = 760 mm of H_g = 101.325 kN/m^2 = 1.01325 bar =1.01325 × 10^5 Pa = 0.101 MPa = 1.113 kg f/cm^2.

The gauge pressure is measured from the gauge and instrument such as Bourdon and manometer is called the gauge pressure.

The absolute pressure measured from the level of absolute zero pressure is called absolute pressure. Absolute zero occurs when the molecular momentum of fluid is zero. Such a situation can only arise if there is a perfect vacuum.

$$P_{abs} = P_{atm} + P_g \tag{3.3}$$

If the pressure of fluid to be measured is less than atmospheric pressure, the gauge will read the negative side of the atmospheric pressure is known as vacuum, rarefaction, or negative pressure.

$$P_{abs} = P_{atm} + P_g = P_{atm} - P_{vac} \qquad (3.4)$$

The pressure measured by instrument without considering the velocity of the fluid. The gauge pressure is the static pressure.

Pressure due to velocity of fluid $= \rho c^2 / 2$ (N/m^2). The kinetic head expressed in force per unit area is called dynamic pressure. KE $= (1/2) mc^2$, J (or N m) and therefore

$$\text{Kinetic pressure} = \frac{1}{2} \frac{mc^2}{V} = \frac{\rho c^2}{2} \ (\text{N/m}^2) \qquad (3.5)$$

Total or stagnation pressure is the sum of static and dynamic pressure.

$$P_{staganation} = P_{static} + P_{dynamic} = P_{static} + \frac{\rho c^2}{2} \qquad (3.6)$$

The average density of a system is the ratio of its total mass to its total volume. The density of a continuum of a particle is given in eq 3.1.

If the value of the property depends upon the mass of the system it is called an extensive property. It is more quantitative in nature, for example, volume, surface area, internal energy, PE, and KE.

3.7 PRIMARY AND SECONDARY THERMODYNAMIC PROPERTIES

There are five basic or primary thermodynamic properties and they are temperature, pressure, volume, entropy, and internal energy.

1. Temperature, T: It is a thermal potential and measure of relative hotness or coldness of a system.
2. Pressure, P: It is a mechanical potential, normal force per unit area.
3. Volume, V: It is the mechanical displacement and the quantity of space possessed by a system.
4. Entropy, S: It is the thermal displacement and the quantity of disorder possessed by a system.

5. Internal energy, U: The kinetic and potential energies of its constituents (atoms and molecules, usually) is the internal energy of a system. The energy associated in a closed system is internal energy.

The secondary properties are two and they are enthalpy and specific heat.

1. Enthalpy, H: The total of internal energy and product of pressure–volume is enthalpy. It is the total energy of a system. The energy associated in an open system is enthalpy. In the open system, the fluid flow with some velocity from inlet to outlet. Some push or work is required for the flow of the working fluid. This energy required for the flow of fluid form the inlet to exit in the open system is called flow work. Therefore in an open system, the energy comprises the internal energy with added flow work.
2. Specific heat capacity, c_p or c_v: Energy needed to raise the temperature of unit mass of system by 1°.

Gases are compressible fluids. So, the specific heats are different in constant pressure process and constant volume process. But the liquids and gases are incompressible fluids. There is no difference between the specific heat at constant pressure and constant volume as the volume is fixed during the processes. The specific heat of solids and liquids are expressed as "c" without refereeing constant pressure or constant volume. The specific heats play an important role in developing the thermodynamic properties of fluid. Without having the information of properties, it is not possible to solve the thermodynamic system. For example, a steam power plant cannot be modeled without the use of steam table or properties. At fundamental level of thermodynamic solution, the specific heats are kept constants. Next level of work, the specific heats can be calculated from the temperature functions developed from the experimental data as an equation fitting. After this level, using the real gas model, the specific heats can be determined from the function of temperature and pressure.

3.8 THERMODYNAMIC STATE, PROCESS, AND CYCLE

The combination of states is called thermodynamic process and the combination of processes with a closed loop is called thermodynamic cycle. A process plant consists many processes with mass and energy interactions. The systems maybe opened or closed for mass transfer. Sometimes, the system work as an open system as well as closed system intermittently. For example, in the fuel firing of an engine system, the inlet valve and out valves remains closed and the system works as a closed system. The same engine during suction and exhaust, the inlet and out valves open, respectively, and open system applies. A complete set of processes which are repetitive in nature can be shown as a thermodynamic cycle in Figure 3.2. Thermodynamic cycle consists states and processes with direction. In a typical thermal power cycle, the cycle is represented in a clock-wise direction. In a refrigeration cycle, the same or different cycle may in anticlock direction. The thermodynamic cycle converts the information from the material flow diagram in to the property information for easy understanding and solution. $P–v$, $T–s$, and $P–h$ diagrams are some of the popular diagrams used to represent the cycle. The thermodynamic state can be defined as a condition defined by the properties. Broadly speaking state is the condition of a system at an instant of time described by its properties. A thermodynamic system passing through a series of states constitutes a path. A process is defined as a transition in which a system changes from one initial state to a final state. In a thermodynamic cycle, the system undergoes through a series of processes from one state to other and finally reaches to the initial state. The properties at the beginning and end of the cycle are the same (Fig. 3.2).

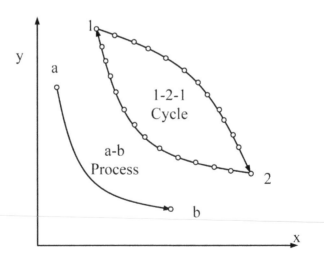

FIGURE 3.2   Thermodynamic process and cycle.

Figure 3.3 shows a typical thermodynamic cycle (Rankine cycle) developed for a thermal power plant with steam a working fluid. Refer Figure 3.1

for the schematic of the Rankine cycle to study the properties at the points. The cycle starts from state 1 and ends with state 1. The diagram shows two thermodynamic properties of temperature and specific entropy respectively extensive property and intensive property. The processes in the cycles are 1–2, 2–3, 3–4, and 4–5–6–1. Some of the processes occur without phase change, that is, 3–4 (pumping in liquid region) and some are with the phase change such as condensation or boiling. The phase change of the fluid is shown from 2–3 (condensation).

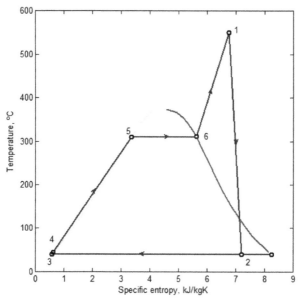

FIGURE 3.3 (See color insert.) Thermodynamic cycle with state, processes, and cycle with reference to Figure 3.1.

The processes can be categorized into three groups as (1) quasi-static process, (2) reversible process, and (3) irreversible process.

3.8.1 QUASI-STATIC PROCESS

Process gives information about initial state and final state only. A quasi-static process is one in which the system deviates from one equilibrium state by only infinitesimal amounts throughout the entire process. In other words, a process closely approximating to a succession of equilibrium states is known as quasi-static process.

3.8.2   REVERSIBLE PROCESS

A reversible process passes through a series of equilibrium states, which can be restored back by reversal of energy interactions. It occurs with infinite slow pace without any dissipative effects such as friction, heat transfer with temperature difference, etc. Practically, all the processes occur with the dissipative effects. A reversible process is a benchmark process to find the degree of deviation from the ideal process. The concept of reversibility is useful to find the effectiveness of a real process so that this comparison is useful to modify or improve the system conditions. Actually, the heat will flow with temperature difference and the rotating elements will move with friction. A reversible process can be shown with a continuous line or curve. The reversible processes are motion of bodies without friction, slow expansion or compression without friction, slow isothermal compression/expansion of a gas, electrolysis of water, and flow of electric current through inductors and capacitors. Reversible process occurs infinitely slow. It is a slow process and closed to the reversible process. In a fast process, more entropy is generated and is an irreversible process.

3.8.3   IRREVERSIBLE PROCESS

An irreversible process passes through nonequilibrium states, which cannot be restored back by reversing the energy interactions between system and surroundings. The complete path of irreversible is not defined due to nonexistence of unique state. Really, all the thermodynamic processes are irreversible. For the sake of simplicity in thermodynamic evaluation and benchmark comparison, the reversible processes are introduced. The human life and age is irreversible as the childhood cannot be obtained back once it is finished. We cannot avoid the friction and potential difference (temperature, pressure, voltage, height, concentration, etc.) in a process. The degree of irreversibility can be estimated by comparing with the reversible process. An irreversible process is expressed by a dotted line. But to gain the clarity in thermodynamic study, irreversible processes also draw in continuous lines or curves but with definition of degree of irreversibility or conditions. The irreversibilities are either internal or external in nature. The internal irreversibilities happen within the system with friction and some dissipative effects. The internal irreversibilities are happens in the viscous flow dissipation, free expansion, throttling, mixing, separation, magnetization, hysteresis, heat generation with current resistance, etc. The external irreversibilities are the heat transfer through a finite temperature differences between hot fluid and cold fluid such as boiler, condenser, etc.

Is a quasi-static process reversible or irreversible? To answer this question, we have to analyze the characteristics of quasi-static process. A single weight is replaced by small weights so that the instantaneous states can be identified. The removing of weights one by one allows the fluid expansion through series of states. The process is not a fictitious process and it is happened in a laboratory setup. Since it is real, we cannot omit the real effects such as friction between piston and cylinder and heat transfer between system and surroundings. Therefore, it is an irreversible process. If a process is assumed completely with neglected dissipative effects and designed completed based on hypothesis without conduction of experiment, then the process becomes a reversible process.

3.9   LAWS OF THERMODYNAMICS

The set of rules and regulation in the energy interactions (heat and work) are framed in laws of thermodynamics. Thermodynamics mainly deals with energy interactions, that is, heat and work. Heat is low-grade energy and work is high-grade energy. The low-grade energy has higher disorder and mostly readily available. Combustion is required to convert the chemical energy into heat for fuels. The work can be produced through many energy conversions and processes and so it is more voluble than heat. Heat can be used to produce the work. But use of work to generate heat is not recommended due to destruction of energy level. Thermodynamic laws are framed to define, solve, and analyze the energy interactions in a systematic way. Generally, there are four thermodynamic laws used to solve the energy interactions in a systematic way. They are zeroth law, first law, second law, and third law of thermodynamics. Apart from these four laws, another law, applied to irreversible thermodynamics is called fourth law of thermodynamics which is in developmental stage.

The zeroth law of thermodynamics defines the property temperature. The first and the second laws are used in the system analysis and solutions quantitatively and qualitatively, respectively. First law defines energy and the second law gives the base for entropy. Third law of thermodynamics gives the idea of 0 K. Thermodynamic laws are more logical without having the mathematical proof. So far, there are no statements to violate the correctness of thermodynamic laws. If it violates the laws it is fictitious and called perpetual motion machine (PMM). The idea of PMM is useful to understand the nature of thermodynamic laws.

Previously, there were no roads and vehicles for the travel. They used animals and traveled through forest and passages. Nowadays, roads and highways are developed which simplifies the travel and journey. Similar to roads, thermodynamics laws help the engineer to understand, solve the properties, model, simulate, analyze, and optimize the system having energy transfers.

3.10   ZEROTH LAW OF THERMODYNAMICS

When two bodies are equilibrium thermally with each other and also with the third body separately, then it can be concluded that the three bodies are thermally equilibrium with each other.

If body 1 and 3 are in thermal equilibrium with 2, then 1 is in thermal equilibrium with 3. Thermal equilibrium can be measured with the temperature property. The three bodies are thermally equilibrium with each other and indicate that all the three bodies are at the same temperature. This concept gives the base for the measurement of temperature. If these three bodies are brought to physical contact, there is no heat transfer among them due to uniform and same temperature. It is so named because it logically precedes the first and second laws of thermodynamics. Zeroth law gives the base for thermometry, art of device of measuring temperature with scientific precision. Zeroth law gives the definition of temperature. This law was framed by Ralph Fowler and Guggenheim. Fowler was a British physicist and astronomer. He named zeroth law of thermodynamics in 1920.

Temperature is the only property which differentiates thermodynamics from other sciences. The first law describes the energy conversion phenomena and its solutions. Second law indicates the probability of happening of a cycle or process and indicates the direction of energy flow. Further, second law distinguishes the quality of energy. The third law gives the interpretation of a system while marching toward absolute zero temperature. Application of both the first and second law to a process gives rise to some properties (internal energy and enthalpy from first law, entropy from the second law).

Thermodynamic law conveys a great philosophy to understand the nature of life. It also helps to understand the thermodynamics. According to zeroth law, thermodynamics is applied to human life; if somebody helps others, the person will save in future from others as the receiver also changes to helping mode. That means that the helping nature is to be habituated to help each other.

3.11 FIRST LAW OF THERMODYNAMICS

Thermodynamics obeys certain rules and regulations, called thermodynamic laws. These laws provide some guidelines to solve, analyze, and refine the systems. The first law of thermodynamics deals with the energy interactions in a system and gives the base for energy balance. The performance can be solved using first law of thermodynamics. In our daily life, we will take the food regularly and performs the works. If the daily exercises and activities are balanced with the food eaten, it results healthy increase in strength (energy). If the activities are not enough, and the person is lazy, the food will convert into fat due to insufficient work.

3.12 FIRST LAW OF THERMODYNAMICS APPLIED TO CYCLE— JOULE'S EXPERIMENT

First law of thermodynamics focuses on energy. It says that energy can be converted from one from to another. It cannot be destroyed or created, just converted only. So, the energy in universe is constant.

The first law of thermodynamics provides the base for measurement of energy through the properties. But it has certain limitations as it is not complete. It is focused on transfer of energy quantities only; it is not studied based on quality of transformation. The rules and regulations in the energy conversion direction are neglected. The description of the process is not complete as it can explain the feasibility or not.

Since the first law explains the energy conversion, it defines the energy conversion efficiency or first law efficiency of a process or a cycle. It is also called thermal efficiency which is the ratio of output to input.

$$\eta = \frac{\text{Output}}{\text{Input}} \tag{3.7}$$

The net output maybe work.

Joule conducted an experiment to demonstrate the first law of thermodynamics applicable to a cycle. Let a certain amount of work W_{1-2} be done upon the system by the paddle wheel. The quantity of work can be measured by the product of weight and the vertical height through which the weight descends. The work input to the insulated vessel causes a rise in the temperature of the fluid. For the process 1–2,

$$Q + U_1 = W + U_2 \tag{3.8}$$

He conducted the experiment in a water container with insulation. The temperature of water is increased by the stirrer work. The work done on the system, $U_1 + W = U_2$. The temperature of water increases from T_1 to T_2. The work is used to raise the internal energy of the system, that is, $U_2 - U_1$.

In second step, the insulation is removed and the water bath is inserted into another water container. The heat is transfer to the water and the system gains its original state.

For the process 2–1,

$$U_2 + Q = U_1 \tag{3.9}$$

Since there is no work, the heat rejection is equal to decrease in internal energy, that is, $U_2 - U_1$. Now, the system undergoes two process, one is forward and the second is backward and completes a cycle. In the forward process, the water temperature is raised and the energy quantity can be determined from the specific heat of water, fluid mass, and temperature rise. Joule conducted many such experiments involving different work interaction in a variety of systems. For example, Joule used work in the form of electrical energy, after measuring the heat transfer by the heating coil, completing the cycle by restoring the initial state.

Joule identified that in all the experiments he conducted; same amount of work is spent to produce the same amount of heat. In all the cases, 4186 J of work is used to raise the temperature of 1 kg of water through 1°C rise. He proved that the ratio of W/Q is constant.

It has been found that W_{1-2} is proportional to Q_{1-2}

$$\oint dW \alpha \oint dQ \tag{3.10}$$

$$\oint dW = J \oint dQ \tag{3.11}$$

In SI units, $J = 1$ N m/J and in MKS units, $J = 427$ kg fm/kcal or it is equal to 4.18 kJ/kcal.

Therefore, first law of thermodynamics says "when a system executes a cyclic process, the algebraic sum of work transfers is proportional to algebraic sum of heat transfers."

For SI units,

$$\oint dQ - \oint dW = 0, \quad \Rightarrow \oint (dQ - dW) = 0 \tag{3.12}$$

The philosophy says that nothing is free in nature. The result will come with the efforts only. Without putting efforts we cannot achieve in life. Humans will expect many things in their lives. But they would not come with free of efforts. With the life struggle and hard work only, the requirements can be fulfilled.

3.13 FIRST LAW OF CLOSED SYSTEM UNDERGOING A CHANGE OF STATE (PROCESS)

When a system undergoing a change of state, the net energy transfer will be stored or accumulated within the system. This stored energy is called internal energy.

If Q is heat supplied and W is the work done by the system,

$$Q - W = \Delta E \tag{3.13}$$

$$Q + E_1 = E_2 + W \tag{3.14}$$

3.14 INTERNAL ENERGY—A PROPERTY OF SYSTEM

Figure 3.4 shows a thermodynamic cycle consisting of processes A and B or A and C.

According to first law,

$$\oint (dQ - dW) = 0 \tag{3.15}$$

For cycle, 1–A–2–B–1

$$\int_1^2 (dQ - dW)_A + \int_2^1 (dQ - dW)_C = 0 \tag{3.16}$$

For cycle 1–A–2–C–1,

$$\int_1^2 (dQ - dW)_A + \int_2^1 (dQ - dW)_C = 0 \tag{3.17}$$

$$\therefore \int_2^1 (dQ - dW)_B = \int_2^1 (dQ - dW)_C = \int_2^1 dE = E_1 - E_2 \tag{3.18}$$

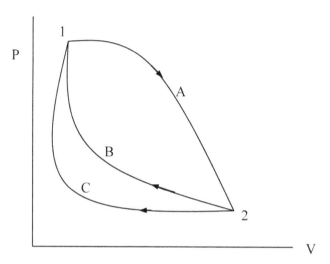

FIGURE 3.4   Processes A, B, and C.

Therefore, internal energy is independent of path and it is point function, so it is called property.

3.15   DIFFERENT FORMS OF STORED ENERGY (INTERNAL ENERGY)

There are two forms of internal energy and they are as follows:

1. *Macroscopic energy*

    i) $KE = (1/2)mC^2 = E_K$ (3.19)

    ii) $PE = mgZ = E_P$ (3.20)

2. *Microscopic energy, molecular energy (U)*

    Energy of one molecule, $\varepsilon = \varepsilon_{trans} + \varepsilon_{rota} + \varepsilon_{vibra} + \varepsilon_{electronic} + \varepsilon_{nuclear}$ (3.21)

    $U = N\varepsilon,$ $N =$ number of molecules

In an ideal gas, there is no intermolecular forces of attraction and repulsion and molecular internal energy depends only on temperature, $U = f(T)$ only.

$$E = (E_K + E_P) + U = (\text{macro}) + \text{micro}$$ (3.22)

In the absence of motion and gravity, $E_K = E_p = 0$

$$E = U \rightarrow Q = \Delta U + W \qquad (3.23)$$

Change in internal energy of an isolated system is zero because there is no heat and work transfer.

3.16 SPECIFIC HEAT AT CONSTANT VOLUME

Specific heat is the thermal property of substance. For gases, the specific heat at constant volume and specific heat at constant pressure can be defined from the first law of thermodynamics.

Specific heat at constant volume is the amount of heat transferred to raise the temperature of gas through 1°C to a unit quantity in a constant volume process.

$$c_v = \left(\frac{dq}{dT} \right)_v \qquad (3.24)$$

while the volume is held constant. For a constant volume process, first law of thermodynamics gives

$$dU = dQ \qquad \text{or} \qquad du = dq \qquad (3.25)$$

$$\therefore c_v = \left(\frac{du}{dT} \right)_v \qquad (3.26)$$

where u is the specific internal energy of the system. If c_v varies with temperature, one can use mean specific heat at constant volume.

$$\bar{c}_v = \frac{\int_{T_1}^{T_2} c_v \, dT}{T_2 - T_1} \qquad (3.27)$$

The total quantity of energy transferred during a constant volume process when the system temperature changes from T_1 to T_2

$$Q = m \int_{T_1}^{T_2} c_v \, dT = U_2 - U_1 \qquad (3.28)$$

The unit for c_v is kJ/kg K. The unit of molar specific heat is kJ/kmol K.

3.17   SPECIFIC HEAT AT CONSTANT PRESSURE

Specific heat at constant pressure is defined as amount of heat required to raise the temperature of gas through 1°C to a unit mass at constant pressure process.

$$c_p = \left(\frac{dq}{dT} \right)_p \tag{3.29}$$

while the pressure is held constant. For a constant pressure process, first law of thermodynamics gives

$$Q + U_1 = W + U_2 \tag{3.30}$$

$$\text{where } W = P(V_2 - V_1) \tag{3.31}$$

therefore,

$$Q = P(V_2 - V_1) + U_2 - U_1 = U_2 + P_2 V_2 - (U_1 + P_1 V_1) = H_2 - H_1 \tag{3.32}$$

$$\text{since } H = U + PV$$

The specific heat at constant pressure, c_p, is defined as the rate of change of enthalpy with respect to temperature when the pressure is held constant.

$$dH = dQ \qquad \text{or} \qquad dh = dq \tag{3.33}$$

$$\therefore c_p = \left(\frac{dh}{dT} \right)_p \tag{3.34}$$

where h is the specific enthalpy of the system. If c_p varies with temperature, one can use mean specific heat at constant pressure.

$$\bar{c}_p = \frac{\int_{T_1}^{T_2} c_p dT}{T_2 - T_1} \tag{3.35}$$

The total quantity of energy transferred during a constant pressure process when the system temperature changes from T_1 to T_2

$$Q = m \int_{T_1}^{T_2} c_p dT = H_2 - H_1 \tag{3.36}$$

The unit for c_p is kJ/kg K. The unit of molar specific heat is kJ/kmol K.

3.18   FIRST LAW APPLIED TO A POLYTROPIC PROCESS, CLOSED SYSTEM

Polytropic process gives the flexibility in the representation of thermodynamic process by the generalized thermodynamic formulations. The polytropic index plays an important role in this expression as it changes from 0 to ∞, facilitates any thermodynamic process in compression region and expansion region. This section relates the heat and work with polytropic index to understand the energy interactions in the process.

We have that the work in polytropic process,

$$W_{1-2} = \frac{P_1V_1 - P_2V_2}{n-1} = \frac{mR(T_1 - T_2)}{n-1}$$

From first law of thermodynamics,

$$Q = W + U_2 - U_1$$

After simplification,

$$Q = \frac{mR(T_1 - T_2)}{(n-1)}\left(\frac{\gamma - n}{\gamma - 1}\right) \tag{3.37}$$

$$Q = W\left(\frac{\gamma - n}{\gamma - 1}\right) \tag{3.38}$$

$$\Delta U = Q - W = W\left(\frac{\gamma - n}{\gamma - 1}\right) - W = W\left(\left(\frac{\gamma - n}{\gamma - 1}\right) - 1\right)$$

$$= W\left(\frac{\gamma - n - \gamma + 1}{\gamma - 1}\right)$$

$$\Delta U = W\left(\frac{1 - n}{\gamma - 1}\right) \tag{3.39}$$

i)   If $n = 0$ (constant pressure process), $Q = W(\gamma/(\gamma - 1))$, since $\gamma > \gamma - 1$, $Q > W$. The heat transfer is more than the work and the increment is equal to the change in internal energy.

ii) If $n = 1$, $Q = W$, it is possible with constant internal energy, that is, isothermal process.

iii) If $n = \gamma$, $Q = 0$, that is, it is an adiabatic process.

iv) At $n = \infty$ (constant volume process), since $W = 0$ and the second term is equal to infinity and $Q \neq 0$.

3.19   POLYTROPIC PROCESS IN OPEN SYSTEM

We know that,

$$h = u + Pv \tag{3.40}$$

$$dh = du + Pdv + vdP \tag{3.41}$$

From the first law,

$$dq = du + Pdv \tag{3.42}$$

$$dh = dq + vdP \tag{3.43}$$

Since the area under Ts diagram is heat similar to work which is the area under Pv diagram,

$$dh = Tds + vdP \tag{3.44}$$

$dq = Tds = dh - vdP$, it is the first law of thermodynamics applied to open system.

In place of W, we can find $-vdP$.

Therefore, in an open system, $\dot{w} = \int -vdP$ \hfill (3.45)

So the work in polytropic process,

$$w_{1-2} = \int_{1}^{2} -v\,dP$$

$$= -\int_{1}^{2} \left(\frac{C}{P}\right)^{1/n} dP$$

$$\because Pv^n = C$$

$$= -C^{1/n}\left(\frac{P^{-(1/n)+1}}{-(1/n)+1}\right)\Bigg|_{1}^{2}$$

$$= -C^{1/n}\left(\frac{P_2^{(n-1)/n} - P_1^{(n-1)/n}}{(n-1)/n}\right)$$

$$= \left(\frac{-\left(P_2 v_2^n\right)^{1/n} P_2^{(n-1)/n} + \left(P_1 v_1^n\right)^{1/n} P_1^{(n-1)/n}}{(n-1)/n}\right)$$

$$= \frac{n\left(P_1 v_1 - P_2 v_2\right)}{n-1}$$

$$W_{\text{open system}} = nW_{\text{closed system}}$$

From this equation, it is possible to compare the work in a closed system and open system at the same working conditions. Between these two systems is the fluid flow, that is,

$$\frac{W_{\text{open system}}}{W_{\text{closed system}}} = n \qquad (3.46)$$

If $n < 1$, $W_{\text{open system}} > W_{\text{closed system}}$

$\quad n = 1$, $W_{\text{open system}} = W_{\text{closed system}}$

$\quad n > 1$, $W_{\text{open system}} < W_{\text{closed system}}$

3.20 FIRST LAW APPLIED TO FLOW PROCESS

In flow processes, attention is focused upon a certain fixed region in space called a control volume. Steady flow process means that the rates of flow of mass and energy across the control surface are constant.

Mass balance, $\dot{m}_1 = \dot{m}_2$

$$\frac{A_1 C_{1'}}{v_1} = \frac{A_2 C_{2'}}{v_2} \quad \text{or} \quad \rho_1 A_1 C_1 = \rho_2 A_2 C_2 - \text{continuity equation} \qquad (3.47)$$

Energy balance: Energy inflow = Energy outflow + Accumulation

In flow process, the work transfer is two types:

1. External work, W
2. Flow work, Pv

$$\dot{m}_1 e_1 + \dot{m}_1 P_1 v_1 + Q = \dot{m}_2 e_2 + \dot{m}_2 P_2 v_2 + W \qquad (3.48)$$

$$e = \text{energy per unit mass} = e_K + \dot{e}_P + u = \frac{C^2}{2} + gZ + u \qquad (3.49)$$

Substituting e

$$\dot{m}_1 \left(\frac{C_1^2}{2} + gZ_1 + u_1 \right) + \dot{m}_1 P_1 v_1 + Q = \dot{m}_2 \left(\frac{C_2^2}{2} + gZ_2 + u_2 \right) + \dot{m}_2 P_2 v_2 + W$$

Since $h = u + Pv$

$$\dot{m}_1\left(h_1 + \frac{C_1^2}{2} + gZ_1\right) + Q = \dot{m}_2\left(h_2 + \frac{C_2^2}{2} + gZ_2\right) + W, \tag{3.50}$$

The above equation is called steady flow energy equation (SFEE).
For SFEE, per unit mass, $\dot{m}_1 = \dot{m}_2$

$$\left(h_1 + \frac{C_1^2}{2} + gZ_1\right) + \frac{Q}{\dot{m}} = \left(h_2 + \frac{C_2^2}{2} + gZ_2\right) + \frac{W}{\dot{m}} \tag{3.51}$$

3.21 SOME EXAMPLES OF STEADY FLOW PROCESS

1. *Nozzles and diffuser*

$$\dot{m}\left(h_1 + \frac{C_1^2}{2} + gZ_1\right) + Q = \dot{m}\left(h_2 + \frac{C_2^2}{2} + gZ_2\right) + W$$

$Z_1 = Z_2$ and C_1 can be neglected.

Since adiabatic, $Q = 0$, and also $W = 0$

$$h_1 = h_2 + \frac{C_2^2}{2}$$

$$C_2 = \sqrt{2(h_1 - h_2)} \tag{3.52}$$

2. *Throttling device*
    The flow through an insulated passage with pressure drop is called throttling. If change in PE is neglected, $Q = W = 0$

$$h_1 + \frac{C_1^2}{2} = h_2 + \frac{C_2^2}{2}$$

The pipe velocities in throttling are so slow; therefore, KE terms can be neglected.

$$h_1 = h_2 \tag{3.53}$$

3. *Turbine and compressors*
    For turbine, neglecting PE and KE

$$\dot{m}h_1 = \dot{m}h_2 + W$$

$$W \doteq m(h_1 - h_2) \tag{3.54}$$

For compressor

$$\dot{m}h_1 + W = \dot{m}h_2$$

$$W \doteq m(h_2 - h_1) \tag{3.55}$$

4. *Heat exchanger*
   For a typical steam condenser,
   Neglecting PE and KE, $Q = 0$ (to surroundings)
   Energy balance

$$m_s h_1 + m_w h_3 = m_s h_2 + m_w h_4$$

$$\text{Heat transfer} = m_s(h_1 - h_2) = m_w(h_4 - h_3) \tag{3.56}$$

Figure 3.1 shows the schematic flow diagram of a thermal power plant, where heat is supplied to the boiler by burning the coal and the rejection of heat at the condenser. The chemical energy of coal is released into thermal energy in the furnace of the boiler. The feedwater converts into superheated steam by absorbing the heat from the hot flue gas. The superheated steam expands in steam turbine and generates power. The expanded steam is condensed completely into water by rejecting heat to the circulating water. This condensed water is pumped back to boiler to complete the cycle.

Now, let us apply the SFEE to the subsystems of the thermal power plant to find the unknown capacity or flow rates. Consider m_s and m_w are the steam and circulating water flow rates in the plant and condenser, respectively. Assume there are no heat losses in the equipment other than the heat transfer between hot fluid and cold fluid. Similarly, the PE and KE values are also neglected compared to the magnitude of enthalpies.

Applying the SFEE to the steam turbine,

$$\dot{m}_1\left(h_1 + \frac{C_1^2}{2} + gZ_1\right) + Q = \dot{m}_2\left(h_2 + \frac{C_2^2}{2} + gZ_2\right) + W$$

Neglecting PE, KE, and heat losses,

$$\dot{m}_s h_1 = \dot{m}_s h_2 + W$$

Therefore, the power generation from the turbine,

$$W \doteq m_s(h_1 - h_2) \tag{3.57}$$

Similarly, the SFEE can be applied to water-cooled condenser to find the circulating water at know steam flow rate,

The PE, KE, and heat losses are neglected. There is no work from the condenser as it is a heat-exchanging device.

By the energy balance, the heat loss by hot fluid is equal to the heat gain by the cold fluid,

$$Q_2 \doteq m_s \left(h_2 - h_3 \right) = \dot{m}_w \left(h_6 - h_5 \right) \tag{3.58}$$

Or, the total energy at the inlet is equal to the total energy at the exit at the condition of no accumulation of energy.

$$\dot{m}_s h_2 + \dot{m}_w h_5 = \dot{m}_s h_3 + \dot{m}_w h_6$$

$$\dot{m}_w = \frac{\dot{m}_s \left(h_2 - h_3 \right)}{\left(h_6 - h_5 \right)} = \frac{\dot{m}_s \left(h_2 - h_3 \right)}{c_{pw} \left(T_6 - T_5 \right)} \tag{3.59}$$

Therefore, the energy balance or heat balance of condenser results the circulating water demand to condenser the steam into complete water.

Now the SFEE is applied to pump,

$$m_3 h_3 + \dot{W}_p = m_4 h_4$$

$$\dot{W}_p = m_s \left(h_4 - h_3 \right) \tag{3.60}$$

In the above equation h_4, enthalpy at the end of the pump is unknown and can be determined from the first law of thermodynamics.

$\dot{W}_p = -\int v \, dP = v(P_4 - P_3) = v(P_1 - P_2) = m_s (h_4 - h_3)$, assuming the specific volume of water as constant.

$$h_4 = h_3 + \frac{v \left(P_1 - P_2 \right)}{m_s} \tag{3.61}$$

Finally after finding the exit enthalpy, the pump capacity can be determined. Applying the SFEE to the last component, that is, boiler,

Similar to the energy balance in condenser,

$$Q_1 \doteq m_g \left(h_7 - h_8 \right) = \dot{m}_s \left(h_1 - h_4 \right) \tag{3.62}$$

Or, the total energy at the inlet is equal to the total energy at the exit at the condition of no accumulation of energy.

$$\dot{m}_g = \frac{\dot{m}_s\left(h_1 - h_4\right)}{\left(h_7 - h_8\right)} = \frac{\dot{m}_s\left(h_1 - h_4\right)}{c_{pg}\left(T_7 - T_8\right)} \tag{3.63}$$

Therefore, the energy balance or heat balance of boiler results the hot gas required or coal feed in case of combustion to generate m_s flow of steam.

Finally, the first law is applied to the cycle shows that

$$Q_1 - Q_2 = W_{st} - W_p \tag{3.64}$$

3.22   SECOND LAW OF THERMODYNAMICS

The first law of thermodynamics is essential for the fundamental energy study and performance evaluation. It describes the energy conversions but with few limitations. To overcome these difficulties and get the completeness of energy conversion study, one more dimension to the thermodynamics is required, which is entropy. The entropy (randomness) of system with primary dimensions of pressure, volume, and temperature defines the state completely. The second law of thermodynamics adds more data to energy conversion behavior and makes a significant study. It results a proper direction to process with the feasibility information. The first law of thermodynamics results energy analysis and the second law is the base for exergy (maximum potential) analysis. Currently, many researchers are focusing second law analysis to refine the system through minimization of entropy generation. Actually, the first law together with second law completes the thermodynamic study.

3.23   LIMITATIONS OF FIRST LAW OF THERMODYNAMICS

The occurrence of a spontaneous process is due to the finite potential difference. For example, water flows from higher altitude to lower altitude, mass with concentration difference, heat with temperature difference, etc. The reverse of these spontaneous process never happens unless the supply of external agency. The second law limits the direction and controls the occurrence of process.

The first law of thermodynamic does not address the following:

1.   The direction of heat flow with temperature.

2. The limitations in energy conversion are not described (practically, it is not possible to convert all the heat into work).
3. The feasibility of the process is not cleared.

The effectiveness of process is not described in first law of thermodynamics, which has significant importance in the design of thermal system. The imperfection of thermodynamic process can be completely scanned using second law of thermodynamics which is not possible with first law. Therefore, second law can be used in the refinement of the thermodynamic system rather just a focus on study.

3.24   QUALITATIVE MEASUREMENT OF ENERGY

In the mathematics, if $x = y$, then $y = x$. The second law of thermodynamics deals the quality of energy and it states that even though $x = y$, $y \neq x$ in a different way. As per the second law of thermodynamics, energy can be grouped into high-grade energy and low-grade energy. High-grade energy is regular and orderly formed energy. In the low-grade energy, more randomness can be observed, that is, no orderly formed energy. Quality of energy can be ascertained by applying the second law of thermodynamics to a process or a system.

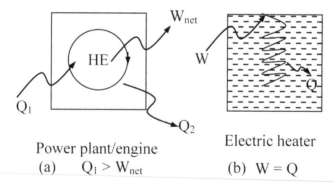

FIGURE 3.5 Qualitative difference between heat and work: (a) heat engine (HE) and (b) electrical heater.

Through a cyclic process or in a thermodynamic cycle, all the heat addition cannot be converted into equal amount of work, that is, it is not possible to convert the low-grade energy into high-grade energy completely. Compulsory, some amount of heat must be rejected to the surroundings. Therefore, work is said to be high-grade energy and heat as low-grade energy. The complete conversion of low-grade energy into high-grade energy in a cycle is impossible. For example, heat is supplied to a thermal power plant by burning coal in a furnace (Fig. 3.5a). This heat is used to generate the steam and the turbine coverts the heat energy into work by expanding the steam. But it will not convert completely into work as most of the heat is rejected in the condenser and exhaust gas through chimney. The heat rejection from the plant or the engine is an unavoidable phenomenon. On the other hand, if water in an insulated container is heated by an electrical heater, the electrical energy completely converts into heat energy assuming minor electrical losses (Fig. 3.5b). Since electricity is high-grade energy, the complete conversion into low-grade energy is possible. It can be concluded that for low-grade applications, use of low-grade energy is better than the high grade as it is more expensive. For example, to produce hot water, direct fuel firing or solar thermal is better than use of electricity.

3.25   CYCLIC HEAT ENGINE

Cyclic heat engine executes a series of repeated thermodynamic processes. It involves heat transfers and work interactions. There is a link between the net heat transfer and net work in a cyclic heat engine.

For example, in a cyclic heat engine Q_1 is the heat addition and Q_2 is the heat rejection, the net heat transfer is

$$Q_{net} = Q_1 - Q_2 \qquad (3.65)$$

Similarly W_T is the turbine expansion positive work and W_p is the pump negative work.

$$W_{net} = W_T - W_P \qquad (3.66)$$

$$\sum_{cycle} Q = \sum_{cycle} W = Q_1 - Q_2 = W_T - W_P \qquad (3.67)$$

$$\eta_{cycle} = \frac{W_{net}}{\text{Heat supplied}} = \frac{W_T - W_P}{Q_1} = \frac{Q_1 - Q_2}{Q_1} = 1 - \frac{Q_2}{Q_1} \qquad (3.68)$$

3.26   ENERGY RESERVOIRS

While solving thermodynamic problems particularly for cycles involves pump, engine, turbine, heater, cooler, etc. the problem is expressed in the symbolic representation. The main resource and sink are expressed in a rectangle symbol. The machines such as pump, prime mover, refrigerator and engines are represented by a circle with direction. The engine cycle is in a clock-wise direction and the refrigerator or heat pump is in anticlock-wise direction.

The energy reservoirs can be categorized into two and they are

1. thermal energy reservoir (TER)
2. mechanical energy reservoir (MER)

TER is an imaginary body having infinite heat capacity that can absorb or reject the heat without changing the thermodynamic properties. TER may be a source or sink. Boiler furnace, combustion chamber, nuclear reactor, sun, etc. maybe treated as heat source while atmospheric air, river, and ocean water maybe treated as heat sink. The temperature of source or sink is assumed to be constant in the study. For example, by drawing a glass of water from ocean its water level will not decrease; similarly, the by rejecting heat from a refrigerator condenser, the atmospheric temperature would not change.

A MER is also an imaginary body or machine that receives or generates the work steadily at uniform rate. It is assumed that the delivery or supply of work to the thermal machine is continuous and fixed.

3.27   KELVIN–PLANK STATEMENT OF SECOND LAW

Kelvin and Max Plank are the two scientists who defined the second law of thermodynamics based on heat-engine reference. The Kelvin–Plank (KP) theory states that it is impossible to construct a heat engine that executes heat with single TER.

The philosophy says that if a person is earning money for his or her life, he/she has to spend some portion toward society after his savings and expenses.

If an engine working with single TER, that engine is called PMM of second kind (PMM 2). Therefore, PMM 2 is impossible.

Figure 3.6a shows the impossibility of heat engine construction with single reservoir and Figure 3.6b is the possibility by adding the heat sink to the engine.

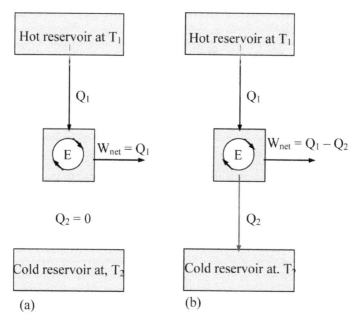

FIGURE 3.6   (a) PMM 2, impossible and (b) possible.

From Figure 3.6b,

$$Q_1 = W + Q_2 \tag{3.69}$$

$$\eta_{cycle} = \frac{W}{Q_1} = \frac{Q_1 - Q_2}{Q_1} \tag{3.70}$$

$$\eta_{cycle} = 1 - \frac{Q_2}{Q_1} \rightarrow \eta < 1 \tag{3.71}$$

If $Q_2 = 0$, $\eta = 100\%$. It is impossible.

From the efficiency relation and Carnot theorem,

$$\eta_{cycle} = 1 - \frac{Q_2}{Q_1} = 1 - \frac{T_2}{T_1}$$

$$\frac{Q_1}{T_1} - \frac{Q_2}{T_2} = 0$$

In a cycle,

$$\sum \frac{Q}{T} = 0$$

It is a property called entropy.

3.28 CLAUSIUS STATEMENT OF SECOND LAW

Clausius conceptualized the second law of thermodynamics based on the heat pump working. As per the Clausius statement, it is impossible to construct a heat pump or refrigerator that removes the heat from a body at lower temperature to a body at high temperature without using work.

The philosophy says that to convert from ordinary to extraordinary, some extra efforts are required and without these additional efforts attaining that stage is impossible.

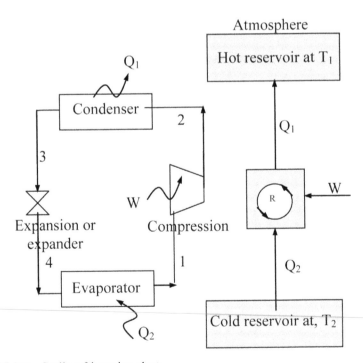

FIGURE 3.7 Cyclic refrigerating plant.

Water always flows from a higher level to lower level naturally. A pump is required to draw the water from a lower level to higher level. Similarly to draw the heat from the lower temperature to higher temperature, a pump (refrigerator or heat pump) is required.

A heat pump or refrigerator is a device working on reversed thermodynamic cycle which removes heat from a low-temperature body and delivers it to a high-temperature body taking external work. A refrigerator is a device which operating in a cycle maintains a body at a temperature lowers than the temperature of surroundings (Fig. 3.7).

$$\oint Q = \oint W$$

$$Q_2 - Q_1 = W_E - W_C$$

$$Q_1 - Q_2 = W_C - W_E \tag{3.72}$$

$$\text{Coefficient of performance, } COP_R = \frac{\text{desired effect}}{\text{Network input}} \tag{3.73}$$

$$\frac{Q_2}{W_C - W_E} = \frac{Q_2}{Q_1 - Q_2} \tag{3.74}$$

Heat pump is a device which operating in a cycle maintains a body at a temperature higher than the temperature of the surroundings (Fig. 3.8).

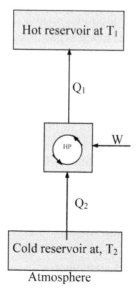

FIGURE 3.8 Cyclic heat pump.

$$COP_{HP} = \frac{Q_1}{W_{net}} = \frac{Q_1}{Q_1 - Q_2}$$

$$COP_{HP} - COP_R = \frac{Q_1}{Q_1 - Q_2} - \frac{Q_2}{Q_1 - Q_2} = 1 \qquad (3.75)$$

$$COP_{HP} = COP_R + 1 \qquad (3.76)$$

$$Q_1 = COP_{HP} W_{net} = (COP_R + 1)W_{net} \qquad (3.77)$$

In heat pump, Q_1 will always be more than W_{net}. Therefore, heat pump provides a thermodynamic advantage over direct heating since high-grade energy can be completely converted into low-grade energy.

3.29 EQUIVALENCE OF KELVIN–PLANK AND CLAUSIUS STATEMENT

The KP's statement demonstrates a heat engine and Clausius's statement defines a refrigerator or a heat pump. These two applications differ but the statements are equal to each other. These two statements are same. To prove this equivalence, it can be shown that the violation of Clausius statement leads to violation of KP statement and similarly the violation of KP leads to violation of Clausius. In the earlier discussion, it has been stated that even $x = y$, $y \neq x$ in the focus of second law of thermodynamics. Now in this section, we are proving that if $x \neq y$, then $y = x$. How it is possible? The Clausius statement and KP statement, respectively, results different applications such as heat pump and heat engine. The refrigerator and heat engine are different areas. But these two statements, that is, Clausius and KP state second law of thermodynamics only.

The reservoir at T_1 can be eliminated so that the heat rejection of heat pump is supplied to the heat engine to produce the work and followed by a heat rejection. These two systems together form a heat engine but with a single heat reservoir. So, the violation of Clausius statement leads to violation of KP statement.

The work generated by the heat engine is supplied to heat pump to draw heat from low temperature to a high temperature. The combined engine and pump together form a heat pump which works without use of external work. It violates the Clausius statement. Therefore, violation of KP statement results in violation of Clausius statement.

On the other hand, Clausius statement will not violate the KP statement and vice-versa. If Clausius statement is not violated, the heat pump and engine together form a new heat engine which converts the work into heat and work with one MER and one TER. Similarly, not violating the KP statement does not violate the Clausius statement.

3.30 CONDITIONS FOR REVERSIBILITY

To study a reversible cycle, all the process involved in it is to be assumed as reversible process. The reversibility is a hypothetical concept used to compare the actual process or cycle with the benchmark reversible process or cycle. In the actual process or irreversible process, heat transfers from the high-temperature body to low-temperature body with finite temperature difference. In the machines, the process occurs with friction causes dissipation of energy. Really, it is impossible to decrease these dissipative effects to zero level. But by properly designing a system, it is possible to minimize these effects.

In the reversible process,

1. Mechanical, thermal, and chemical equilibrium must be satisfied.
2. There should be no dissipative effects.

3.31 CARNOT CYCLE

A benchmark heat engine or a heat pump is required to estimate the maximum amount of gain from these thermal machines. Carnot machine is such an imaginary or hypothetical machine shown as a master product. Carnot cycle is a reversible cycle consisting of four processes as shown in Figure 3.9.

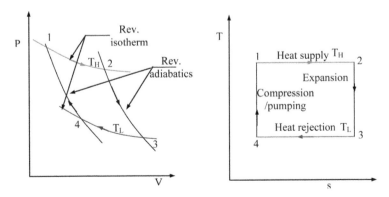

FIGURE 3.9 **(See color insert.)** Carnot engine (a) P–V diagram and (b) T–s diagram.

1. *Reversible isothermal process (1–2)*
   Heat is added at constant temperature process

   $$Q_1 = (U_2 - U_1) + W_{1-2} = W_{1-2} \text{ (since } U_1 = U_2) \tag{3.78}$$

   $$Q_1 = W_{1-2} = mRT_H \ln\left(\frac{V_2}{V_1}\right) \tag{3.79}$$

2. *Isentropic expansion process (2–3)*
   $$0 = (U_3 - U_2) + W_{2-3}$$

   $$-\int_{T_2}^{T_3} mc_v \, dT = \int_{V_2}^{V_3} P \, dV$$

   $$-\int_{T_2}^{T_3} mc_v \, dT = \int_{V_2}^{V_3} \frac{mRT}{V} dV$$

   $$-mc_v \int_{T_2}^{T_3} \frac{dT}{T} = mR \int_{V_2}^{V_3} \frac{dV}{V}$$

   $$-mc_v \ln\left(\frac{T_3}{T_2}\right) = mR \ln\left(\frac{V_3}{V_2}\right)$$

   $$-mc_v \ln\left(\frac{T_L}{T_H}\right) = mc_v \ln\left(\frac{T_H}{T_L}\right) = mR \ln\left(\frac{V_3}{V_2}\right) \tag{3.80}$$

3. *Isothermal process (3–4)*
   Heat is rejected at isothermal process
   $$Q_2 = (U_4 - U_3) + W_{3-4} = W_{3-4} \text{ (since } U_3 = U_4)$$

   $$Q_2 = W_{3-4} = mRT_L \ln\left(\frac{V_4}{V_3}\right)$$

   Since Q_2 is negative and $V_3 > V_4$

   $$Q_2 = mRT_L \ln\left(\frac{V_3}{V_4}\right) \tag{3.81}$$

4. *Isentropic compression process (4–1)*
   $$0 = (U_1 - U_4) - W_{4-1}$$

$$-\int_{T_4}^{T_1} mc_v\, dT = \int_{V_4}^{V_1} P\, dV$$

$$-\int_{T_4}^{T_1} mc_v\, dT = \int_{V_4}^{V_1} \frac{mRT}{V}\, dV$$

$$-mc_v \int_{T_4}^{T_1} \frac{dT}{T} = mR \int_{V_4}^{V_1} \frac{dV}{V}$$

$$-mc_v \ln\left(\frac{T_1}{T_4}\right) = mR \ln\left(\frac{V_1}{V_4}\right)$$

$$-mc_v \ln\left(\frac{T_H}{T_L}\right) = mR \ln\left(\frac{V_1}{V_4}\right)$$

$$mc_v \ln\left(\frac{T_H}{T_L}\right) = mR \ln\left(\frac{V_4}{V_1}\right) \tag{3.82}$$

Combining eqs 3.18 and 3.20,

$$mc_v \ln\left(\frac{T_H}{T_L}\right) = mR \ln\left(\frac{V_3}{V_2}\right) = mR \ln\left(\frac{V_4}{V_1}\right)$$

Therefore,

$$\frac{V_3}{V_2} = \frac{V_4}{V_1}$$

$$\frac{V_3}{V_4} = \frac{V_2}{V_1} \tag{3.83}$$

$$\sum_{cycle} Q_{net} = \sum_{cycle} W_{net}$$

$$Q_1 - Q_2 = W_{net}$$

For Carnot power cycle,

$$\eta_{Carnot\ cycle} = \frac{W_{net}}{Q_1} = \frac{Q_1 - Q_2}{Q_1} = 1 - \frac{Q_2}{Q_1} = 1 - \frac{mRT_L \ln(V_3/V_4)}{mRT_H \ln(V_2/V_1)} = 1 - \frac{T_L}{T_H} \tag{3.84}$$

Or

$$\eta_{\text{Carnot cycle}} = \frac{W_{\text{net}}}{Q_1} = \frac{Q_1 - Q_2}{Q_1} = 1 - \frac{Q_2}{Q_1} = 1 - \frac{T_2(S_3 - S_4)}{T_1(S_2 - S_1)} = 1 - \frac{T_2}{T_1}$$

Similarly for Carnot refrigeration cycle,

$$\text{COP}_R = \frac{Q_2}{W_{\text{net}}} = \frac{Q_2}{Q_1 - Q_2} = \frac{T_2(S_3 - S_4)}{T_1(S_2 - S_1) - T_2(S_3 - S_4)} = \frac{T_2}{T_1 - T_2} \qquad (3.85)$$

3.32   CARNOT THEOREM

It has been stated that the Carnot cycle has the highest performance and so it is the master cycle that can be used as a benchmark. But before preceding this concept, there is a need to prove this theorem. As per the Carnot theorem, the Carnot engine shows the highest thermal efficiency than the other heat engines operating between the same temperature limits.

No heat engine working in a cycle between two constant temperature reservoirs can be more efficient than a reversible engine working between the same two reservoirs.

It can be proved by assuming practical heat engine have more efficient than reversible heat engine and proving the negative.

Let us consider two heat engines E_A and E_B operating between the same temperatures of source and sink. E_A is the actual engine (not reversible) and E_B us the reversible or hypothetical engine. The comparison of these two engines should be made under the same temperature limits and the same energy supply.

$$\text{Let } Q_{1A} = Q_{1B} = Q_1 \qquad (3.86)$$

Assume $\eta_A > \eta_B$ (which is not actually true and the reverse to be proved)

$$\frac{W_A}{Q_{1A}} > \frac{W_B}{Q_{1B}} \rightarrow W_A > W_B \qquad (3.87)$$

Reverse the heat reversible heat engine, B. Now the heat engine works as a heat pump with the same temperature limits and energy transformation.

Since $W_A > W_B$, some part of W_A equal to W_B maybe fed to drive the reversed heat engine E_B. Since $Q_{1A} = Q_{1B} = Q_1$, the heat discharged by E_B maybe supplied to E_A. The source maybe eliminated.

The net result is that E_A and reversed E_B together constitute a heat engine which violates KP statement of second law. Hence the initial assumption of $\eta_A > \eta_B$ is wrong.

Hence $\eta_B \geq \eta_A$.

3.33 COROLLARY OF CARNOT'S THEOREM

The efficiency of reversible engine or Carnot cycle is the quality and depends on temperature only. It is independent of nature and quantity of the working fluids.

The efficiencies of all reversible heat engines operating between the same heat reservoirs are the same.

To prove this statement, it is assumed like in previous section

i)   Assume $\eta_A > \eta_B \rightarrow$ PMM 2
ii) Assume $\eta_B > \eta_A \rightarrow$ PMM 2

Therefore, $\eta_A = \eta_B$.

3.34 ENTROPY

The word entropy was first used by Clausius, taken from the Greek word "trope" meaning "transformation." Entropy is the fourth property (dimension) in thermodynamics after pressure, volume, and temperature. Pressure, volume, and temperature are the properties that can be measured with the instruments but entropy cannot be measured using instrument, so it is a special property. Actually it brings a new look to thermodynamics as it adds the quality to the energy. Entropy is well used to determine or estimate the decay or dissipative effects in the process or cycle to find the degree of reversibility or irreversibility.

3.35 ENTROPY IN A REVERSIBLE ADIABATIC PROCESS

Figure 3.10 shows two reversible adiabatic processes. Two reversible adiabatic paths cannot intersect each other. Assume reverse of this statement. In this cycle, net work is being produced by exchanging heat with a single reservoir in isotherm. It is PMM 2. Therefore two reversible adiabatic paths cannot intersect each other. That means two or more reversible adiabatic processes cannot have a common state. Two constant temperature processes cannot have a common state because of property of matter temperature remaining constant. There must be a property of matter that remains constant during reversible adiabatic process. This new property is called entropy.

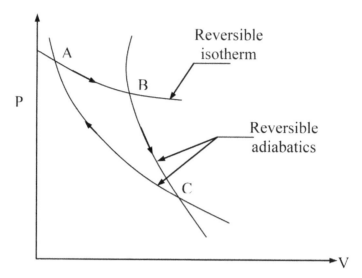

FIGURE 3.10   Assumption of reversible adiabatics intersecting each other.

3.36   CLAUSIUS THEOREM—DEFINITION OF ENTROPY

Let a system from an equilibrium state "*i*" (initial) to another equilibrium state "*f*" (final) by following the reversible path "*i–f*" as shown in Figure 3.11.

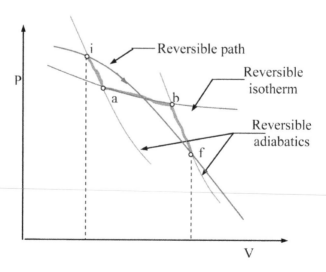

FIGURE 3.11   (See color insert.) Reversible path substituted by two reversible adiabatics and a reversible isotherm.

Let a reversible adiabatic "*i–a*" be drawn through "*i*" and another reversible adiabatic "*b–f*" be drawn through "*f*." Then, a reversible isotherm "*a–b*" is drawn in such a way that the area under "*i–a–b–f*" is equal to area under "*i–f*."

$$i\text{--}f\text{:} \qquad\qquad Q_{i\text{-}f} = U_f - U_i + W_{i\text{-}f} \qquad\qquad\qquad (3.88)$$

$$i\text{--}a\text{--}b\text{--}f\text{:} \qquad Q_{i\text{-}a\text{-}b\text{-}f} = U_f - U_i + W_{i\text{-}a\text{-}b\text{-}f} \qquad\qquad (3.89)$$

Since the initial and final states are same in both the processes, the changes in internal energies in these two processes are same.

$$\text{Since} \quad W_{i\text{-}f} = W_{i\text{-}a\text{-}b\text{-}f} \qquad\qquad\qquad\qquad\qquad (3.90)$$

$$\text{Therefore,} \quad Q_{i\text{-}f} = Q_{i\text{-}a\text{-}b\text{-}f} = Q_{i\text{-}a} + Q_{a\text{-}b} + Q_{b\text{-}f} = Q_a \qquad (3.91)$$

In this comparison, it can be clearly stated that the process "*i–f*" is equal to "*i–a–b–f*" as all the three energies (heat, work, and change in internal energy) are same in both processes.

Any reversible process can be replaced by a reversible zigzag path between the same-end states, constituting a reversible adiabatic followed by a reversible isotherm and then by a reversible adiabatic.

Let us take a smooth loop or cycle to represent a reversible cycle as shown in Figure 3.12.

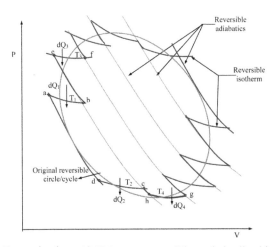

FIGURE 3.12 (**See color insert.**) Unknown reversible cycle is sliced into a large number of Carnot cycle elements.

Let an elementary Carnot cycle, "a–b–c–d–a"

$$\frac{dQ_1}{dQ_2} = \frac{T_1}{T_2}$$

$$\frac{dQ_1}{T_1} = \frac{dQ_2}{T_2} \tag{3.92}$$

If heat supply is in positive direction and rejection is negative (including sign convention)

$$\frac{dQ_1}{T_1} = -\frac{dQ_2}{T_2} \tag{3.93}$$

$$\therefore \frac{dQ_1}{T_1} + \frac{dQ_2}{T_2} = 0 \tag{3.94}$$

Similarly for elementary Carnot cycle, "e–f–g–h–e"

$$\frac{dQ_3}{T_3} + \frac{dQ_4}{T_4} = 0 \tag{3.95}$$

$$\therefore \frac{dQ_1}{T_1} + \frac{dQ_2}{T_2} + \frac{dQ_3}{T_3} + \frac{dQ_4}{T_4} + \cdots = 0 \tag{3.96}$$

$$\oint \frac{dQ}{T} = 0 \text{ is property—Clausius theorem} \tag{3.97}$$

In Figure 3.11, "i–f" is a thermodynamic process with the initial state "i" and final state "f." The forward process from "i" to "f" is shown as R_1 and the return from "f" to "i" as R_2.

According to Clausius theorem

$$\oint_{R_1 R_2} \frac{dQ}{T} = 0 \tag{3.98}$$

$$\int_{i,R_1}^{f} \frac{dQ}{T} + \int_{f,R_2}^{i} \frac{dQ}{T} = 0 \tag{3.99}$$

$$\int_{i,R_1}^{f} \frac{dQ}{T} = -\int_{f,R_2}^{i} \frac{dQ}{T} = \int_{i,R_2}^{f} \frac{dQ}{T} = S_f - S_i \tag{3.100}$$

Therefore, dQ/T is independent of path called entropy and it is a property.

If the system is taken from an initial equilibrium state "i" to a final equilibrium state "f" by an irreversible path, since entropy is a point or state function, and the entropy change is independent of the path followed, the nonreversible path is to be replaced by a reversible path to integrate for the evaluation of entropy change in the reversible process.

$$S_f - S_i = \int_{i,1}^{f} \frac{dQ_{rev}}{T} = (\Delta S)_{irrev} \tag{3.101}$$

The area under P–v diagram is work. Similarly, the area under T–s diagram is called heat transfer. It can be shown from the entropy definition of Clausius theorem.

$$\frac{dQ_{rev}}{T} = dS \tag{3.102}$$

$$dQ_{rev} = TdS \tag{3.103}$$

$$Q_{rev} = \int T\, dS = \text{Area under } T - s \text{ diagram} \tag{3.104}$$

3.37 CLAUSIUS INEQUALITY THEOREM

It demonstrates the difference between dQ/T and dS. When a system undergoes a complete cyclic process, the integral of dQ/T around the cycle is less than zero or equal to zero. The equality and inequality signs hold for reversible and irreversible cycles, respectively.

$$\oint \frac{dQ}{T} \leq 0 \tag{3.105}$$

From Clausius theorem, for reversible process, we proved that $\oint dQ/T = 0$. In the case of an irreversible engine, it has been shown that $\eta_I < \eta_R$, hence

$$\left(1 - \frac{dQ_2}{dQ_1}\right)_I < \left(1 - \frac{dQ_2}{dQ_1}\right)_R \tag{3.106}$$

$$\left(1 - \frac{dQ_2}{dQ_1}\right)_I < 1 - \frac{T_2}{T_1}, \quad \frac{dQ_2}{dQ_1} > \frac{T_2}{T_1} \tag{3.107}$$

$$\frac{dQ_2}{T_2} > \frac{dQ_1}{T_1} \tag{3.108}$$

$$\frac{dQ_1}{T_1} - \frac{dQ_2}{T_2} < 0 \text{ Considering sign} \tag{3.109}$$

$$\frac{dQ_1}{T_1} + \frac{dQ_2}{T_2} < 0 \tag{3.110}$$

Similarly for other elementary irreversible cycle,

$$\frac{dQ_1}{T_1} + \frac{dQ_2}{T_2} + \frac{dQ_3}{T_3} + \frac{dQ_4}{T_4} + \cdots < 0 \tag{3.111}$$

$$\oint dQ/T < 0 \text{ for irreversible cycle}$$

Thus, for a cycle of processes—reversible and irreversible

$$\oint \frac{dQ}{T} \leq 0 \tag{3.112}$$

$$\oint \frac{dQ}{T} > 0 \text{ Impossible} \tag{3.113}$$

3.38   ENTROPY CHANGE IN AN IRREVERSIBLE PROCESS

Let A and B are reversible processes and C is an irreversible process as shown in Figure 3.13.

For the reversible cycle consisting of A and B

$$\oint_R \frac{dQ}{T} = {}_A\int_1^2 \frac{dQ}{T} + {}_B\int_2^1 \frac{dQ}{T} = 0 \tag{3.114}$$

$$\text{or } {}_A\int_1^2 \frac{dQ}{T} = -{}_B\int_2^1 \frac{dQ}{T} \tag{3.115}$$

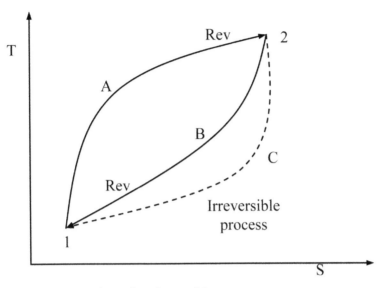

FIGURE 3.13   Entropy change in an irreversible process.

For the irreversible cycle consisting of A and C by the inequality of Clausius,

$$\oint \frac{dQ}{T} =_A \int_1^2 \frac{dQ}{T} +_C \int_2^1 \frac{dQ}{T} < 0 \tag{3.116}$$

From the above equation,

$$-_B \int_2^1 \frac{dQ}{T} +_C \int_2^1 \frac{dQ}{T} < 0 \tag{3.117}$$

$$_C \int_2^1 \frac{dQ}{T} <_B \int_2^1 \frac{dQ}{T} \tag{3.118}$$

Since the path B is reversible

$$_B \int_2^1 \frac{dQ}{T} =_B \int_2^1 dS \tag{3.119}$$

Since entropy is a property, entropy change for the paths B and C would be the same,

$$B\int_{2}^{1}dS =_C \int_{2}^{1}dS \tag{3.120}$$

$$C\int_{2}^{1}\frac{dQ}{T} < \int_{B\,2}^{1}dS <_C \int_{2}^{1}dS < S_1 - S_2 \tag{3.121}$$

$dQ/T < dS$ for irreversible process;

$dQ/T < dS$ for reversible process.

3.39   COMBINED FIRST AND SECOND LAW OF THERMODYNAMICS

Second law, $dQ_{\text{rev}} = TdS$ (3.122)

First law, $dQ = dU + PdV$ (3.123)

$TdS = dU + PdV$ (3.124)

Also, $H = U + PV$ (3.125)

$dH = (dU + PdV) + VdP = TdS + VdP$ (3.126)

$TdS = dH - VdP$ (3.127)

$TdS = dU + PdV = dH - VdP$ (3.128)

3.40   FORMULAE FOR ENTROPY CHANGE

For closed system,

$$TdS = dU + PdV$$

$$dS = \frac{dU}{T} + P\frac{dV}{T}$$

$$\because PV = mRT \Rightarrow \frac{P}{T} = \frac{mR}{V}$$

$$dS = mc_v\frac{dT}{T} + mR\frac{dv}{v} \tag{3.129}$$

Integrating both sides, $\int dS = mc_v\int dT/T + mR\int dv/v$

$$S_2 - S_1 = mc_v \ln \frac{T_2}{T_1} + mR \ln \frac{V_2}{V_1} \qquad (3.130)$$

It can be also expressed in terms of pressure ratio and volume ratio.

$$S_2 - S_1 = mc_v \ln \frac{T_2}{T_1} + m\left(c_p - c_v\right) \ln \frac{V_2}{V_1}$$

$$= mc_v \ln \frac{T_2}{T_1} - mc_v \ln \frac{V_2}{V_1} + mc_p \ln \frac{V_2}{V_1}$$

$$= mc_v \ln \frac{T_2}{T_1} \times \frac{V_1}{V_2} + mc_p \ln \frac{V_2}{V_1} \qquad \because \frac{P_1 V_1}{T_1} = \frac{P_2 V_2}{T_2}$$

$$= mc_v \ln \frac{P_2}{P_1} + mc_p \ln \frac{V_2}{V_1} \qquad (3.131)$$

In terms of temperature ratio and pressure ratio

For open system,

$$TdS = dH - VdP$$

$$dS = \frac{dH}{T} - V\frac{dP}{T} \qquad \because PV = mRT \Rightarrow \frac{V}{T} = \frac{mR}{P}$$

$$dS = mc_p \frac{dT}{T} - mR\frac{dP}{P}$$

Integrating both sides, $\int dS = mc_p \int \frac{dT}{T} - mR \int \frac{dP}{P}$

$$S_2 - S_1 = mc_P \ln \frac{T_2}{T_1} - mR \ln \frac{P_2}{P_1} \qquad (3.132)$$

Therefore,

$$S_2 - S_1 = mc_v \ln \frac{T_2}{T_1} + mR \ln \frac{V_2}{V_1} = mc_P \ln \frac{T_2}{T_1} - mR \ln \frac{P_2}{P_1} = mc_v \ln \frac{P_2}{P_1} + mc_p \ln \frac{V_2}{V_1} \quad (3.133)$$

The temperature–entropy diagram (Fig. 3.14) shows the relation between polytropic index and slope in this chart. The changes can be grouped into two categories the first variation is the increase in polytropic index from 0 to 1, that is, from constant pressure to isothermal process. In this range, the increase in index shows a corresponding drop in the slope. Similarly in the second variation, that is, the increase of index from γ to ∞, the slope in

T–s diagram drops continuously. Finally, the slop in this diagram inversely varies with the polytropic index.

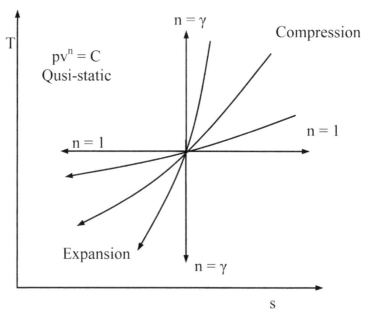

FIGURE 3.14   Temperature–entropy diagram for thermodynamic processes in compression and expansion regions.

KEYWORDS

- **thermodynamics**
- **thermal system development**
- **energy**
- **entropy**
- **exergy**

CHAPTER 4

THERMODYNAMIC PROPERTIES OF AMMONIA–WATER MIXTURE

ABSTRACT

Thermodynamic properties of working fluid in a thermal system play an important role in its evaluation. Evaluation of properties for binary fluid mixture is complex compared to single fluid system due to additional property of concentration. The regions of phase with reference to bubble-point temperature and dew-point temperature are described to understand the state of fluid. The Gibbs function is derived to generate the properties and the derived properties are presented with the equation fitting coefficients. A methodology has been developed to identify the region of fluid from the randomly supplied properties, that is, pressure, temperature, and concentration. The resulted properties are arranged in tabular form and charts in Appendix section.

4.1 INTRODUCTION

Ammonia–water mixture properties play a critical role in thermodynamic evaluation of vapor absorption systems. Currently, researchers are focusing on the vapor absorption systems in the area of Kalina cycle system, vapor absorption refrigeration, and cooling cogeneration cycle which is also called as combined power and cooling cycle. Compared to the assessment of thermodynamic properties of pure substance, evaluation of binary mixture properties is complicated due to addition of an extra property, that is, mixture concentration. Ammonia–water mixture is a well-known and established zeotropic mixture for vapor absorption systems. In this mixture, ammonia has low boiling point and the water has high boiling point comparatively. This difference in the boiling point allows the separation and distillation of the mixture which is required in the cycles. Unlike pure substance, the binary mixture changes its temperature in phase-change heat transfer for bubble-point temperature (BPT) to dew-point temperature (DPT). In this work, the thermodynamic properties of ammonia–water mixture have been generated from the reported correlations, derivations, and the iterations using MATLAB programing. The function programs or subroutines are called in the main program of vapor absorption cycles for modeling, simulation, and analysis. The temperature–concentration (T–x), specific volume–concentration (v–x), specific enthalpy–concentration (h–x), specific entropy–concentration (s–x), and specific exergy–concentration (e–x) graphs for ammonia–water mixtures are plotted up to 50 bar pressure. The tabulated properties are furnished in the Appendix.

4.2 BUBBLE-POINT AND DEW-POINT TEMPERATURES

To evaluate the mixture properties, first the BPT and DPT are to be determined at given pressure and concentration. Pressure, temperature, and concentration are required to find the properties of mixture without knowing the region or phase. The mixture temperature along with the BPT and DPT decides the region of mixture, that is, liquid mixture (subcooled or compressed liquid mixture), saturated liquid, liquid–vapor mixture, saturated vapor, and superheated vapor.

Figure 4.1 is developed to describe the regions in binary fluid mixture. The regions are grouped with reference to the BPT and DPT which are compared to the mixture pressure, temperature, and concentration. The temperature–concentration plot is generated at 40 bar pressure. The BPT curve represents all the starting states of phase change and DPT curve the completed states of phase change of the mixture at different concentrations. The BPT and DPT curves are also known as saturated liquid curve and saturated vapor curve, respectively. Between these two curves, that is, BPT and DPT, the mixture exists in the liquid–vapor mixture. Below the BPT, the mixture is completely liquid and subcooled condition. Above the DPT, the mixture is vapor condition in superheated condition. The phase change starts from BPT and ends at DPT at a particular concentration. At zero concentration and 100% concentration, it is a pure substance and shows the respective saturation temperature at the given pressure. The drawing shows five regions, namely, subcooled liquid, saturation liquid, liquid vapor mixture, saturated vapor, and superheated vapor. BPT and DPT are determined from the curve-fit equations as shown in eqs A.1 and A.2 and corresponding coefficients are tabulated at Tables 4.1 and 4.2, respectively, for BPT and DPT.

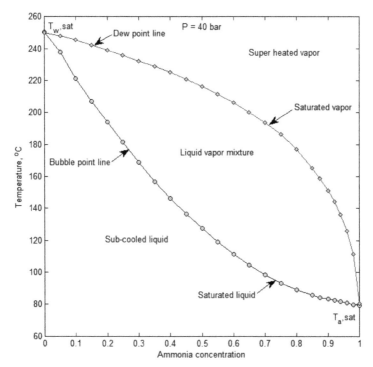

FIGURE 4.1 Temperature–concentration diagram with five regions at 40 bar.

TABLE 4.1 Constants for BPT Used in Eq 4.1.

i	m_i	n_i	a_i
1	0	0	$+0.322302 \times 10^1$
2	0	1	-0.384206×10^0
3	0	2	$+0.460965 \times 10^{-1}$
4	0	3	-0.378945×10^{-2}
5	0	4	$+0.135610 \times 10^{-3}$
6	1	0	$+0.487755 \times 10^0$
7	1	1	-0.120108×10^0
8	1	2	$+0.106154 \times 10^{-1}$
9	2	3	-0.533589×10^{-3}
10	4	0	$+0.785041 \times 10^1$
11	5	0	-0.115941×10^2
12	5	1	-0.523150×10^{-1}
13	6	0	$+0.489596 \times 10^1$
14	13	1	$+0.421059 \times 10^{-1}$

$$T_{\text{bp}} = T_0 \sum_i a_i \left(1-x\right)^{m_i} \left[\ln\left(\frac{P_0}{P}\right)\right]^{n_i} \tag{4.1}$$

$$T_{\text{dp}} = T_0 \sum_i a_i \left(1-y\right)^{m_i} \left[\ln\left(\frac{P_0}{P}\right)\right]^{n_i} \tag{4.2}$$

Pressure, temperature, and concentration are the three properties required to find the thermodynamic properties of the mixture. The given temperature is compared with the BPT and DPT to decide the location of mixture. If the temperature is less than BPT, it is subcooled liquid or compressed liquid. The temperature equal to BPT is saturated liquid. Between BPT and DPT is the mixture of liquid and vapor. Temperature equal to DPT shows the saturated vapor. The temperature above the DPT is superheated vapor.

TABLE 4.2 Constants for DPT Used in Eq 4.2.

i	m_i	n_i	a_i
1	0	0	$+0.324004 \times 10^1$
2	0	1	-0.395920×10^0
3	0	2	$+0.435624 \times 10^{-1}$
4	0	3	-0.218943×10^{-2}
5	1	0	-0.143526×10^1
6	1	1	$+0.105256 \times 10^1$
7	1	2	-0.719281×10^{-1}
8	2	0	$+0.122362 \times 10^2$
9	2	1	-0.224368×10^1
10	3	0	-0.201780×10^2
11	3	1	$+0.110834 \times 10^1$
12	4	0	$+0.145399 \times 10^2$
13	4	2	$+0.644312 \times 10^0$
14	5	0	-0.221246×10^1
15	5	2	-0.756266×10^0
16	6	0	-0.135529×10^1
17	7	2	$+0.183541 \times 10^0$

4.3 SPECIFIC ENTHALPY AT LIQUID PHASE

The properties are determined from Gibbs free energy function. In liquid phase, the Gibbs free energy for both liquid and gas phases has been determined from eqs 4.3 and 4.4, respectively.

$$h_l^m = x h_{NH_3}^l + (1-x) h_{NH_3}^l + h_E \tag{4.3}$$

The following eqs 4.4–4.16 are derived to frame expression for the liquid enthalpy. The reduced temperature and reduced pressure are $T_r = T/T_B$ and $P_r = P/P_B$, respectively.

$$h = -RT_B T_r^2 \left[\frac{\partial}{\partial T_r} \left(\frac{G_r}{T_r} \right) \right]_{P_r} \tag{4.4}$$

where the base temperature and base pressure are $T_B = 100$ K and $P_B = 10$ bar.

$$G = h_0 - Ts_0 + \int_{T_0}^{T} c_p dT + \int_{P_0}^{P} v dp - T \int_{T_0}^{T} \left(\frac{c_p}{T} \right) dT \tag{4.5}$$

$$G_r^l = \left\{ \begin{array}{l} h_{ro}^l - T_r s_{ro}^l + B_1 (T_r - T_{ro}) + \dfrac{B_2}{2} (T_r^2 - T_{ro}^2) + \dfrac{B_3}{3} (T_r^3 - T_{rp}^3) \\[2mm] -B_1 T_r \ln \left(\dfrac{T_r}{T_{ro}} \right) - B_2 T_r (T_r - T_{ro}) - \dfrac{B_3}{2} (T_r^2 - T_{ro}^2) T_r \\[2mm] + \left(A_1 + A_3 T_r + A_4 T_r^2 \right) (P_r - P_{ro}) + \dfrac{A_2}{2} (P_r^2 - P_{ro}^2) \end{array} \right\} \tag{4.6}$$

$$\frac{G_r^l}{T_r} = \left\{ \begin{array}{l} \dfrac{h_{ro}^l}{T_r} - s_{ro}^l + B_1 \left(1 - \dfrac{T_{ro}}{T_r} \right) + \dfrac{B_2}{2} \left(T_r - \dfrac{T_{ro}^2}{T_r} \right) + \dfrac{B_3}{3} \left(T_r^2 - \dfrac{T_{ro}^3}{T_r} \right) \\[2mm] -B_1 \ln \left(\dfrac{T_r}{T_{ro}} \right) - B_2 (T_r - T_{ro}) - \dfrac{B_3}{2} (T_r^2 - T_{rp}^2) \\[2mm] + \left(\dfrac{A_1}{T_r} + A_3 + A_4 T_r \right) (P_r - P_{ro}) + \dfrac{A_2}{2T_r} (P_r^2 - P_{ro}^2) \end{array} \right\} \tag{4.7}$$

$$\frac{\partial}{\partial T_r} \left(\frac{G_r^l}{T_r} \right) = \left\{ \begin{array}{l} -\dfrac{h_{ro}^l}{T_r^2} + B_1 \left(\dfrac{T_{ro}}{T_r^2} \right) + \dfrac{B_2}{2} + B_2 \left(\dfrac{T_{ro}^2}{2T_r^2} \right) - \dfrac{B_1}{T_r} - B_2 + \dfrac{B_3}{3} (2T_r) \\[2mm] + \dfrac{B_3}{3} \left(\dfrac{T_{ro}^3}{T_r^2} \right) - B_3 T_r + \left(A_4 - \dfrac{A_1}{T_r^2} \right) (P_r - P_{ro}) - \dfrac{A_2}{2T_r^2} (P_r^2 - P_{ro}^2) \end{array} \right\} \tag{4.8}$$

$$T_r^2 \frac{\partial}{\partial T_r} \left(\frac{G_r}{T_r} \right) = \left\{ \begin{array}{l} -h_{ro}^l + B_1 (T_{ro} - T_r) + \dfrac{B_2}{2} (T_{ro}^2 - T_r^2) + \dfrac{B_3}{3} (T_{ro}^3 - T_r^3) \\[2mm] + \left(A_4 T_r^2 - A_1 \right) (P_r - P_{ro}) - \dfrac{A_2}{2} (P_r^2 - P_{ro}^2) \end{array} \right\} \tag{4.9}$$

$$
h^L = -RT_B \left\{
\begin{array}{l}
\left[-h_{\mathrm{ro}}^L + B_1 \left(T_{\mathrm{ro}} - T_r \right) + \dfrac{B_2}{2} \left(T_{\mathrm{ro}}^2 - T_r^2 \right) + \dfrac{B_3}{3} \left(T_{\mathrm{ro}}^3 - T_r^3 \right) \right] \\[2mm]
+ \left(A_4 T_r^2 - A_1 \right) \left(P_r - P_{\mathrm{ro}} \right) - \dfrac{A_2}{2} \left(P_r^2 - P_{\mathrm{ro}}^2 \right)
\end{array}
\right\}
\tag{4.10}
$$

Equation 4.10 is used to find the enthalpy of water and liquid ammonia.

The Gibbs excess energy G_r^E for liquid mixtures has been expressed in eq 4.11.

$$
G_r^E = \left\{ F_1 + F_2 (2x - 1) + F_3 (2x - 1)^2 \right\} (1 - x)
\tag{4.11}
$$

$$
F_1 = E_1 + E_2 P_r + \left(E_3 + E_4 P_r \right) T_r + \frac{E_5}{T_r} + \frac{E_6}{T_r^2}
\tag{4.12}
$$

$$
F_2 = E_4 + E_8 P_r + \left(E_9 + E_{10} P_r \right) T_r + \frac{E_{11}}{T_r} + \frac{E_{12}}{T_r^2}
\tag{4.13}
$$

$$
F_3 = E_{13} + E_{14} P_r + \frac{E_{15}}{T_r} + \frac{E_{16}}{T_r^2}
\tag{4.14}
$$

$$
h^E = -RT_B T_r^2 \left\{ \frac{\partial}{\partial T_r} \left(\frac{G_r^E}{T_r} \right) \right\}_{P_r, x}
\tag{4.15}
$$

$$
h^E = -RT_B (1 - x) \left\{
\begin{array}{l}
\left(-E_1 - E_2 P_r - \dfrac{2E_5}{T_r} - \dfrac{3E_6}{T_r^2} \right) \\[3mm]
+ (2x - 1) \left(-E_7 - E_8 P_r - \dfrac{2E_{11}}{T_r} - \dfrac{3E_{12}}{T_r^2} \right) \\[3mm]
+ (2x - 1)^2 \left(-E_{13} - E_{14} P_r - \dfrac{2E_{15}}{T_r} - \dfrac{3E_{16}}{T_r^2} \right)
\end{array}
\right\}
\tag{4.16}
$$

4.4   SPECIFIC ENTHALPY AT VAPOR PHASE

The enthalpy of pure component in the vapor phase can be expressed in the following equation:

$$
h_v^m = x h_{\mathrm{NH}_3}^v + (1 - x) h_{\mathrm{H}_2\mathrm{O}}^v
\tag{4.17}
$$

For the gas phase, Gibbs free energy equation is

$$
G_r^v =
\begin{cases}
h_{\mathrm{ro}}^v - T_r s_{\mathrm{ro}}^v + D_1(T_r - T_{\mathrm{ro}}) + \dfrac{D_2}{2}\left(T_r^2 - T_{\mathrm{ro}}^2\right) + \dfrac{D_3}{3}\left(T_r^3 - T_{\mathrm{ro}}^3\right) \\[2mm]
-D_1 T_r \ln\left(\dfrac{T_r}{T_{\mathrm{ro}}}\right) - D_2 T_r (T_r - T_{\mathrm{ro}}) - \dfrac{D_3}{2}\left(T_r^2 - T_{\mathrm{ro}}^2\right) + T_r \ln\left(\dfrac{P_r}{P_{\mathrm{ro}}}\right) \\[2mm]
+ C_1(P_r - P_{\mathrm{ro}}) + C_2\left(\dfrac{P_r}{T_r^3} - \dfrac{4 P_{\mathrm{ro}}}{T_{\mathrm{ro}}^3} + \dfrac{3 P_{\mathrm{ro}} T_r}{T_{\mathrm{ro}}^4}\right) \\[2mm]
+ C_3\left(\dfrac{P_r}{T_r^{11}} - \dfrac{12 P_{\mathrm{ro}}}{T_{\mathrm{ro}}^{11}} + \dfrac{11 P_{\mathrm{ro}} T_r}{T_{\mathrm{ro}}^{12}}\right) + \dfrac{C_4}{3}\left(\dfrac{P_r^3}{T_r^{11}} - \dfrac{12 P_{\mathrm{ro}}^3}{T_{\mathrm{ro}}^{11}} + \dfrac{11 P_{\mathrm{ro}}^3 T_r}{T_{\mathrm{ro}}^{12}}\right)
\end{cases}
\tag{4.18}
$$

$$
\frac{G_r^v}{T_r} =
\begin{cases}
\dfrac{h_{\mathrm{ro}}^v}{T_r} - s_{\mathrm{ro}}^v + D_1\left(1 - \dfrac{T_{\mathrm{ro}}}{T_r}\right) + \dfrac{D_2}{2}\left(T_r - \dfrac{T_{\mathrm{ro}}^2}{T_r}\right) + \dfrac{D_3}{3}\left(T_r^2 - \dfrac{T_{\mathrm{ro}}^3}{T_r}\right) \\[2mm]
-D_1 \ln\left(\dfrac{T_r}{T_{\mathrm{ro}}}\right) - D_2 (T_r - T_{\mathrm{ro}}) - \dfrac{D_3}{2}\left(T_r - \dfrac{T_{\mathrm{ro}}^2}{T_r}\right) + \ln\left(\dfrac{P_r}{P_{\mathrm{ro}}}\right) \\[2mm]
+ \dfrac{C_1}{T_r}(P_r - P_{\mathrm{ro}}) + C_2\left(\dfrac{P_r}{T_r^4} - \dfrac{4 P_{\mathrm{ro}}}{T_{\mathrm{ro}}^3 T_r} + \dfrac{3 P_{\mathrm{ro}}}{T_{\mathrm{ro}}^4}\right) + C_3\left(\dfrac{P_r}{T_r^{12}} - \dfrac{12 P_{\mathrm{ro}}}{T_{\mathrm{ro}}^{11} T_r} + \dfrac{11 P_{\mathrm{ro}}}{T_{\mathrm{ro}}^{12}}\right) \\[2mm]
+ \dfrac{C_4}{3}\left(\dfrac{P_r^3}{T_r^{12}} - \dfrac{12 P_{\mathrm{ro}}^3}{T_{\mathrm{ro}}^{12} T_r} + \dfrac{11 P_{\mathrm{ro}}^3}{T_{\mathrm{ro}}^{12}}\right)
\end{cases}
\tag{4.19}
$$

$$
\frac{\partial}{\partial T_r}\left(\frac{G_r^v}{T_r}\right) =
\begin{cases}
-\dfrac{h_{\mathrm{ro}}^v}{T_r^2} + D_1\left(\dfrac{T_{\mathrm{ro}}}{T_r^2}\right) + \dfrac{D_2}{2}\left(1 + \dfrac{T_{\mathrm{ro}}^2}{T_r^2}\right) + \dfrac{D_3}{3}\left(2 T_r + \dfrac{T_{\mathrm{ro}}^3}{T_r^2}\right) - D_1\left(\dfrac{1}{T_r}\right) \\[2mm]
-D_2 - \dfrac{D_3}{2}\left(1 + \dfrac{T_{\mathrm{ro}}^2}{T_r^2}\right) - \dfrac{C_1}{T_r^2}(P_r - P_{\mathrm{ro}}) + C_2\left(-\dfrac{4 P_r}{T_r^5} + \dfrac{4 P_{\mathrm{ro}}}{T_{\mathrm{ro}}^3 T_r^2}\right) \\[2mm]
+ C_3\left(\dfrac{-12 P_r}{T_r^{13}} + \dfrac{12 P_{\mathrm{ro}}}{T_{\mathrm{ro}}^{11} T_r^2}\right) + \dfrac{C_4}{3}\left(\dfrac{-12 P_r^3}{T_r^{13}} + \dfrac{12 P_{\mathrm{ro}}^3}{T_{\mathrm{ro}}^{11} T_r^2}\right)
\end{cases}
\tag{4.20}
$$

$$
T_r^2 \frac{\partial}{\partial T_r}\left(\frac{G_{\mathrm{ro}}^v}{T_r}\right)_{P_r} =
\begin{cases}
-h_{\mathrm{ro}}^v + D_1 T_{\mathrm{ro}} + \dfrac{D_2}{2}\left(T_r^2 + T_{\mathrm{ro}}^2\right) + \dfrac{D_3}{3}\left(2 T_r^3 + T_{\mathrm{ro}}^3\right) \\[2mm]
-D_1 T_r - D_2 T_r^2 - \dfrac{D_3}{2}\left(T_r^2 + T_{\mathrm{ro}}^2\right) - C_1(P_r - P_{\mathrm{ro}}) \\[2mm]
+ C_2\left(-\dfrac{4 P_r}{T_r^3} + \dfrac{4 P_{\mathrm{ro}}}{T_{\mathrm{ro}}^3}\right) + C_3\left(\dfrac{-12 P_r}{T_r^{11}} + \dfrac{12 P_{\mathrm{ro}}}{T_{\mathrm{ro}}^{11}}\right) \\[2mm]
+ \dfrac{C_4}{3}\left(\dfrac{-12 P_r^3}{T_r^{11}} + \dfrac{12 P_{\mathrm{ro}}^3}{T_{\mathrm{ro}}^{11}}\right)
\end{cases}
\tag{4.21}
$$

$$h^v = -RT_B T_r^2 \frac{\partial}{\partial T_r}\left(\frac{G_r^v}{T_r}\right)_{P_r} = -RT_B \begin{bmatrix} -h_{\mathrm{ro}}^v + D_1 T_{\mathrm{ro}} + \frac{D_2}{2}\left(T_r^2 + T_{\mathrm{ro}}^2\right) + \frac{D_3}{3}\left(2T_r^3 + T_{\mathrm{ro}}^3\right) \\[2mm] -D_1 T_r - D_2 T_r^2 - \frac{D_3}{2}\left(T_r^2 + T_{\mathrm{ro}}^2\right) - C_1\left(P_r - P_{\mathrm{ro}}\right) \\[2mm] +C_2\left(-\frac{4P_r}{T_r^3} + \frac{4P_{\mathrm{ro}}}{T_{\mathrm{ro}}^3}\right) + C_3\left(\frac{-12P_r}{T_r^{11}} + \frac{12P_{\mathrm{ro}}}{T_{\mathrm{ro}}^{11}}\right) \\[2mm] +\frac{C_4}{3}\left(\frac{-12P_r^3}{T_r^{11}} + \frac{12P_{\mathrm{ro}}^3}{T_{\mathrm{ro}}^{11}}\right) \end{bmatrix} \quad (4.22)$$

4.5   SPECIFIC ENTROPY AT LIQUID AND VAPOR PHASES

The molar entropy of the liquid and vapor phases is specified and simplified from eqs 4.23 to 4.33.

$$s = -R\left(\frac{\partial G_r}{\partial T_r}\right)_{P_r} \tag{4.23}$$

$$\frac{\partial G_r^l}{\partial T_r} = \begin{Bmatrix} -s_{\mathrm{ro}}^l + B_1 + \frac{B_2}{2}\left(2T_r\right) + \frac{B_3}{3}\left(3T_r^2\right) - B_1\left(\ln\frac{T_r}{T_{\mathrm{ro}}}+1\right) \\[2mm] -B_2\left(2T_r - T_{\mathrm{ro}}\right) - \frac{B_3}{2}\left(3T_r^2 - T_{\mathrm{ro}}^2\right) + \left(A_3 + 2A_4 T_r\right)\left(P_r - P_{\mathrm{ro}}\right) \end{Bmatrix} \tag{4.24}$$

$$s^l = -R\left(\frac{\partial G_r^l}{\partial T}\right)_{P_r} = -R\begin{Bmatrix} -s_{\mathrm{ro}}^l + B_1 + \frac{B_2}{2}\left(2T_r\right) + \frac{B_3}{3}\left(3T_r^2\right) - \left(B_1 \ln\frac{T_r}{T_{\mathrm{ro}}}+1\right) \\[2mm] -B_2\left(2T_r - T_{\mathrm{ro}}\right) - \frac{B_3}{2}\left(3T_r^2 - T_{\mathrm{ro}}^2\right) \\[2mm] +\left(A_3 + 2A_4 T_r\right)\left(P_r - P_{\mathrm{ro}}\right) \end{Bmatrix} \tag{4.25}$$

$$s_E = -R\left(\frac{\partial G_r^E}{\partial T_r}\right)_{P_r,x} \tag{4.26}$$

$$\frac{\partial G_r^E}{\partial T_r} = (1-x)\begin{Bmatrix} \left(E_3 + E_4 P_r - \frac{E_5}{T_r^2} - \frac{2E_6}{T_r^3}\right) + (2x-1)\left(E_9 + E_{10} P_r - \frac{E_{11}}{T_r^2} - \frac{2E_{12}}{T_r^3}\right) \\[2mm] +(2x-1)^2\left(-\frac{E_{15}}{T_r^2} - \frac{2E_{16}}{T_r^3}\right) \end{Bmatrix} \tag{4.27}$$

$$s_E = -R\left(\frac{\partial G_r^E}{\partial T_r}\right)_{P_r,x} = -R(1-x)\begin{cases}\left(E_3 + E_4 P_r - \dfrac{E_5}{T_r^2} - \dfrac{2E_6}{T_r^3}\right) \\[2mm] +(2x-1)\left(E_9 + E_{10}P_r - \dfrac{E_{11}}{T_r^2} - \dfrac{2E_{12}}{T_r^3}\right) \\[2mm] +(2x-1)^2\left(-\dfrac{E_{15}}{T_r^2} - \dfrac{2E_{16}}{T_r^3}\right)\end{cases} \quad (4.28)$$

$$s_{\text{mix}} = -R\left(x\ln(x) + (1-x)\ln(1-x)\right) \quad (4.29)$$

$$s_l^m = xs_a^l + (1-x)s_w^l + s_E + s_{\text{mix}} \quad (4.30)$$

$$\frac{\partial G_r^v}{\partial T_r} = \begin{cases}-s_{\text{ro}}^v + D_1 + D_2 T_r + D_3 T_r^2 - D_1\left(1 + \ln\dfrac{T_r}{T_{\text{ro}}}\right) - D_2\left(2T_r - T_{\text{ro}}\right) \\[2mm] -\dfrac{D_3}{2}\left(3T_r^2 - T_{\text{ro}}^2\right) + \ln\left(\dfrac{P_r}{P_{\text{ro}}}\right) + C_2\left(-\dfrac{3P_r}{T_r^4} + \dfrac{3P_{\text{ro}}}{T_{\text{ro}}^4}\right) \\[2mm] +C_3\left(-\dfrac{11P_r}{T_r^{12}} + \dfrac{11P_{\text{ro}}}{T_{\text{ro}}^{12}}\right) + \dfrac{C_4}{3}\left(-\dfrac{11P_r^3}{T_r^{12}} + \dfrac{11P_{\text{ro}}^3}{T_{\text{ro}}^{12}}\right)\end{cases} \quad (4.31)$$

$$s^v = \left(\frac{\partial G_r^v}{\partial T_r}\right)_{P_r} = -R\begin{cases}-s_{\text{ro}}^v + D_1 + D_2 T_r + D_3 T_r^2 - D_1\left(1 + \ln\dfrac{T_r}{T_{\text{ro}}}\right) \\[2mm] -D_2\left(2T_r - T_{\text{ro}}\right) - \dfrac{D_3}{2}\left(3T_r^2 - T_{\text{ro}}^2\right) + \ln\left(\dfrac{P_r}{P_{\text{ro}}}\right) \\[2mm] +C_2\left(-\dfrac{3P_r}{T_r^4} + \dfrac{3P_{\text{ro}}}{T_{\text{ro}}^4}\right) + C_3\left(-\dfrac{11P_r}{T_r^{12}} + \dfrac{11P_{\text{ro}}}{T_{\text{ro}}^{12}}\right) \\[2mm] +\dfrac{C_4}{3}\left(-\dfrac{11P_r^3}{T_r^{12}} + \dfrac{11P_{\text{ro}}^3}{T_{\text{ro}}^{12}}\right)\end{cases} \quad (4.32)$$

$$s_v^m = xs_a^v + (1-x)s_w^v + s_{\text{mix}} \quad (4.33)$$

4.6 SPECIFIC VOLUME AT LIQUID AND VAPOR PHASES

The specific volume of the liquid and vapor phases is simplified from eqs 4.34 to 4.43.

$$v = \frac{RT_B}{P_B}\left(\frac{\partial G_r}{\partial P_r}\right)_{T_r} \tag{4.34}$$

$$\frac{\partial G_r^l}{\partial P_r} = \left(A_1 + A_3 T_r + A_4 T_r^2\right) + A_2 P_r \tag{4.35}$$

$$v^l = \frac{RT_B}{P_B}\left(\frac{\partial G_r^l}{\partial P_r}\right)_{T_r} = \frac{RT_B}{P_B}\left(A_1 + A_3 T_r + A_4 T_r^2\right) + A_2 P_r \tag{4.36}$$

$$v^E = \frac{RT_B}{P_B}\left(\frac{\partial G_r^E}{\partial P_r}\right)_{T_r,x} \tag{4.37}$$

$$\frac{\partial G_r^E}{\partial P_r} = \left(E_2 + E_4 T_r\right) + (2x-1)\left(E_8 + E_{10} T_r\right) + (2x-1)^2 (1-x)(E_{14}) \tag{4.38}$$

$$v^E = \frac{RT_B}{P_B}\left(\frac{\partial G_r^E}{\partial P_r}\right)_{T_r} = \frac{RT_B}{P_B}$$

$$\left[\left(\left(E_2 + E_4 T_r\right) + (2x-1)\left(E_8 + E_{10} T_r\right) + (2x-1)^2 (E_{14})\right)(1-x)\right] \tag{4.39}$$

$$v_m^l = xv_a^l + (1-x)v_w^l + v^E \tag{4.40}$$

$$\frac{\partial G_r^v}{\partial P_r} = \frac{T_r}{P_r} + C_1 + \frac{C_2}{T_r^3} + \frac{C_3}{T_r^{11}} + \frac{C_4 P_r^2}{T_r^{11}} \tag{4.41}$$

$$v^v = \frac{RT_B}{P_B}\left(\frac{\partial G_r^v}{\partial P_r}\right)_{T_r} = \frac{RT_B}{P_B}\left(\frac{T_r}{P_r} + C_1 + \frac{C_2}{T_r^3} + \frac{C_3}{T_r^{11}} + \frac{C_4 P_r^2}{T_r^{11}}\right) \tag{4.42}$$

$$v_m^v = xv_a^v + (1-x)v_w^v \tag{4.43}$$

The coefficients used in eqs 4.6, 4.12–4.14, 4.18, 4.24, and 4.31 are tabulated in Tables 4.3 and 4.4.

Figure 4.2 is a flowchart to describe the method of finding the thermodynamic properties of ammonia–water mixture as per the state condition. As per the earlier discussion, the properties are grouped into five regions. The identified regions are subcooled liquid, saturated liquid, liquid–vapor mixture, saturated vapor, and superheated vapor. To find the properties of mixture, it is required to supply the mixture properties, that is, pressure, temperature, and concentration. After stating the mixture condition, at the supplied pressure and concentration BPT and DPT can be evaluated. Now, the supplied temperature is compared with the BPT. If the temperature is less than the BPT, it is subcooled liquid or equals to BPT, then it a saturated liquid. If the temperature lies between the BPT and DPT, then the state falls between liquid and vapor condition, that is, liquid–vapor mixture. Solving the liquid–vapor mixture properties is complicated compared to either liquid condition only or vapor condition only. In case the given temperature is greater than the DPT, then it can be concluded that the resulted region as a superheated vapor

TABLE 4.3 Constants Used to Find the Properties of Pure Substance (i.e., Ammonia Only and Water Only).

Coefficient	Ammonia	Water
A_1	3.971423×10^{-2}	2.748796×10^{-2}
A_2	-1.790557×10^{-5}	-1.016665×10^{-5}
A_3	-1.308905×10^{-2}	-4.452025×10^{-3}
A_4	3.752836×10^{-3}	8.389246×10^{-4}
B_1	1.634519×10^{1}	1.214557×10^{1}
B_2	-6.508119	-1.898065
B_3	1.448937	2.911966×10^{-2}
C_1	-1.049377×10^{-2}	2.136131×10^{-2}
C_2	-8.288224	-3.169291×10^{1}
C_3	-6.647257×10^{2}	-4.634611×10^{4}
C_4	-3.045352×10^{3}	0.0
D_1	3.673647	4.019170
D_2	9.989629×10^{-2}	-5.175550×10^{-2}
D_3	3.617622×10^{-2}	1.951939×10^{-2}
h^L	4.878573	21.821141
h^v	26.468879	60.965058
s^L	1.644773	5.733498
s^v	8.339026	13.453430
T_{ro}	3.2252	5.0705
P_{ro}	2.0000	3.0000

TABLE 4.4 Constants Used to Find the Gibbs Excess Energy.

Coefficients	Value
E_1	−41.733398
E_2	0.02414
E_3	6.702285
E_4	−0.011475
E_5	63.608967
E_6	−62.490768
E_7	1.761064
E_8	0.008626
E_9	0.387983
E_{10}	0.004772
E_{11}	−4.648107
E_{12}	0.836376
E_{13}	−3.553627
E_{14}	0.000904
E_{15}	21.361723
E_{16}	−20.736547

Figure 4.3 is the plot for BPT and DPT up to 50 bar pressure. Figure 4.4 shows the changes in saturated vapor specific volume with ammonia mass fraction at various pressures. The specific volume has been calculated at BPT for the given pressure and ammonia mass fractions for the liquid region. The trend increases with the increase in pressure. Similarly, Figure 4.5 is the saturated vapor specific volume diagram. It is generated with the BPT and vapor ammonia mass fraction. At lower pressure, the changes in specific volume in both liquid region and vapor region are minor.

Figures 4.4 and 4.5 depict the specific volume of vapor and liquid, respectively. The specific volume of liquid and vapor show opposite trends with the concentration and pressure variations. The variation of specific volume of vapor is minor with a change in concentration. The concentration plays a significant role on specific volume of liquid. The specific volume changes at high pressure are more compared to the low-pressure range in the liquid region. At low pressure, the specific volume changes are significant compared to the changes in specific volume at the higher pressure. The specific volume of liquid increases with increase in pressure and the specific volume of vapor decreases with increase in pressure.

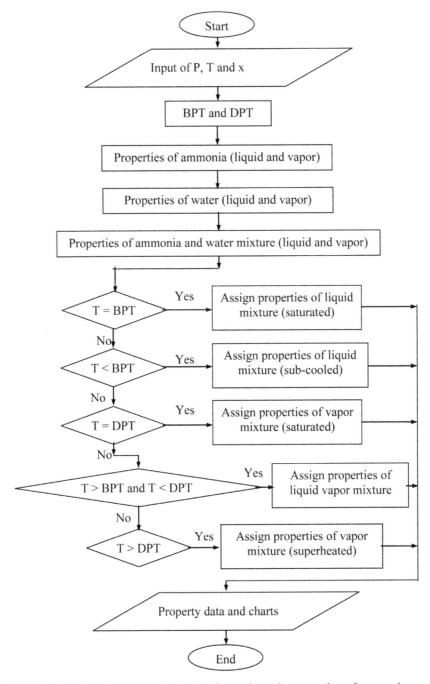

FIGURE 4.2 Flowchart to evaluate the thermodynamic properties of ammonia–water mixture.

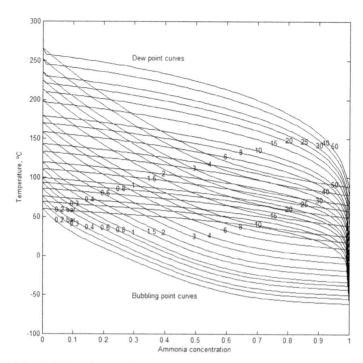

FIGURE 4.3   Bubble and dew point temperatures up to 100 bar.

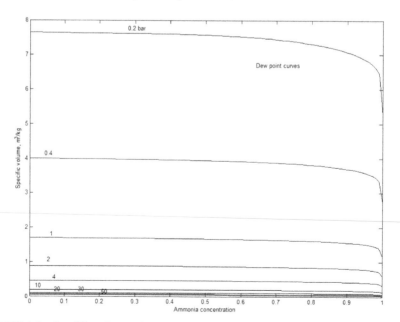

FIGURE 4.4   Specific volume of saturated vapor.

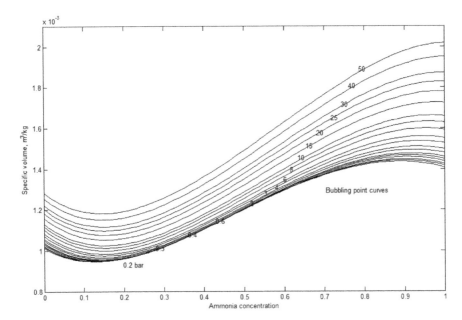

FIGURE 4.5   Specific volume of saturated liquid.

Figures 4.6 and 4.7 summarize the ammonia–water mixture on enthalpy–concentration chart for vapor and liquid, respectively, generated up to 50 bar. It is an enthalpy–concentration plot as function of ammonia concentration and pressure. The vapor curves consist of saturated vapor and auxiliary curves. The liquid curves are the saturated liquid curves with the BPT information. The properties of subcooled mixture can also be read from this chart. The enthalpies of liquid and vapor increase with the increase in pressure. The upper curve is the vapor curve, resulted from liquid concentration and DPT. The lower curve is the liquid enthalpy plot, resulted from the BPT and liquid ammonia concentration. The auxiliary curve is resulted from BPT and vapor ammonia concentration. The curves are generated from 0.2 to 50 bar pressure.

Figures 4.8 and 4.9 show the specific entropy of vapor and liquid, respectively, with a change in concentration from 0 to 1 and the pressure from 0.2 to 50 bar. The specific entropy is the function of ammonia mass concentration and pressure for the saturation curves. The specific entropy values are decreasing with increase in pressure, in vapor region. But in the liquid region, it is opposite, similar to specific volume. The specific entropy of fluid increases with increase in pressure.

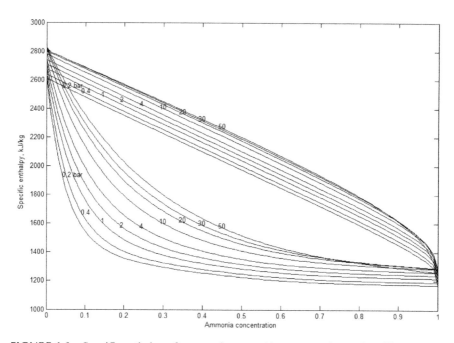

FIGURE 4.6   Specific enthalpy of saturated vapor with concentration and auxiliary curves.

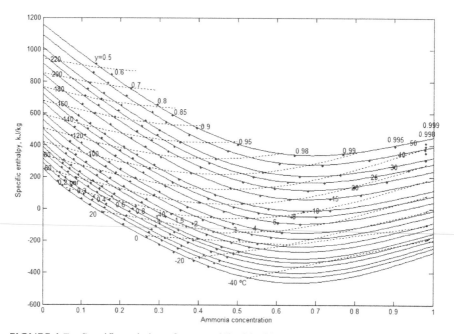

FIGURE 4.7   Specific enthalpy of saturated liquid with concentration and temperature.

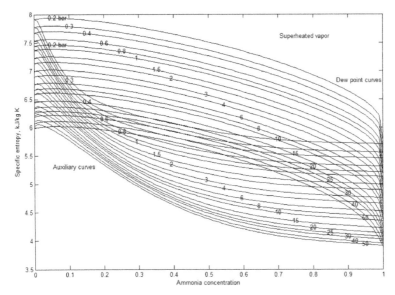

FIGURE 4.8   Specific entropy of saturated vapor with concentration.

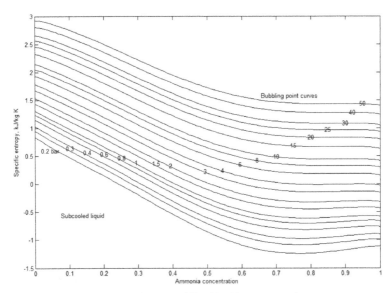

FIGURE 4.9   Specific entropy of saturated liquid with concentration.

The specific exergy is the maximum useful work obtained from enthalpy and entropy values.

$$e = h - T_0 s \qquad\qquad (4.44)$$

The exergy–concentration plot for ammonia–water mixture at various pressures is shown in Figures 4.10 and 4.11, respectively, for vapor and liquid. In liquid region, the curves are widened at high-pressure side. The specific exergy of vapor increases with increase in pressure. Exergy is the function of enthalpy and entropy. The liquid exergy has two treads of increasing and decreasing with changes in pressure and concentration.

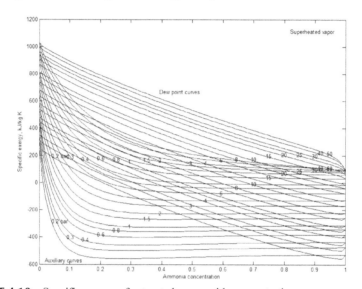

FIGURE 4.10   Specific exergy of saturated vapor with concentration.

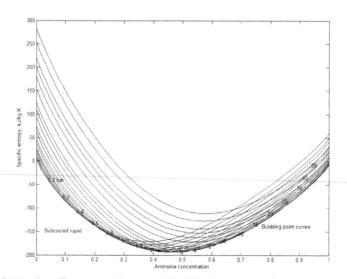

FIGURE 4.11   Specific exergy of saturated liquid with concentration.

4.7   ALGORITHM TO FIND THE PROPERTIES OF AMMONIA–WATER MIXTURE

The algorithm has been prepared to identify the region for thermodynamic properties generation using MATLAB codes.

```
%DIFFERENT PHASES OF AMMONIA-WATER MIXTURE
if(((T-Tbp)>=-0.05)&&((T-Tbp)<=0.05))||(T==Tbp)

%1. saturated liquid mixture
df=0;
RGN=1;
elseif(T<Tbp&&T<Tdp)

%2. sub cooled liquid mixture
df=0;
RGN=2;
elseif(((T-Tdp)>=-0.05)&&((T-Tdp)<=0.05))||(T==Tdp)

%3. saturated vapor mixture
df=1;
RGN=3;
elseif(T>Tbp&&T<Tdp)
%4. liquid-vapor mixture
RGN=4;
elseif(T>Tdp)

%5. superheated mixture
df=1;
RGN=5;
end
switch(RGN)

case 1
hm=(1-df)*hf+df*hg2;
sm=(1-df)*sf+df*sg2;
vm=(1-df)*vf+df*vg2;
Tdp=Tdp-273.15;
Tbp=Tbp-273.15;
```

```
case 2
if(((Tbp-Tdp)>=-1)&&((Tbp-Tdp)<=1))
df=1;
end
hm=(1-df)*hf+df*hg2;
sm=(1-df)*sf+df*sg2;
vm=(1-df)*vf+df*vg2;
Tdp=Tdp-273.15;
Tbp=Tbp-273.15;
hg1=0;
hg2=0;

case 3
hm=(1-df)*hf+df*hg2;
sm=(1-df)*sf+df*sg2;
vm=(1-df)*vf+df*vg2;
Tdp=Tdp-273.15;
Tbp=Tbp-273.15;

case 4
%AMMONIA-WATER LIQUID MIXTURE
hm=(1-df)*hf+df*hg2;
sm=(1-df)*sf+df*sg2;
vm=(1-df)*vf+df*vg2;
Tdp=Tdp-273.15;
Tbp=Tbp-273.15;

case 5
hm=(1-df)*hf+df*hg2;
sm=(1-df)*sf+df*sg2;
vm=(1-df)*vf+df*vg2;
Tdp=Tdp-273.15;
Tbp=Tbp-273.15;
end
```

4.8 AQUA–AMMONIA TABLES AND CHARTS

The properties tables and charts are developed and presented in the Appendix of this book. The considered variable in the tables and plots are pressure, concentration, and temperature. Tables A.1–A.13 show the properties of ammonia–water mixture and the focused resulted are temperature, specific volume, specific enthalpy, specific entropy, liquid concentration, and vapor concentration. The same properties are also plotted on enthalpy–concentration charts from Figures A.1 to A.13.

KEYWORDS

- ammonia–water mixture
- thermodynamic evaluation
- combined power
- cooling cycle
- zeotropic mixture

CHAPTER 5

BINARY MIXTURE THERMODYNAMIC PROCESSES

ABSTRACT

A thermal system is the combination of many subsystems. Each subsystem undergoes thermodynamic process. Similar to binary fluid properties, binary fluid processes are also complex in nature compared to single fluid system. It is easy to evaluate the binary fluid power system, cooling system, or combined system with a study on basic binary fluid processes and their solutions. The processes are formulated using mass balance and energy balance equations. The developed processes are explained with properties charts drawn to scale. Some typical numerical solutions are outlined to understand the process solutions.

5.1 INTRODUCTION

Thermodynamic evaluation of binary fluid system is somewhat complicated compared to the single-fluid system due to the extra property of concentration in addition to pressure and temperature. The understanding of processes evaluation helps to develop the computer models to iterate the results from the functions of properties. Therefore, thermodynamic processes can be solved with the binary mixture properties. The essential processes in the vapor absorption cycles (power, cooling, and combined power and cooling) are absorption, condensation, mixing, pumping, heat recovery, boiling, separation, distillation, throttling, preheating, superheating, and expansion. In the process evaluation, the systematic arrangement of mass balance and energy balance links plays an important role. The resulted structured formulation generates the state properties, that is, pressure, volume, temperature, concentration, entropy, density, energy, etc. The processes can be solved either by graphical method (which is suitable for manual method) or analytical approach. The graphical method is the manual method suitable for understanding and practicing of the processes solutions. The analytical method is best suited to develop a computer models. The computational models can be further can be used in the plant simulation and optimization. The process's solutions are described with the property charts drawn to scale using the computation tools. All the process's solutions can be integrated to develop a solution to a whole cycle or plant. Therefore, the theoretical solutions are framed from the properties generation, processes solutions, and cycle or plant modeling.

5.2 THERMODYNAMIC PROCESSES WITH BINARY MIXTURES

The heat-transfer processes in the vapor absorption cycles occur in boiler, condenser, subcooler, solution heat exchanger, absorber, distillation, super-heater, high-temperature regenerator, and low-temperature regenerator. Other than heat-transfer processes, the other processes involved in this subject are throttling, expansion, pumping, adiabatic mixing, and separation. In the mixing, separation, dephlegmator or distillation solutions, the application of lever rule to generate the mass and energy balance formulation. The following section details the thermodynamic processes with formulae and solutions to understand the simulation and analysis of binary vapor system for power or cooling or for both. For heat exchanger's solution, the regular practice can be adopted. The processes are outlined using enthalpy–concentration chart for ammonia–water mixture working fluid.

5.2.1 SEPARATION

Separator is a physical cylindrical shaped cylinder that allows the separation of liquid by gravity from the bottom from the liquid–vapor mixture. The vapor is tapped from the top of the cylinder. The concentration of the liquid and vapor in the separation chamber will be different.

Figure 5.1 shows the separation process after the vapor generator (boiler). In some designs, separation is the integral part of the boiler where heating and separation takes place in a common place. The liquid–vapor mixture in the boiler enters into separator where the vapor is collected at the upper portion and the liquid at the bottom. The liquid has low concentration and thus called weak solution. The vapor is saturated vapor at the pressure and temperature. Here, three concentrations exists, namely, inlet mixture concentration, weak solution concentration, and saturated vapor concentration. It is assumed that the separation process occurs at constant pressure.

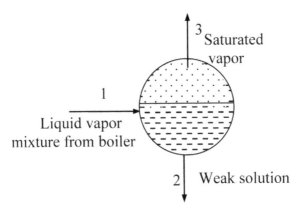

FIGURE 5.1   Separation of liquid–vapor mixture into weak solution and saturated vapor.

The separator parameters are key parameters which influence the performance of the system. The parameters in the separator are pressure, temperature, dryness fraction (vapor fraction), and vapor concentration. The separation can be solved either at a given pressure and temperature or at known temperature and vapor concentration. At known pressure and temperature, the unknown vapor concentration can be determined. Similarly at known temperature and vapor concentration, unknown pressure can be determined. To solve the separation process, the separator inlet condition is required, that is, either separator inlet concentration or separator dryness fraction (mass of vapor to the total mass). Figure 5.2 is the enthalpy–concentration chart that shows separation process solved at given separator pressure (13 bar), dryness fraction (30%), and vapor concentration (0.88). The solved results are separator temperature, concentrations, and enthalpies. The liquid–vapor mixture (1) is separated into liquid (2) and vapor (3).

The mass balance in separation process is

$$m_1 = m_2 + m_3 \tag{5.1}$$

The total ammonia portion before separation and after separation is the same.

$$m_1 x_1 = m_2 x_2 + m_3 x_3 \tag{5.2}$$

Eliminating m_2,

$$m_1 x_1 = (m_1 - m_3)x_2 + m_3 x_3$$

$$m_1 x_1 = m_1 x_2 - m_3 x_2 + m_3 x_3$$

$$m_1(x_1 - x_2) = m_3(x_3 - x_2)$$

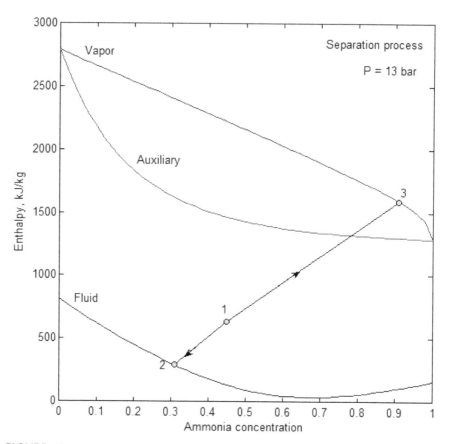

FIGURE 5.2   Separation process at known pressure, dryness fraction, and vapor concentration.

Therefore, the dryness fraction (dryness fraction) is defined from the simplification:

$$DF = \frac{m_3}{m_1} = \frac{x - x_l}{x_v - x_l} = \frac{x_1 - x_2}{x_3 - x_2} \tag{5.3}$$

Similarly by eliminating m_1 in the ammonia balance eq 5.3, wetness fraction or liquid fraction:

$$WF = \frac{m_2}{m_1} = \frac{x_v - x}{x_v - x_l} = \frac{x_3 - x_1}{x_3 - x_2} \tag{5.4}$$

Equation 5.3 shows that the dryness fraction or vapor fraction is inversely proportional to the distance between liquid and mixture. Similarly, eq 5.4 indicates that liquid fraction or wetness fraction is inversely proportional to the distance between the vapor and mixture. It is called lever rule.

Similar to ammonia balance, the energy before separation is equal to the total energy after the separation neglecting the heat losses from the process.

$$m_1 h_1 = m_2 h_2 + m_3 h_3 \tag{5.5}$$

Eliminating m_2,

$$m_1 h_1 = (m_1 - m_3)h_2 + m_3 h_3$$
$$m_1 h_1 = m_1 h_2 - m_3 h_2 + m_3 h_3$$
$$m_1(h_1 - h_2) = m_3(h_3 - h_2)$$

The dryness fraction also can be written as a function of enthalpies. After the simplification of the above equations,

$$DF = \frac{m_3}{m_1} = \frac{h - h_l}{h_v - h_l} = \frac{h_1 - h_2}{h_3 - h_2} \tag{5.6}$$

Similarly by eliminating m_1 in the energy balance eq 5.5, wetness fraction or liquid fraction,

$$WF = \frac{m_2}{m_1} = \frac{h_v - h}{h_v - h_l} = \frac{h_3 - h_1}{h_3 - h_2} \tag{5.7}$$

Therefore, strong solution concentration,

$$x_1 = x_2 + DF(x_3 - x_2) \tag{5.8}$$

For example, to find the separator temperature, liquid concentration, solution concentration, and solution enthalpy in separator is located after the vapor generator in a vapor absorption refrigeration system at the known following data. The required dryness fraction is 30% at a pressure of 13 bar. The ammonia concentration in vapor at the exit of separator is 0.88.

Analytical solution involves use of property tables and formulae.

The dryness fraction for saturated vapor phase is 1. Therefore, from the property table, at 13 bar pressure, the solution temperature can be iterated between 131.32°C at 0.8 concentration and 112.54°C at 0.9 concentration. It is approximately 117°C.

Similarly, the dryness fraction is zero for saturated liquid. At zero dryness fraction, the liquid concentration can be iterated at 117°C, that is, between 0.2 (135.65°C) and 0.3 (112.33°C). It is approximately 0.28.

OR

At 100% dryness fraction (complete vapor), 13 bar, the liquid concentration at 0.8 is 0.22 and at 0.9, it is 0.3. It can be iterated between 0.22 and 0.3 for 0.88 concentration and it is approximately 0.28. At 0% dryness fraction (complete liquid), the fluid temperature is 135.65°C at 0.2 and 112.33°C at 0.3. Similarly, the temperature also can be iterated and it is 117°C.

The dryness fraction is

$$DF = \frac{x - x_l}{x_v - x_l} = \frac{x_1 - x_2}{x_3 - x_2}$$

Therefore, solution concentration,

$$x = x_1 = x_2 + DF\,(x_3 - x_2)$$
$$0.28 + 0.3(0.88 - 0.28) = 0.46$$

From the properties tables,

$$h_2 = 334 \text{ kJ/kg}$$
$$h_3 = 1638 \text{ kJ/kg}$$
$$h_1 = (1 - DF)h_2 + DFh_3 = (1 - 0.3)334 + 0.3 \times 1638 = 725.2 \text{ kJ/kg}$$

A scale and pencil can be used with enthalpy–concentration chart in graphical solution. For graphical solution, locate a State 3 on vapor curve at 0.88 concentration and 13 bar. Draw a horizontal line till the auxiliary curve and extend to a vertical line from this intersection. The vertical line meets the saturated liquid line at liquid concentration. At State 2 (liquid), both liquid concentration and temperature can be read as 0.28 and 117°C. From the lever rule or locating point with length inversely proportional to the dryness fraction, the point 1 can be solved as 0.46. The solution enthalpy from the graph is 770 kJ/kg. Figure 5.3 shows the solution for separation process on enthalpy–concentration chart.

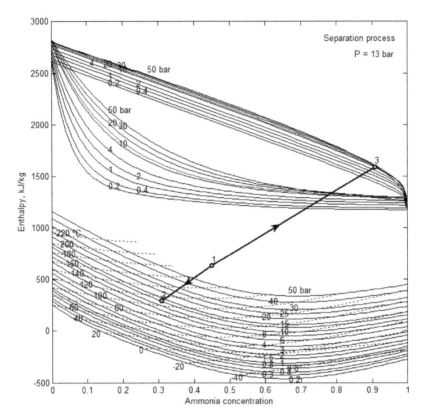

FIGURE 5.3   Separation process on concentration chart.

5.2.2   ADIABATIC MIXING

Mixing is a process with or without heat transfer. The mixing without heat transfer is called adiabatic mixing and the mixing with heat transfer is called diabatic mixing. Two or more fluids are mixed in a mixing and it is required to find the state properties after completion of mixing. In mixing, the pressure assumed as constant. For example, at the inlet of absorber, vapor is absorbed into liquid solution and mixed with each other. This absorption is considered as an adiabatic mixing process. Later, it is condensed by rejecting heat to surroundings. This total process is termed as adiabatic process with heat transfer (addition or rejection, rejection in this case) or diabatic process. The final temperature depends on the temperature of the fluids before mixing process. Figure 5.4 shows the adiabatic mixing of two fluids into a single fluid.

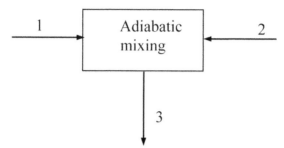

FIGURE 5.4   Adiabatic mixing.

Figure 5.5 outlines this adiabatic mixing process on enthalpy–concentration diagram. Two streams 1 and 2 are mixed at State 3, which lies in two-phase region.

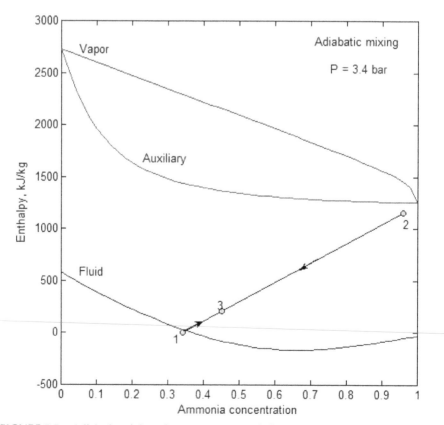

FIGURE 5.5   Adiabatic mixing of two streams on enthalpy–concentration diagram.

The mass and energy balances for the mixing chamber are outlined.

$$m_1 + m_2 = m_3 \tag{5.9}$$

$$m_1 x_1 + m_2 x_2 = m_3 x_3 \tag{5.10}$$

$$m_1 h_1 + m_2 h_2 = m_3 h_3 \tag{5.11}$$

From eq 5.9, the mass at the outlet is calculated, and eqs 5.10 and 5.11 result the mixture concentration and mixture specific enthalpy at the outlet of the mixing chamber. From the enthalpy–concentration diagram of mixing streams (Fig. 5.5), mixture concentration and mixture enthalpy have been calculated from the following equations.

From eq 5.10, eliminating m_1,

$$x_3 = \frac{m_1 x_1 + m_2 x_2}{m_3} = \frac{(m_3 - m_2)x_1 + m_2 x_2}{m_3} = \frac{m_3 x_1 - m_2 x_1 + m_2 x_2}{m_3}$$

$$\therefore x_3 = x_1 + \frac{m_2}{m_3}(x_2 - x_1)$$

$$\therefore \frac{x_3 - x_1}{x_2 - x_1} = \frac{m_2}{m_3} \tag{5.12}$$

Similarly,

$$h_3 = h_1 + \frac{m_2}{m_3}(h_2 - h_1)$$

$$\therefore \frac{h_3 - h_1}{h_2 - h_1} = \frac{m_2}{m_3} \tag{5.13}$$

The outlet concentration and enthalpy can also be calculated using the graphical method. The mixture temperature can be iterated from the enthalpy h_3.

Alternatively by eliminating, m_2

$$x_3 = \frac{m_1 x_1 + m_2 x_2}{m_3} = \frac{m_1 x_1 + (m_3 - m_1)x_2}{m_3} = \frac{m_1 x_1 + m_3 x_2 - m_1 x_2}{m_3}$$

$$x_3 = x_2 + \frac{m_1}{m_3}(x_1 - x_2)$$

$$x_3 = x_2 - \frac{m_1}{m_3}(x_2 - x_1)$$

$$\therefore \frac{x_3 - x_2}{x_2 - x_1} = \frac{m_1}{m_3} \tag{5.14}$$

Similarly,

$$\frac{h_3 - h_2}{h_2 - h_1} = \frac{m_1}{m_3} \tag{5.15}$$

It shows that the length of 1–3 in concentration–enthalpy chart is proportional to m_2/m_3, that is, m_2/m_3 times of the total length 1–2. Similarly 3–2 length is proportional to m_1/m_3.

From the mass balance equation of the mixed solution can be found as

$$x_3 = x_1 + \frac{m_2}{m_3}(x_2 - x_1) \tag{5.16}$$

Similarly from the energy balance,

$$h_3 = h_1 + \frac{m_2}{m_3}(h_2 - h_1) \tag{5.17}$$

For example, a subcooled liquid of 0.82 kg/s of aqua–ammonia ($x_1 = 0.34$) at 47°C and 3.4 bar mixes adiabatically with another liquid–vapor mixture with flow rate of 0.18 kg/s and 12°C at the same pressure. To establish the state points on h–x diagram and result, the mixture concentration, mixture enthalpy, isotherm through the mixture state, and amount of liquid and vapor after mixing either analytical or graphical solution can be adopted. The mixture obtained in the two-phase region is a blend of compressed liquid and liquid–vapor mixture. As per the lever rule, the distance between 1 and 3 is the ratio of mass at 2 to mass at 3 and between 2 and 3 is the ratio of mass at 1 to mass at 3. The solution is shown in enthalpy–concentration chart in Figure 5.6.

Pressure $= P_1 = P_2 = P = 3.4$ bar

$$m_1 = 0.82 \text{ kg/s}$$
$$x_1 = 0.34$$
$$T_1 = 47°C$$
$$m_2 = 0.18 \text{ kg/s}$$
$$x_2 = 0.96$$
$$T_2 = 12°C$$

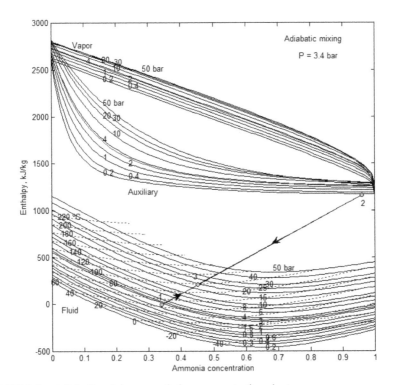

FIGURE 5.6 Adiabatic mixing on enthalpy–concentration chart.

Using chart:

State 1 is established on the concentration enthalpy chart as the intersection of concentration and temperature 47°C. With reference to the saturation liquid curve, State 1 can be defined. If the state is below the curve, it is subcooled or compressed liquid on the curve shows the saturation state and above the state reveals the liquid–vapor mixture. Since State 1 lies below the saturation curve, it is subcooled state. State 2 can be obtained by drawing isotherm at 12°C from saturated liquid at 3.4 bar, extension vertically up till auxiliary, followed by horizontal line till vapor curve and finally joint this saturation vapor point to the initial saturation liquid state. This inclined line is called isotherm. The isotherm will intersect the 0.96 concentration vertical line and it is termed as State 2. State 3 is located on the line joining 1–2 at the distance $m_2/m_3 = 0.18/1$, that is, 18% of the length of point 1. Then, one obtains,

$$x_3 = 0.45 \text{ and } h_3 = 206 \text{ kJ/kg}$$

The trial-and-error procedure gives the 49°C isotherm passing through State 3.

Since the mixture state is above the saturation liquid curve, it is liquid and vapor mixture and consists of vapor whose concentration corresponds to state, x_v is 0.98 and the liquid concentration, x_l is 0.37).

In the mixture,

$$\frac{m_v}{m_3} = \frac{x_3 - x_l}{x_v - x_l} = \frac{0.45 - 0.37}{0.98 - 0.37} = 0.136$$

$$m_v = 0.136 \text{ kg/s}$$

The liquid mass is then,

$$m_l = m_3 - m_v = 1 - 0.136 = 0.86 \text{ kg/s}$$

Using tables:

At 3.4 bar, 0.34 concentration, and 47°C, the liquid condition is subcooled, since the bubble-point temperature at this condition is 54.5°C. At 3.4 bar pressure curve and 0.34 concentration line, State 1 can be identified.

$$h_1 = -2.55 \text{ kJ/kg}$$

Similarly at 3.4 bar and 12°C liquid–vapor mixture and the concentration of 0.96,

$$h_2 = 1157.5 \text{ kJ/kg}$$

$$x_3 = x_1 + \frac{m_2}{m_3}(x_2 - x_1) = 0.34 + \frac{0.18}{1}(0.96 - 0.34) = 0.45$$

$$h_3 = h_1 + \frac{m_2}{m_3}(h_2 - h_1) = -2.55 + \frac{0.18}{1}(1157.5 - (-2.55)) = 206.25 \text{ kJ/kg}$$

From the tables, the DF lies between 0.1 and 0.2 and it is 0.136.

5.2.3 MIXING WITH HEAT TRANSFER (DIABATIC MIXING)

The diabatic mixing process may be either mixing with heat addition or mixing with heat rejection. Figure 5.7 shows a diabatic mixing process with heat rejection. It happens in an absorber where two fluids mix with each other and condenses by rejection heat to the surroundings. In absorber, these two processes happen together. For simplicity, two thermodynamic processes are shown separately to understand the adiabatic mixing and heat-transfer process. The formulation helps to find the heat-transfer load in the absorber to condense the vapor mixture completely into saturated liquid mixture.

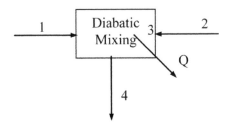

FIGURE 5.7 Diabatic mixing.

The mixing with heat transfer occurs in absorber and boiler. In absorber, heat is rejected after mixing to condenser the liquid–vapor mixture. In the boiler, heat is added to convert to liquid into liquid–vapor mixture. Later, the liquid and vapor are separated at the exit of boiler. Figure 5.8 depicts a mixing process with heat rejection in a typical absorber. State 1 is slightly subcooled liquid and State 2 is liquid–vapor mixture. After the mixing of these two fluids, the resultant fluid is in the state of liquid–vapor mixture. This liquid–vapor mixture is cooled by rejecting heat to the surroundings to the saturated liquid condition. The process is exothermic, so heat (Q) is released in the mixing process. The released heat is rejected.

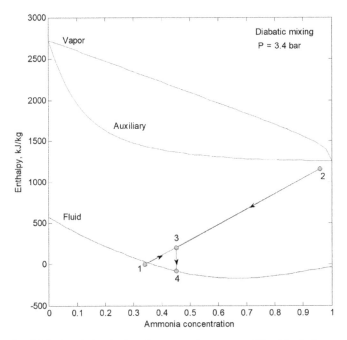

FIGURE 5.8 Diabatic mixing in absorber to result a saturated liquid solution.

Following are the mass balance and energy balance formulations and its simplifications.

$$m_1 + m_2 = m_3 = m_4 \tag{5.18}$$

$$m_1 x_1 + m_2 x_2 = m_3 x_3 + m_4 x_4 \tag{5.19}$$

$$m_1 h_1 + m_2 h_2 = m_4 h_4 + Q \tag{5.20}$$

From eq 5.18, the mass at the outlet is calculated, and eqs 5.19 and 5.20 result the mixture concentration and mixture specific enthalpy at the outlet of the mixing chamber. From the enthalpy–concentration diagram of mixing streams (Fig. 5.8), mixture concentration and mixture enthalpy have been calculated from the following equations.

Similar to the earlier discussion, the concentration of the mixture can be evaluated. For example, the process with heat rejection like absorber can be formulated as follows (Fig. 5.8).

From eq 5.20, eliminating m_1,

$$(m_4 - m_2)h_1 + m_2 h_2 = m_4 h_4 + Q$$

$$m_4 h_1 - m_2 h_1 + m_2 h_2 = m_4 h_4 + Q$$

$$m_4 h_1 + m_2 (h_2 - h_1) = m_4 h_4 + Q$$

$$h_4 = \frac{m_4 h_1 + m_2 (h_2 - h_1) - Q}{m_4}$$

$$= h_1 + \frac{m_2}{m_4}(h_2 - h_1) - \frac{Q}{m_4}$$

Following equation results, the heat-transfer duty in absorber:

$$\frac{Q}{m_3} = h_1 + \frac{m_2}{m_3}(h_2 - h_1) - h_4 \tag{5.21}$$

Since the mixing-process and heat-transfer processes are solved separately, the mixture concentration is same during the heat-transfer process. Therefore, in h–x diagram is a vertical line. The distance of heat-transfer line in h–x diagram is equal to the Q/m_3 as per eq 5.21. In case of adiabatic mixing, the exit state lies between State 1 and 2.

In diabatic process, it is required to find the enthalpy of the mixture after mixing and heat transfer. Equation 5.21 can be rewritten for enthalpy of the mixture after the heat transfer, that is, h_4.

$$h_4 = h_1 + \frac{m_2}{m_3}(h_2 - h_1) + \frac{Q}{m_3} \tag{5.22}$$

The concentration remains the same and the temperature can be iterated from the saturated liquid condition from known pressure and concentration.

For example, a subcooled liquid of 0.82 kg/s of aqua–ammonia (x_1 = 0.34) at 47°C and 3.4 bar mixes adiabatically with another liquid–vapor mixture with flow rate of 0.18 kg/s and 12°C at the same pressure. If heat rejected from the system to condense the vapor into saturated liquid, the following section outlined the solution. The diabatic process is depicted on enthalpy–concentration chart (Fig. 5.9).

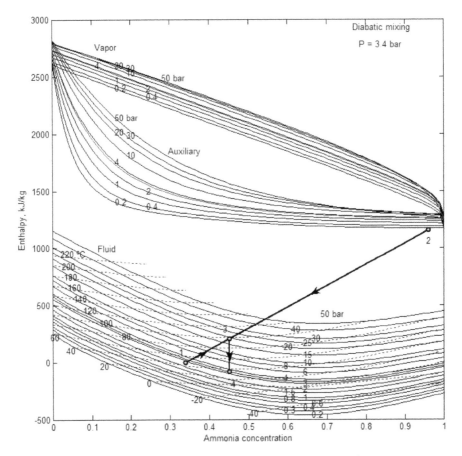

FIGURE 5.9 Adiabatic mixing with heat transfer (diabatic mixing) shown on enthalpy–concentration chart.

h_4 is the enthalpy at the bubble-point temperature of 34.5°C at 3.4 bar. The heat transfer for condensation

$$Q = m_4 h_1 + m_2(h_2 - h_1) - m_4 h_4$$
$$= 1 \times (-2.55) + 0.18 (1157.5 - (-2.55)) - 1 \times (-83.39) = 289.64 \text{ kJ/kg}$$

Since the concentration of the mixture is unchanged, State 4 is located below 3, with enthalpy of −83.38 kJ/kg.

Let us check some other model for diabatic process as shown in Figure 5.10. Consider a stream of 15 kg/s of aqua–ammonia ($x_1 = 0.8$) at 0°C and 5 bar mixes adiabatically with another saturated liquid stream with the flow rate 10 kg/s and 100°C at the same pressure. To establish first the state points on h–x diagram and obtain mixture concentration, mixture enthalpy, isotherm through the mixture state, and amount of liquid and vapor after mixing, the following procedure is adopted and then extended to diabatic process.

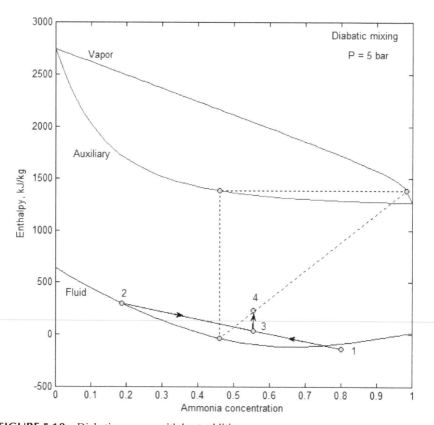

FIGURE 5.10   Diabatic process with heat addition.

Pressure $= P_1 = P_2 = P = 5$ bar

$$m_1 = 15 \text{ kg/s}$$
$$x_1 = 0.8$$
$$T_1 = 0°C$$
$$m_2 = 10 \text{ kg/s, saturated liquid}$$
$$T_2 = 100°C$$

Using chart:

State 1 is established on the concentration enthalpy chart as the intersection of concentration and temperature 0°C. State 2 lies on the saturation line at 5 bar and 100°C temperature. The concentration ($x_1 = 0.8$) at State 2 is 0.18. State 3 is located on the line joining 1–2 at the distance $m_2/m_3 = 0.4$, that is, 40% of the length of point 1. Then, one obtains,

$$x_3 = 0.55 \text{ and } h_3 = 36.16 \text{ kJ/kg}$$

and trial-and-error procedure gives the 36°C isotherm passing through State 3.

Since the mixture state is above the saturation temperature line, it consists of vapor whose concentration corresponds to state B ($x_v = 0.99$ and the liquid concentration is $x_l = 0.52$). Therefore, the mass of vapor with $m = m_l + m_v = 25$ kg/s.

$$\frac{m_v}{m} = \frac{x_3 - x_l}{x_v - x_l} = \frac{0.55 - 0.52}{0.99 - 0.52} = 0.076$$

$$m_v = 1.9 \text{ kg/s}$$

The liquid mass is then,

$$m_l = m_3 - m_v = 25 - 1.9 = 23.1 \text{ kg/s}$$

Using tables:

At 5 bar, 0.8 concentration, and 0°C, the liquid condition is subcooled since, the bubble-point temperature at this condition is 11.09°C. Approximately at on 3.6 bar pressure curve and 0.8 concentration line, State 1 can be identified.

$$h_1 = -140 \text{ kJ/kg}$$

Similarly at 5 bar and 100°C saturated liquid solution (DF = 0), the concentration is 0.18. At this condition,

$$h_2 = 301 \text{ kJ/kg}$$

$$x_3 = x_1 + \frac{m_2}{m_3}(x_2 - x_1) = 0.8 + \frac{10}{25}(0.8 - 0.18) = 0.55$$

$$h_3 = h_1 + \frac{m_2}{m_3}(h_2 - h_1) = -140 + \frac{10}{25}(301 - (-140)) = 36 \, \text{kJ/kg}$$

From the tables, the DF lies between 0 and 0.1 with reference to the concentration ($x_3 = 0.55$) and enthalpy ($h_3 = 36$ kJ/kg). It results as 0.076.

A heat transfer to the system at a rate of 5000 kJ/s is considered for diabatic process. The state of mixture after heat transfer, liquid concentration, and vapor concentrations can be determined. If the required exit condition is saturated liquid after the heat rejection, the required heat transfer also can be obtained.

Using eq 5.22, one gets the enthalpy of the leaving streams as

$$h_4 = h_1 + \frac{m_2}{m_3}(h_2 - h_1) + \frac{Q}{m_3}$$

$$= -149.42 + \frac{10}{25}(301.04 - (-149.42)) + \frac{5000}{25} = 236.16 \, \text{kJ/kg}$$

Since the concentration of the mixture is unchanged, State 4 is located above 3 with enthalpy value 236.16 kJ/kg. Using the trial-and-error procedure, it is found that 46°C isotherm passes through 4. The concentrations are $x_v = 0.99$ and $x_l = 0.46$

Then from the equation in the present notations:

$$\frac{m_v}{m_4} = \frac{x_4 - x_l}{x_v - x_l} = \frac{0.55 - 0.46}{0.99 - 0.46} = 0.18$$

$$m_v = 4.5 \, \text{kg/s}$$

The liquid mass is then,

$$m_l = m_3 - m_v = 25 - 4.5 = 20.5 \, \text{kg/s}$$

Using tables:

$$h_4 = h_1 + \frac{m_2}{m_3}(h_2 - h_1) + \frac{Q}{m_3}$$

$$= -149.42 + \frac{10}{25}(301.04 - (-149.42)) + \frac{5000}{25} = 236.16 \, \text{kJ/kg}$$

At $h_4 = 236.16$ and $x_4 = 0.55$, the dryness fraction should be between 0.1 and 0.2. It can be iterated and the result is 0.18.

For saturated liquid condition:

If the whole mixture is saturated corresponding to 5 bar and concentration 0.55, the enthalpy is scaled as -94.64 kJ/kg. Now making the energy balance for the control volume, it is seen

$$h_4 = h_1 + \frac{m_2}{m_3}(h_2 - h_1) - \frac{Q}{m_3}$$

$$Q = m_3\left(h_1 + \frac{m_2}{m_3}(h_2 - h_1) - h_4\right)$$

$$Q = 25\left(-140.42 + \frac{10}{25}(301.04 - (-140.42)) - (-94.64)\right) = 3270\,\text{kJ/kg}$$

5.2.4   THROTTLING

Throttling is an important process in a refrigeration plant to generate the cooling. It is an irreversible process where the fluid temperature decreases after the throttling. For throttling, expansion valves or capillary tube are used. The throttling valves separated the high-pressure lines and low-pressure lines. In a vapor absorption refrigeration, refrigerant line and weal solution line needs throttling. It is assumed that the work and heat transfer in throttling are neglected. The resulted process is an isenthalpic process.

In a typical vapor absorption system, throttling of subcooled ammonia–water liquid solution from 13 to 4 bar is considered for solution. The initial temperature is 31°C with ammonia mass concentration of 0.96. Iteration is made to find the temperature after throttling and 0.8°C final temperature has been resulted (Fig. 5.11). In graphical solution, locate a point for State 1 at 31°C temperature line with 0.96 concentration line. Using the trial-and-error procedure, it is found that 0.8°C isotherm passes through 2. The concentrations are $x_v = 0.99$ and $x_l = 0.95$.

$$\frac{m_v}{m_2} = \frac{x_2 - x_l}{x_v - x_l} \tag{5.23}$$

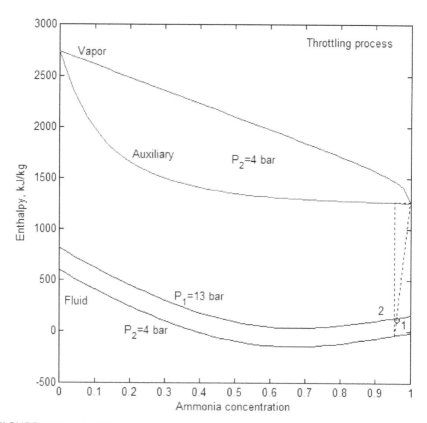

FIGURE 5.11   Throttling process in a typical vapor absorption system.

Then from eq 5.23 in the present notations,

$$\frac{m_v}{m_2} = \frac{x_2 - x_l}{x_v - x_l} = \frac{0.96 - 0.95}{0.99 - 0.96} = 0.17$$

$$m_v = 0.17 \text{ kg/s}$$

The liquid mass is then,

$$m_l = m_2 - m_v = 1 - 0.17 = 0.83 \text{ kg/s}$$

Out of 1 kg/s of throttled mixture, the vapor portion is 0.17 kg/s and the liquid mass is 0.83 kg/s.

In the above solution, let us check the exit temperature of throttling if the concentration is shifted from 0.96 to 0.75 without affecting the initial pressure and temperature.

Figure 5.12 is plotted for the throttling solution, with the solution concentration 0.75 instead of 0.96 and the other conditions are same as in earlier description.

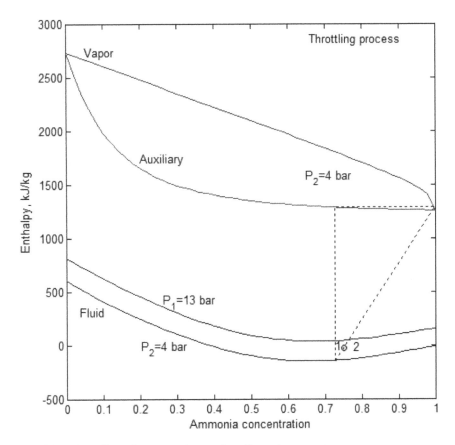

FIGURE 5.12   Effect of concentration on throttling exit temperature.

Similar to earlier solution, locate a point for State 1 at 31°C temperature line with 0.96 concentration line. Using the trial-and-error procedure, it is found that 8.6°C isotherm passes through 2. The concentrations are $x_v = 0.99$ and $x_l = 0.73$. To refer the exit temperature from the throttling, the temperature curve passing through State 2 can be referred in Figure 5.13.

$$\frac{m_v}{m_2} = \frac{x_2 - x_l}{x_v - x_l} = \frac{0.75 - 0.72}{0.99 - 0.95} = 0.085$$

$$m_v = 0.085 \text{ kg/s}$$

The liquid mass is then,

$$m_l = m_2 - m_v = 1 - 0.085 = 0.915 \text{ kg/s}$$

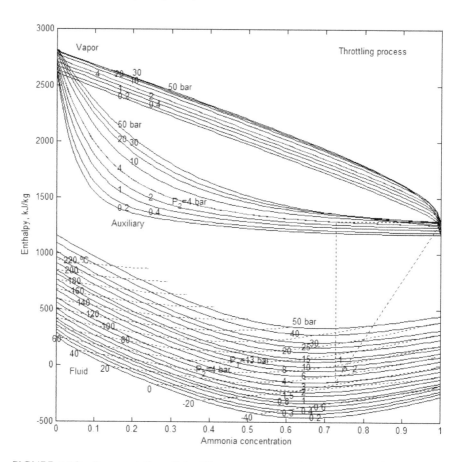

FIGURE 5.13 Representation of throttling process on enthalpy–concentration chart with temperature curves.

Out of 1 kg/s of throttled mixture, the vapor portion is 0.085 kg/s and the liquid mass is 0.915 kg/s.

In this problem, the change in the concentration from 0.96 to 0.75 increases the temperature of the fluid at the exit of the throttling. Therefore, a higher concentration decreases the temperature of the fluid at the exit of the throttling. In cooling plant, the refrigerant concentration should be high to result low temperature at the exit of throttling. Dephlegmator is used to increase the concentration of fluid. A high concentration drops the refrigerant flow rate and drops the power due to increased distillation. Therefore, if concentration is increased with dephlegmator in power plant, the output will suffer due to suppressed fluid in turbine.

Similarly in a typical another throttling, the saturated aqua solution leaves the generator at 90°C and at 12 bar, expanded through a throttling device. The pressure after throttling is found to be 4 bar.

The saturated liquid concentration before throttling is found as 0.395 from the intersection of 90°C isotherm and 12 bar saturation liquid line. This is also the point after throttling. With 4 bar pressure curves, an isotherm can be generated by hit and trial method to get result liquid concentration as 0.343 and vapor concentration as 0.97 and isotherm.

5.2.5 DEPHLEGMATOR

The refrigerant concentration is maintained high to result a low temperature in a refrigeration plant. In a low-concentration mixture, high content of moisture will exists. To increase the concentration, this moisture needs to be condensed for removal from the mixture. This process increases the concentration of the working fluid. The process of distillation of the working fluid to increase the concentration is called dephlegmation. The equipment is called dephlegmator. Dephlegmator is a heat exchanger in which circulating water flow to cool the working fluid. The cooling of the working fluid removes the moisture and so increased concentration. The heat from the working fluid is transferred to the circulating water or some other fluid to gain the performance of the refrigerator. The boiling or condensing temperature of the water is higher than the ammonia. Therefore, first the moisture in dephlegmator condenses without condensing the ammonia. By gravity, the condensed moisture is dripped to the boiler. To condense ammonia, very low temperature is required and therefore ammonia remains in vapor form in the dephlegmator. The process in dephlegmator is steady state process and the process pressure is assumed as constant. Dephlegmator favors the refrigeration plant. But it is not favoring the vapor absorption power plant, that is, Kalina power plant's performance. The use of dephlegmator in a power plant facilitates to operate the boiler at the low pressure relatively without its association. Therefore, a low capacity or low-cost vapor absorption power plant can be designed with the use of dephlegmator. The resulting two-phase mixture is then fed to adiabatic separator, where the saturated liquid and saturated vapors are separated. It is assumed that the entire process takes place at a constant pressure.

The dephlegmator is required for ammonia–water mixture but not to LiBr–water mixture. Since LiBr is in solid as well as in liquid form, it will not evaporate. The refrigerant water is pure in vapor form.

Figure 5.14 shows the arrangement of components with dephlegmator in a vapor absorption system. The process details are shown in enthalpy–concentration diagram (Fig. 5.15).

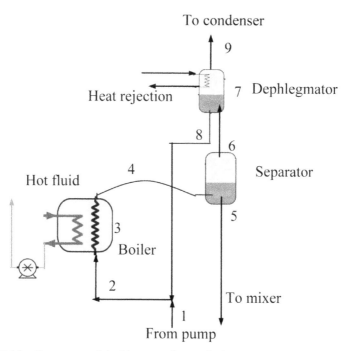

FIGURE 5.14   Generator and dephlegmator in a typical vapor absorption system.

The dry fraction (mass of vapor to the total mass) in separator after the generator,

$$VF_1 = \frac{x_4 - x_5}{x_6 - x_5}$$

(5.24)

Vapor separated in generator from m_4,

$$m_6 = DF_1 m_4$$

(5.25)

Liquid separated in generator from m_4,

$$m_5 = (1 - DF_1) m_4$$

(5.26)

Similar to separator, vapor–mass ratio in dephlegmator,

$$DF_2 = \frac{x_7 - x_8}{x_9 - x_8} \qquad (5.27)$$

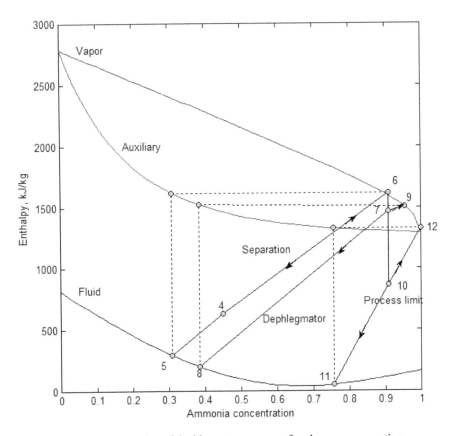

FIGURE 5.15 Representation of dephlegmator process after the vapor separation.

Vapor separated in dephlegmator from m_7,

$$m_9 = DF_2 m_7 \qquad (5.28)$$

Liquid separated in dephlegmator from m_7,

$$m_8 = (1 - DF_2)m_7 \qquad (5.29)$$

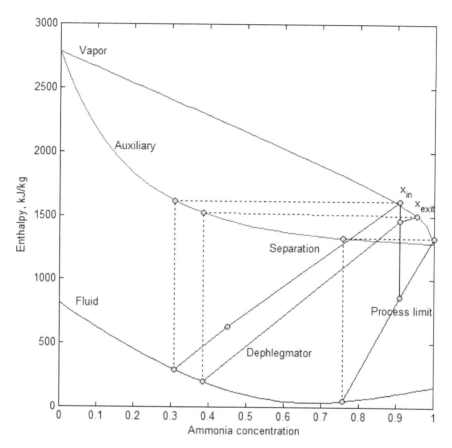

FIGURE 5.16   Actual dephlegmator process and maximum dephlegmator process to define the effectiveness of process.

The heat rejection from the vapor to surroundings or solution results increases in vapor concentration. As shown in Figure 5.16, the removal of water from the vapor increases the concentration from x_{in} at the inlet of dephlegmator to x_{exit}. It also depicts the maximum limit of heat rejection to generate 100% pure ammonia vapor. Based on this, the effectiveness of dephlegmator is defined as the ratio of actual rise in concentration to the maximum possible rise.

Effectiveness of dephlegmator is function of inlet and outlet concentrations of heat exchanger.

$$\varepsilon_{dephlegmator} = \frac{x_{exit} - x_{in}}{1 - x_{in}} \qquad (5.30)$$

$$h_7 = h_6 - \frac{Q_d}{m_6} \tag{5.31}$$

where Q_d is the heat loss in dephlegmator. The heat loss can be determined from the required status (concentration) at the exit and the enthalpies.

For example, in a typical vapor absorption system with ammonia–water solution, a solution enters the boiler at 12 bar, 70°C, and 0.45 concentration. It is heated up to 120°C. The separated vapor is connected to a dephlegmator and it is cooled by rejecting 100 kW of heat to the surroundings. The return liquid solution from the dephlegmator is mixed at the inlet of generator. Solve the dryness fractions, concentrations, and mass details in generator and dephlegmator at 1 kg/s of mass at the inlet of generator (Fig. 5.17).

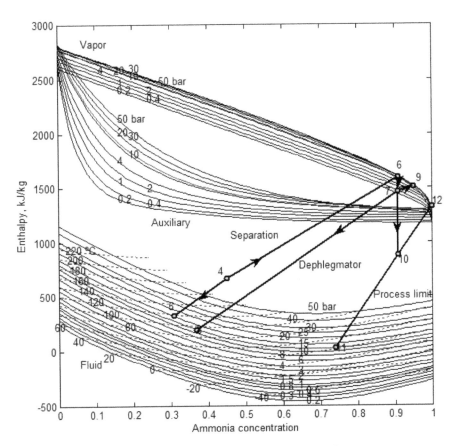

FIGURE 5.17 Dephlegmator process on enthalpy–concentration chart.

Neglecting the mixing effect between 1, 2, and 4, the strong solution concentration x_1 and x_2 can be assumed same.

$$x_1 = x_2 = x_{ss} = 0.45$$

During heating at constant concentration,

$$x_3 = x_4 = x_{ss} = 0.45$$

At 12 bar and 120°C, the weak solution concentration, that is, $x_5 = x_{ws}$ can be obtained from the h–x chart. A vertical line up to auxiliary line and horizontal extension up to vapor curve results a vapor concentration (x_6). Therefore, from the chart

$$x_5 = 0.25$$
$$x_6 = 0.85$$

The dryness fraction in separator after the generator,

$$VF_1 = \frac{x_4 - x_5}{x_6 - x_5} = \frac{0.45 - 0.25}{0.85 - 0.25} = 0.33$$

Mass of vapor separated from $m_4, m_6 = DF_1 m_4 = 0.33 \times 1 = 0.33$ kg/s
Mass of liquid separated from $m_4, m_5 = (1 - DF_1)m_4 = (1 - 0.33) \times 1 = 0.67$ kg/s

Now from the chart, all the enthalpies can be obtained as

$$h_4 = 810 \text{ kJ/kg}$$
$$h_5 = 357.6 \text{ kJ/kg}$$
$$h_6 = 1675.5 \text{ kJ/kg}$$

$$h_7 = h_6 - \frac{Q_d}{m_6} = 1675.5 - \frac{100}{0.33} = 1372.3 \text{ kJ/kg}$$

$$x_7 = x_6 = 0.85$$
$$m_7 = m_6 = 0.33$$

State 7 can be located from the enthalpy and concentration. Using trial-and-error method, the liquid concentration and vapor concentrations can be obtained.

$$x_8 = 0.36$$

$$x_9 = 0.95$$

The dryness fraction in dephlegmator,

$$VF_2 = \frac{x_7 - x_8}{x_9 - x_8} = \frac{0.85 - 0.36}{0.95 - 0.36} = 0.84$$

Mass of vapor separated from $m_7, m_9 = DF_2 m_7 = 0.84 \times 0.33 = 0.28$ kg/s

Mass of liquid separated from $m_7, m_8 = (1 - DF_2)m_7 = (1 - 0.84) \times 0.33 = 0.05$ kg/s

In case of the solution with the maximum possible dephlegmator effect, the vapor concentration at the exit of dephlegmator is 0.999. It means approximately a complete removal of water content.

$$x_{10} = x_6 = 0.85$$

$$m_{10} = m_6 = 0.33$$

A vertical line can be drawn from State 6 in downward direction but the end point, that is, State 10 can be obtained from the trial-and-error method started from the saturated liquid state, 11 to nearly pure ammonia vapor state, 12. Finally the saturated vapor state, 12 (99.9% ammonia or nearly zero water), and saturated liquid state, 11, can be joined to result the required state, 10. From the graph,

$$x_{11} = 0.74$$

$$x_{12} = 0.999$$

$$h_6 = 1675.5 \text{ kJ/kg}$$

$$h_{10} = 608.5 \text{ kJ/kg}$$

The heat to be rejected from State 6 to get 99.9% ammonia vapor from dephlegmator,

$$Q_{ds} = m_6(h_6 - h_{10}) = 0.33 (1675.5 - 608.5) = 351.8 \text{ kJ/s}$$

The dryness fraction in dephlegmator 99% distillation,

$$VF_3 = \frac{x_{10} - x_{11}}{x_{12} - x_{11}} = \frac{0.85 - 0.74}{0.99 - 0.74} = 0.44$$

Mass of vapor separated from m_{10},

$$m_{12} = DF_3 m_{10} = 0.44 \times 0.33 = 0.114 \text{ kg/s}$$

Mass of liquid separated from m_{10},

$$m_{11} = (1 - DF_3)m_{10} = (1 - 0.44) \times 0.33 = 0.185 \text{ kg/s}$$

5.2.6 HEAT EXCHANGER

Vapor absorption system consists of different types of heat exchangers such as economizer, evaporator, superheater, low-temperature regenerator, high-temperature regenerator, and condenser. In the evaporator section, the boiling takes place over a range of temperature. By virtue of varying boiling point, two-component working fluid is able to match or run parallel to the hot fluid temperature line while recovering energy and hence the exit hot fluid temperature can be low. Ammonia–water mixture is a zeotropic mixture and nonisothermal. A pure substance boils at constant temperature and pressure, whereas zeotropic mixture boils at increasing temperature. With the constant temperature boiling nature in the Rankine cycle, the structural loss will be high. The temperature gap between the working fluid and heat source is high. Due to nonisothermal phase change of aqua–ammonia, a closer match between fluids has been observed. When the structural loss is reduced, the driving force for the heat transfer is reduced and reduces exergy loss.

Heat recovery vapor generator (HRVG) is a boiler in which heat from the hot source is utilized for generating vapor. HRVG consists of economizer, evaporator, and superheater. A typical heat recovery used in Kalina power plant has been depicted in Figure 5.18. In economizer sensible heat transfers, the temperature of the fluid increases to its bubble-point temperature. In HRVG, the boiling process is incomplete, that is, it results the liquid–vapor mixture. The temperature of the boiler is below the dew-point temperature. The liquid and vapor from the boiler are separated in the same vessel or in a separate cylindrical drum. At high-temperature heat recovery, directly superheated vapor can be generated. The heat source fluid may be pressurized water, glycol-based water, steam, hot gas, or any waste heat. In the evaporator, the working fluid is passing through tubes surrounded by heat source stream, gaining heat and converted to vapor. Superheater is a device at which separated vapor or saturated vapor is converted into dry superheated vapor. The superheated vapor is then expanded in the turbine. A small amount of superheat is enough to ensure a dry expansion in turbine.

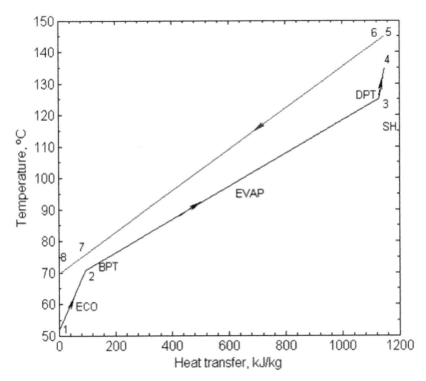

FIGURE 5.18 Temperature–heat-transferred diagram for heat recovery vapor generator. BPT, bubble-point temperature; DPT, dew-point temperature; ECO, economizer; EVA, evaporator; SH, superheater.

5.2.7 CONDENSER

The condensation of two-component working fluid also occurs over a range of temperatures and hence permits additional heat rejection in the condensation system. The condenser pressure is much higher in two-component fluid cycle, and so no need to maintain vacuum at condenser. More power can be generated at lower sink temperature. In KCS, a high condenser pressure have been efficiently achieved which is not possible in the Rankine cycle. Condenser pressure in Kalina cycle depends on sink temperature and concentration of ammonia in the mixture. The condenser concentration can be optimized as it is a function of condenser pressure and saturated liquid temperature. The saturated liquid temperature is maintained above the circulating cooling water inlet temperature with the terminal temperature difference.

5.2.8   TURBINE EXPANSION

The ammonia vapor turbine is a component for expanding ammonia vapor which is coupled to generator for electric work output. A mixture turbine can be used in decentralized power generation, cogeneration, trigeneration, etc. Expansion in mixture turbine results relatively dry or a saturated vapor at exit compared to wet steam in Rankine cycle, which requires protection of blades in the last few stages. Due to this reason, reheater is not essential for Kalina power turbine. With high pressure of vapor and low specific volume at exit of turbine, the size of the exhaust system could be comparably smaller than steam Rankine cycle.

The net electrical work output from the turbine,

$$W_t = m(h_1 - h_2)\eta_{m,t}\eta_{ge} \tag{5.32}$$

where

$$h_2 = h_1 - \eta_t \left(h_1 - h_2'\right) \tag{5.33}$$

Subscripts 1 and 2 are the inlet and outlet states and h_2' are the isentropic enthalpy at the exit condition. The exit enthalpy has been calculated from isentropic efficiency. The actual temperature of the turbine exit can be iterated from the enthalpy.

5.2.9   SOLUTION PUMP

The pump in the plant pressurizes the liquid stream and delivers to the subsequent components. The pump exit condition can be calculated in a similar way of turbine exit condition. From the energy balance, the pump work can be predicted.

$$W_P = \frac{m_1 \left(h_2 - h_1\right)}{\eta_{m,p}} \tag{5.34}$$

where $\eta_{m,p}$ is pump mechanical efficiency.

5.3   RELATION BETWEEN CIRCULATION RATIO AND DRYNESS FRACTION

Some situations, instead of dryness fraction, circulation ratio will be used in the analysis of the vapor absorption system. Figure 5.19 shows the weak solution concentration, strong solution concentration, and refrigerant concentration in a typical vapor absorption system.

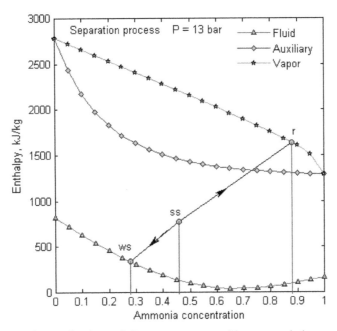

FIGURE 5.19 **(See color insert.)** Separator process with strong solution concentration, weak solution concentration, and refrigerant concentration.

Generally, the mass details in a typical vapor absorption refrigeration system are expressed in terms of circulation ratio, λ. The circulation ratio is defined as the ratio of weak solution to refrigerant flow rate, that is,

$$\lambda = \frac{m_{ws}}{m_r}$$

The mass balance at the separator is

$$m_{ss} = m_{ws} + m_r$$

since $\lambda = m_{ws} = m_r$, divide both sides by m_r

$$\frac{m_{ss}}{m_r} = \frac{m_{ws}}{m_r} + 1 = \lambda + 1$$

$$m_{ss} = (1 + \lambda)m_s$$

The dryness fraction is defined as the ratio of vapor to strong solution mass flow rate.

$$VF = \frac{m_r}{m_{ss}} = \frac{m_r}{(1+\lambda)m_r} = \frac{1}{(1+\lambda)}$$

$$\lambda = \frac{1}{VF} - 1$$

Using lever rule, they can be expressed in concentrations,

$$\lambda = \frac{m_{ws}}{m_r} = \frac{x_r - x_{ss}}{x_{ss} - x_{ws}}$$

$$VF = \frac{m_r}{m_{ss}} = \frac{x_{ss} - x_{ws}}{x_r - x_{ws}}$$

5.4 SUMMARY

All the processes involved in vapor absorption system are detailed with mass balance and energy balance formulae. The detailed processes are separation, adiabatic mixing, diabatic mixing, throttling, dephlegmator process, heat transfer in economizer, evaporator, and superheater, condensation, and expansion in turbine. The processes are shown in enthalpy–concentration chart with scale to understand clearly. The sample solutions are given with analytical and graphical approach. These solutions with pictorial representation and formulation assist the design engineer in developing and evaluating a vapor absorption system easily without any complexity.

KEYWORDS

- thermodynamic evaluation
- binary fluid system
- computer models
- absorption
- condensation

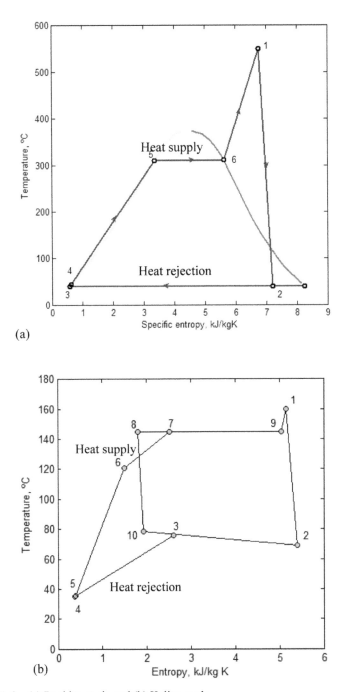

FIGURE 1.4 (a) Rankine cycle and (b) Kalina cycle.

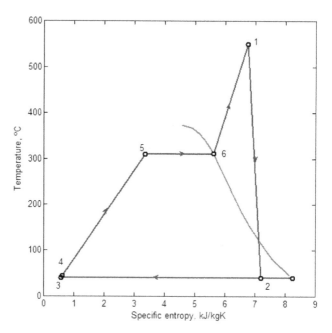

FIGURE 3.3 Thermodynamic cycle with state, processes, and cycle with reference to Figure 3.1.

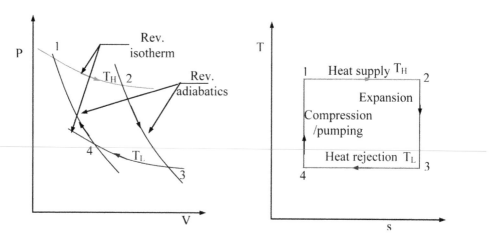

FIGURE 3.9 Carnot engine (a) *P–V* diagram and (b) *T–s* diagram.

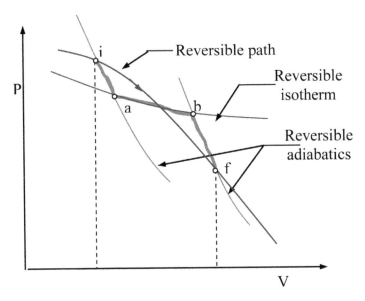

FIGURE 3.11 Reversible path substituted by two reversible adiabatics and a reversible isotherm.

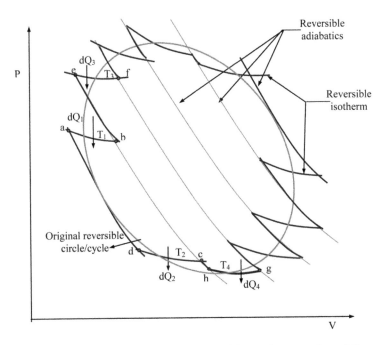

FIGURE 3.12 Unknown reversible cycle is sliced into a large number of Carnot cycle elements.

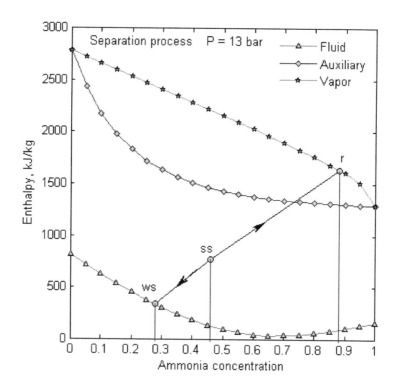

FIGURE 5.19 Separator process with strong solution concentration, weak solution concentration, and refrigerant concentration.

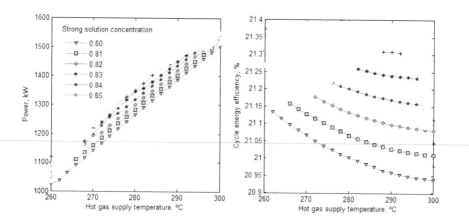

FIGURE 7.6 Effect of hot gas inlet temperature with strong solution concentration on (a) plant's power generation and (b) cycle thermal efficiency.

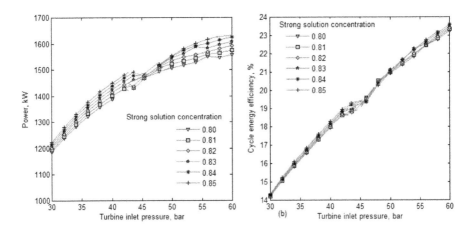

FIGURE 7.7 Effect of turbine inlet pressure with strong solution concentration on (a) plant's power generation and (b) cycle thermal efficiency.

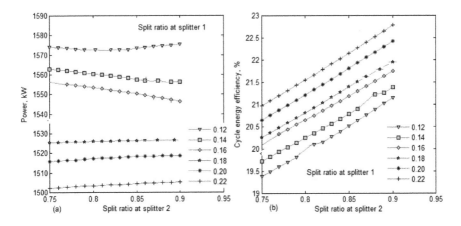

FIGURE 7.8 Effect of separations in the cycle on (a) plant's power generation and (b) cycle thermal efficiency.

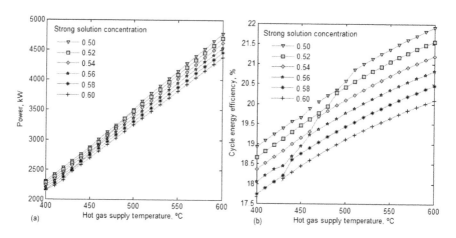

FIGURE 8.5   Influence of hot gas inlet temperature with strong solution concentration on (a) power generation and (b) cycle thermal efficiency.

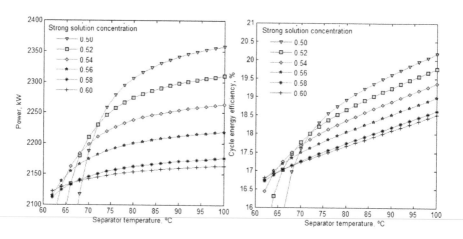

FIGURE 8.6   Influence of separator temperature with strong solution concentration on (a) power generation and (b) cycle thermal efficiency.

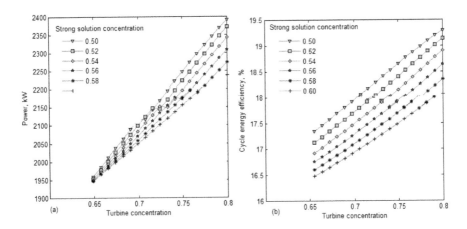

FIGURE 8.7 Influence of turbine concentration with strong solution concentration on (a) plant's power generation and (b) cycle thermal efficiency.

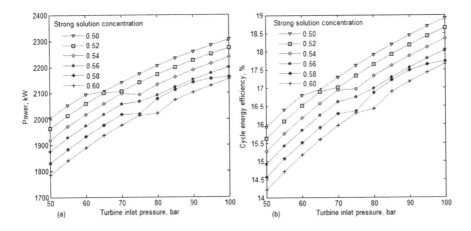

FIGURE 8.8 Role of turbine inlet pressure with strong solution concentration on (a) power generation and (b) thermal efficiency of cycle.

CHAPTER 6

KALINA CYCLE SYSTEM WITH LOW-TEMPERATURE HEAT RECOVERY

ABSTRACT

Kalina cycle system (KCS) designed for low-temperature heat recovery has a simple layout compared to the intermediate- and high-temperature heat recoveries. The basic thermal power plant consists of four processes, namely, heat addition, expansion, heat rejection, and pumping (compression). In addition to these components/processes, KCS has some additional essential and optional equipment. The role of every component on overall performance of total system is investigated. The hot gas considered in this study is the waste heat reject from a typical process plant. The combustion model is developed to evaluate the properties of source, that is, hot exhaust gas. After the modeling and simulation, the results are analyzed to find the optimum conditions for the efficient operation of the plant.

6.1 INTRODUCTION

Kalina cycle system (KCS) has a greater flexibility to suit wide range of heat source temperatures. The source may be a direct fuel burning, solar thermal collectors, or water heat recovery. To suit the source temperature, KCS configuration needs to be updated and modified. Most of heat losses such as engine exhaust and chimney gas, etc. are at the low-temperature range from 100°C to 200°C. This section is focused on development and investigation on low-temperature heat recovery-operated KCS systems. The cycle features are diagnosed individually to highlight the significance of its role. Thermodynamics plays a key role on deciding the operation conditions by modeling, simulation, analysis, and parametric optimization before sizing. Development of thermodynamic properties of ammonia–water mixture simplifies the design and assessment of the KCS plant. Processes solutions are developed after the thermodynamic properties generation to simplify the tedious theoretical work and analyze the basic processes in KCS plants. In modeling, the equations are derived to find the unknown properties and energy values. The performance of the plants has been predicted using mass, energy, and exergy balance equations. The key parameters and its operational range that influences on the performance of the cycles have been identified and optimized. The work is useful in designing and cost estimation of the plant equipment. In this chapter, the addition of superheater (SH), low-temperature regenerator (LTR), high-temperature regenerator (HTR), and dephlegmator (with external heat rejection and internal heat recovery) are elaborated. The full KCS without dephlegmator, series arrangement of heaters, and parallel heaters in KCS are compared.

6.2   BASIC KCS

Figure 6.1 shows the schematic diagram of the basic KCS and its fluid flow lines. The properties are determined with the thermodynamic formulation and they are shown in Figure 6.2a enthalpy–concentration and Figure 6.2b temperature–entropy diagram. Basically, KCS needs heat source (boiler), turbine, absorber (sink), pump, and throttling. The boiler also integrates a separator where the liquid and vapor divide. So compared to a single fluid Rankine cycle, the minimum additional components are separator and throttling. In single fluid system with ammonia as a working fluid, the condensation of ammonia vapor at the exit of turbine is difficult as its condensation temperature is very low at higher concentration. It demands high pressure to condenser at the standard sink temperature. To avoid this difficulty, in KCS, the turbine exit vapor is diluted by mixing with the weak solution before condensation. The concentration of the fluid is decreased by mixing the weak solution with turbine vapor. The resulted fluid is known as strong solution. The concentration is defined as the ratio of mass of ammonia to the total mass. The concentration of strong solution is more than the weak solution concentration. The absorption process is continued by rejecting the heat in the absorber. The function of absorber is the same as the condenser.

The condensed liquid (4) is pumped to the boiler. The boiler consists of two sections, the first section is the economizer and the second part is the evaporator. In the evaporator, the liquid solution is converted into liquid (8) and vapor (1) mixture. It indicates the incomplete evaporation or boiling. In the Rankine cycle, the liquid is completely converted into vapor. In KCS, the boiling starts at bubble-point temperature (BPT) (6) and stops at a temperature below the dew-point temperature. The dew-point temperature is the temperature where the liquid converts completely into saturated vapor, that is, complete evaporation or boiling. In the boiler, after the heat addition, the liquid and vapor are separated. These two outlets, that is, saturated liquid (8) and saturated vapor (1), are connected to throttling (8–9) and turbine, respectively. The liquid solution is the weak solution as the strong solution is distilled in the generator and separator assembly by removing the vapor. The throttling allows transferring the liquid from the high pressure to low pressure before entering into absorber. After throttling, the weak solution (9) is mixed with the turbine vapor (2) and the resultant concentration (3) is above the weak solution concentration and below the turbine vapor concentration, that is, between weak solution and turbine vapor concentrations. The concentration of the condensed liquid solution is high compared to the weak solution concentration. So, this solution flowing in pump and boiler is called strong solution. In the boiler, the liquid is converted to partial vapor by adding heat from the hot gas (12–14) as shown in Figure 6.1.

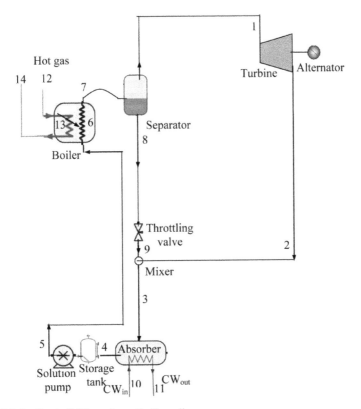

FIGURE 6.1   Basic KCS—schematic flow diagram.

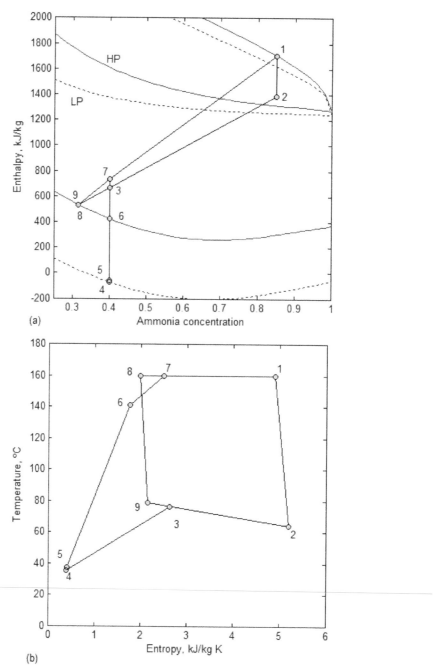

FIGURE 6.2 Simple KCS on (a) enthalpy–concentration chart and (b) temperature–entropy chart.

The BPT is evaluated determined from the local hot gas temperature and the pinch-point (PP) temperature. Thermodynamic processes are referred on property chart (Fig. 6.2). Figure 6.2a and b shows the heat addition (5–6–7) and heat rejection (3–4) with change in temperature. Figure 6.2a and b clears the separation of fluid (7) into saturated liquid (8) and saturated vapor (1). The figure also details the mixing process of 2 and 9 into 3.

The solution of basic KCS involves the handling of assumptions and development of mass and energy balance equations. Assumptions used in KCS analysis are as follows. Terminal temperature difference (TTD) at evaporator of heat recovery vapor generator (HRVG) inlet with hot gas is taken at 15°C. PP in boiler is 20°C. Approach point (AP) in the boiler is 2°C. Vapor concentration at exit of separator, that is, generator, is 0.85. Strong solution concentration is 0.4. The isentropic efficiency of vapor turbine and solution pump is considered as 80% and 75%, respectively. The mechanical efficiency of the solution pump ($\eta_{m,p}$) and mixture turbine ($\eta_{m,t}$) is taken at 96%. Electrical generator efficiency (η_{ge}) is taken as 98%. The condensate leaving the condenser is assumed as complete saturated liquid. Pressure drop and heat loss in pipelines are neglected. The PP is the minimum temperature difference in the HRVG maintained to ensure the positive heat transfer between the hot fluid and cold fluid. It is the temperature difference between the hot gas exit temperature in the evaporator with the working fluid saturation temperature. AP is maintained at the inlet of the evaporator. If the liquid mixture is entering at the BPT in the evaporator, there is a sudden transition of fluid from liquid to vapor and it may also happen in the economizer itself. To avoid this sharp transition from liquid to vapor, AP is maintained so that there is no phase change in the economizer and it ensures a liquid entry to evaporator.

The turbine inlet temperature is

$$T_1 = T_{12} - \text{TTD} \tag{6.1}$$

The separator temperature is found from the turbine inlet temperature and degree of superheat (DSH). DSH is the temperature difference between turbine inlet and separator. In basic KCS, SH is not included and so the separator temperature is equal to the turbine inlet temperature, that is, $T_7 = T_1$. The high pressure (P_1) is calculated from the separator temperature and vapor concentration (T_7, x_7), since it is the function of temperature and concentration at the saturated vapor state. The separator temperature, T_8, is on liquid curve which is a BPT at the high pressure and concentration, x_8. From this relation, the liquid solution concentration, that is, weak solution concentration x_8, is found through the iteration. The low pressure is determined from mixture concentration (x_4) and temperature (T_4) at absorber outlet.

The economizer's exit temperature or inlet to evaporator, T_6, is maintained below the BPT to avoid the sharp transition at the evaporator inlet. This temperature difference is equal to AP.

The temperature of strong solution at the evaporator inlet is

$$T_6 = T_{bp} - AP \tag{6.2}$$

where T_{bp} is the BPT at boiler pressure and concentration.

The hot fluid temperature at the evaporator section of boiler is

$$T_{13} = T_{bp} + PP \tag{6.3}$$

For specific results, KCS is solved at unit mass of strong solution at separator inlet (7). In separator, out of one unit mass of mixture, dryness (DF) is the dryness fraction and (1 − DF) of WF is the wetness fraction. After applying lever rule for separation process,

$$VF_{separator} = \frac{x_7 - x_8}{x_1 - x_8} \tag{6.4}$$

The simplifications of mass balance equations result in the unknown mass flow rate in the cycles.

$$m_1 = VF_{separator} m_7 \tag{6.5}$$

The liquid fluid from the separator is

$$m_8 = (1 - VF_{separator}) m_7 \tag{6.6}$$

The concentration after mixing is found from the mass balance equation.

$$x_3 = \frac{m_2 x_2 + m_9 x_9}{m_3} \tag{6.7}$$

The concentration's result, x_3, is cross-checked with the initial assumption of strong solution concentration to check the completeness of mass and energy balance equations. The exit temperature of turbine is determined from the isentropic efficiency, inlet, and outlet process parameters.

The energy balance in the mixing after turbine and before absorber is

$$h_3 = \frac{m_2 h_2 + m_9 h_9}{m_3} \qquad (6.8)$$

From the iteration procedure, T_3 is resulted from the enthalpy, h_3. The circulating water demand in the absorber is

$$m_{10} = \frac{m_3 (h_3 - h_4)}{c_{p,w} (T_{11} - T_{10})} \qquad (6.9)$$

In the first stage, the KCS is solved with unit mass at the absorber exit. Later, the actual working fluid generation is obtained from the source conditions, that is, boiler and SH (in basic KCS SH is not included).

The working fluid in KCS is resulted from the energy balance in boiler

$$m_{wf} = \frac{m_{12} (h_{12} - h_{13})}{m_6 (h_7 - h_6)} \qquad (6.10)$$

Mass of hot flue gas, m_{12}, is determined from the combustion and gas composition solutions.

For coal combustion, the fuel is defined by a general formula as $C_{a_1} H_{a_2} O_{a_3} N_{a_4}$. For single carbon atom fuel, coefficient a_1 becomes one. The coefficients a_2, a_3, and a_4 are H/C, O/C, and N/C mole ratios, respectively. The moisture content in coal has been accounted in thermodynamic evaluation.

The following is the chemical reaction in coal combustion.

$$\left(C_{a_1} H_{a_2} O_{a_3} N_{a_4} \right)_{coal} + \left(a_5 \left(O_2 + 3.76 N_2 \right) \right)_{air} + \left(a_6 H_2 O\ (l) \right)_{moisture\ content}$$
$$\Rightarrow \left(b_1 CO_2 + b_2 H_2 O + b_3 O_2 + b_4 N_2 \right)_{product\ gas} \qquad (6.11)$$

The minimum air fuel ratio (stoichiometric) is 9.22 with the adiabatic flame temperature of 1120°C.

At the combustion temperature of 900°C, the resulted chemical equation is

$$\left(CH_{0.8094} O_{0.1388} N_{0.0274} \right)_{coal} + \left(1.2125 \left(O_2 + 3.76 N_2 \right) \right)_{air} + \left(0.086 H_2 O\ (l) \right)_{moisture\ content}$$
$$\Rightarrow \left(CO_2 + 0.4857 H_2 O + 0.2690 O_2 + 5.29 N_2 \right)_{product\ gas} \qquad (6.12)$$

The gas enthalpy at temperature T is

$$h_{gat,T} = b_1 h_{CO_2,T} + b_2 h_{H_2O,T} + b_3 h_{O_2,T} + b_4 h_{N_2,T} \qquad (6.13)$$

Therefore,

$$
m_{12} = \frac{m_{\text{hot gas in N m}^3/\text{h}} \left(b_1 M_{CO_2} + b_2 M_{H_2O} + b_3 M_{O_2} + b_4 M_{N_2} \right)}{\left(b_1 + b_2 + b_3 + b_4 \right) \times 22.4 \times 3600} \tag{6.14}
$$

The above equation can also be written using mole fractions.

Table 6.1 results from the mass and energy balances applied to the plant. The source temperature (hot gas) 175°C is resulted in 160°C of turbine inlet temperature with the assumption of TTD in boiler. The gas flow rate, 100,000 N m³/h, is equal to 36.89 kg/s of fluid flow. The circulating water demand in the absorber is 37.88 kg/s. At this state of source, the generated working fluid flow in the absorber and boiler is 1.72 kg/s. After the boiler, the fluid is divided into weak solution (1.45 kg/s) and vapor (0.27 kg/s). Due to considered isentropic efficiency of turbine, the entropy is increased from 4.89 to 5.20 kJ/kg K.

TABLE 6.1   Material Flow Details of Basic KCS at 100,000 N m³/h of Hot Gas at 175°C.

State	P (bar)	T (°C)	x	m (kg/s)	h (kJ/kg)	s (kJ/kg K)
1	37.67	160.00	0.85	0.27	1707.04	4.89
2	2.58	64.37	0.85	0.27	1386.13	5.20
3	2.58	76.09	0.40	1.72	663.47	2.63
4	2.58	35.00	0.40	1.72	−72.57	0.39
5	37.67	35.82	0.40	1.72	−65.92	0.40
6	37.67	141.31	0.40	1.72	424.79	1.76
7	37.67	160.00	0.40	1.72	737.82	2.50
8	37.67	160.00	0.32	1.45	530.71	1.99
9	2.58	78.90	0.32	1.45	533.15	2.16
10	1.01	30.00	0.00	37.88	0.55	0.00
11	1.01	38.00	0.00	37.88	1.43	0.00
12	1.01	175.00	0.00	36.89	157.21	0.43
13	1.01	161.31	0.00	36.89	142.61	0.39
14	1.01	139.78	0.00	36.89	119.71	0.34

6.3   KCS WITH SUPERHEATER

SH in a power plant ensures a dry expansion in turbine and also added power generation due to increase in source temperature. In organic Rankine cycle (ORC), the wet fluid (in *T–s* diagram, the vapor dome is symmetrical) demands a mandatory SH to avoid the wet expansion in the turbine and hence saves the life of turbine. In the dry fluids (vapor dome bends toward right in *T–s* diagram), the use of SH may not be required as there is no moisture content. Aqua ammonia is the mixture of wet fluid (water) and dry fluid (ammonia). Therefore, the role of SH on performance needs to be revealed.

Figure 6.3 shows the use of SH at the end of separator and before the turbine. A small temperature rise in SH is shown in properties chart (Fig. 6.4). The SH ensures the dry expansion in the turbine. In this configuration, compared to the earlier description on basic KCS, the additional component included is a SH. The vapor from the generator and separator flows to SH where the vapor temperature increases. The superheated vapor is expanded in the turbine and the remaining processes are same as in earlier plant.

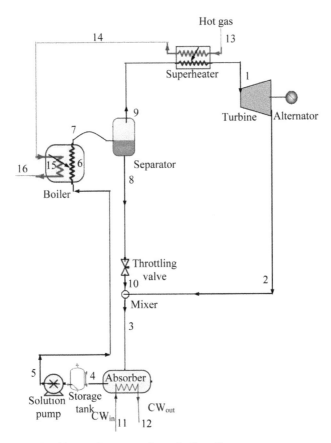

FIGURE 6.3 KCS with superheater—schematic flow diagram.

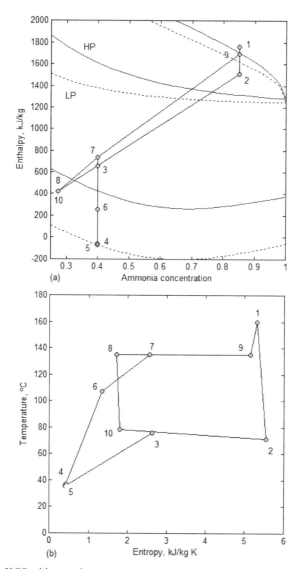

FIGURE 6.4   KCS with superheater on property chart.

The DSH considered is 15°C.

The cycle energy efficiency is determined as follows:

$$\eta_{1\,cycle} = \frac{W_{turbine} - W_{pump}}{Q_{boiler} + Q_{superheater}} = \frac{m_1\left(h_1 - h_2\right)\eta_{turbine}\eta_{gen} - m_4\left(h_5 - h_4\right)/\eta_{turbine}}{m_5\left(h_7 - h_5\right) + m_1\left(h_1 - h_9\right)} \quad (6.15)$$

Figure 6.5 depicts the influence of DSH with source temperature on dryness fraction of fluid at the exit of vapor turbine. The increase in source temperature increases the turbine inlet temperature and also the high pressure. The dryness fraction of exit vapor decreases with increase in temperature at a fixed DSH. In case of an increase in DSH, the dryness fraction increases from 0.95 to 1, that is, almost completely vapor. Increase in DSH at a fixed source temperature indicates a decrease in boiler temperature. Approximately after 25°C, the dryness fraction of vapor becomes to one, that is, there is no liquid portion at the exit of turbine. Therefore, to ensure the complete dry expansion in the turbine, at least 25°C of DSH is required.

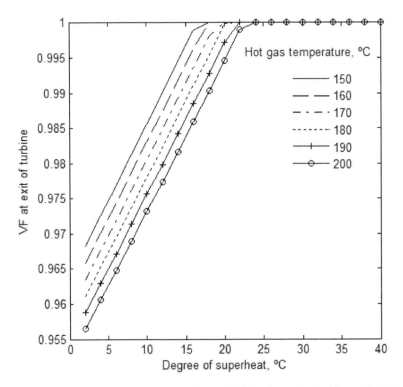

FIGURE 6.5 Dryness fraction (vapor fraction) of fluid at the exit of turbine with DSH and source temperature.

The performance variations of KCS plant with SH are studied in Figure 6.6 with a change in DSH with the source temperature, that is, hot gas inlet temperature. The turbine inlet temperature is less than the hot gas inlet temperature with an amount equal to TTD. The variation in (a) power and (b) cycle energy efficiency has been shown in Figure 6.6. The increase in hot gas temperature increases both the power and efficiency. An increase in power with DSH has been observed up to a small amount of superheat and there is a fall in power with a rise in DSH. But the increase in DSH is not favoring the efficiency of the cycle.

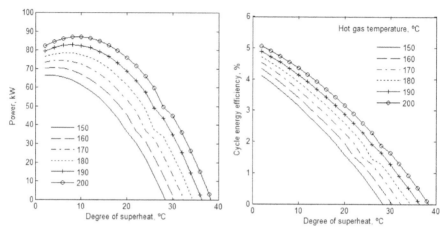

FIGURE 6.6 Influence of degree of superheat with strong solution concentration on performance of KCS (a) specific power and (b) thermal efficiency.

Table 6.2 shows the generated results of plant properties obtained from mass and energy balances. The hot gas 100,000 N m³/h (36.89 kg/s) is used to generate 3.18 kg/s of working fluid. The plant is demanding 69.37 kg/s of circulating water to dissipate the heat to surroundings. Compared to the basic KCS, the SH model increased the condenser load significantly due to increased working fluid. The specific entropy of fluid in turbine is increased from 5.14 to 5.41 kJ/kg K due to irreversibility. The SH decreases the boiler pressure from 37.67 to 25.04 bar compared to the basic KCS. It influences the heat balance in boiler and so increased working fluid. Even though there is no much benefit from SH for dryness fraction at exit of turbine, it improves the heat recovery in the boiler by decreasing the boiler high pressure. Since the turbine inlet temperature is the same in both basic KCS and SH plant, the separator temperature differs in both the models. In basic KCS, due to the absence of SH, separator and turbine inlet are same. But with SH, these two are different. The separator temperature in SH is lower than the basic KCS.

TABLE 6.2 Material Balance Results of KCS with SH.

State	P (bar)	T (°C)	x	m (kg/s)	h (kJ/kg)	s (kJ/kg K)
1	25.04	160.00	0.85	0.62	1744.39	5.14
2	2.58	68.83	0.85	0.62	1459.63	5.41
3	2.58	75.76	0.40	3.18	657.11	2.62
4	2.58	35.00	0.40	3.18	−72.57	0.39
5	25.04	35.64	0.40	3.18	−67.81	0.40
6	25.04	120.67	0.40	3.18	321.22	1.51
7	25.04	145.00	0.40	3.18	736.83	2.53
8	25.04	145.00	0.29	2.56	462.14	1.82
9	25.04	145.00	0.85	0.62	1701.95	5.05
10	2.58	78.51	0.29	2.56	464.57	1.94
11	1.01	30.00	0.00	69.37	0.30	0.00
12	1.01	38.00	0.00	69.37	0.78	0.00
13	1.01	175.00	0.00	36.89	157.21	0.43
14	1.01	174.35	0.00	36.89	156.49	0.42
15	1.01	140.67	0.00	36.89	120.68	0.34
16	1.01	108.91	0.00	36.89	87.15	0.26

6.4 KCS WITH LOW-TEMPERATURE REGENERATOR

Figure 6.7 outlines the schematic layout of KCS with incorporation of LTR as a heat recovery. The properties are depicted in Figure 6.8a and b on enthalpy–concentration and temperature–entropy diagrams, respectively. The property values are tabulated in Table 6.3. The additional feature of KCS compared to conventional cycles, that is, ORC and steam Rankine cycle, is heat recovery. In ORC, partially there is a scope to recover the heat at the exit of turbine. It is not possible with steam turbine. In steam power plant, some steam from the turbine needs to be removed for heat recovery in the regenerator. There are two possible heat recoveries and they are LTR and HTR. At the exit of steam power plant, the steam temperature is very nearer to sink temperature and around 45–50°C. Therefore, heat recovery is not possible at this instant. The fluid temperature is 64.37°C at the exit of KCS turbine. To ensure the easy condensation, the turbine fluid is diluted by mixing with the weak solution (78.90°C). Therefore after mixing, LTR is located at the inlet of condenser. It decreases the condenser duty and also generator heat load. Even though it will not contribute on power generation, the LTR improves the efficiency of KCS. At LTR, the pumped fluid temperature is increased from 35.82°C to 88.57°C with a drop in mixed fluid (strong solution) temperature from 76.09°C to 63.03°C.

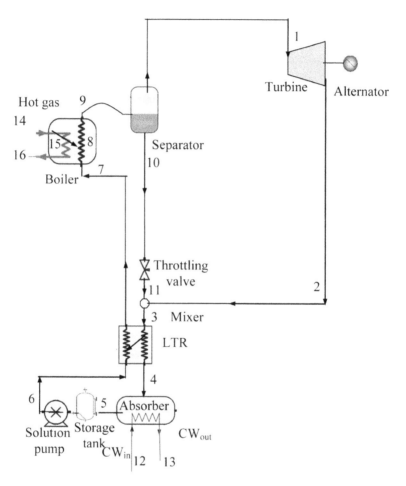

FIGURE 6.7 KCS with LTR—schematic flow diagram.

The amount of heat recovery in LTR depends on the assumption used in thermodynamic simulation. The maximum temperature rise in the LTR is up to the BPT corresponding to the high pressure. So, the actual temperature rise in the LTR is converted into the ratio of these two temperature differences. LTR temperature ratio is defined as the temperature rise in the LTR to the maximum possible temperature rise.

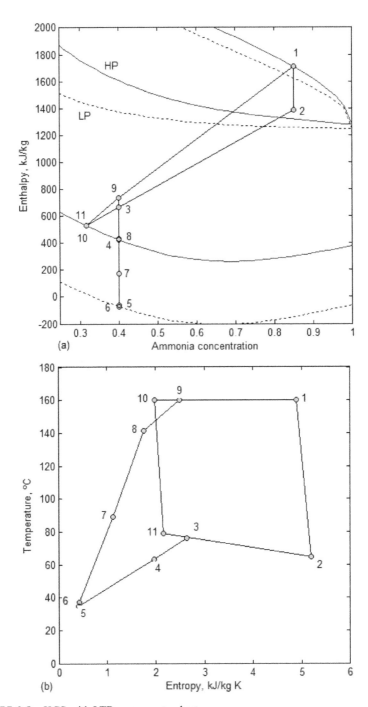

FIGURE 6.8 KCS with LTR on property chart.

$$\theta_{\text{LTR}} = \frac{T_7 - T_6}{T_8 - T_6} \tag{6.16}$$

The above equation is used to find the exit temperature of LTR, T_7 from the temperature ratio assumption. But this temperature should be below BPT (T_8) and mixer temperature (T_3). Now T_7 (88.57°C) is below T_3 (76.09°C) and T_8 (141.31°C). Figure 6.9 indicates the effect of this assumption on LTR temperature ratio on cycle efficiency as it would not affect the power generation. The efficiency increases with increase in source temperature and LTR temperature ratio. It shows that a more heat recovery in LTR indicates the high efficiency. Since the power production is not disturbed with LTR, these results are omitted.

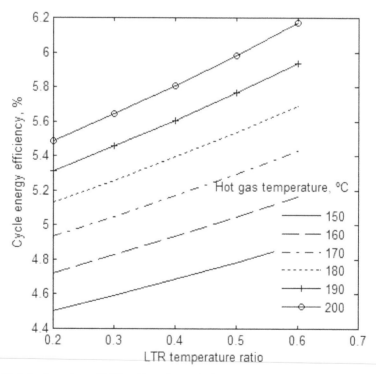

FIGURE 6.9 Location of LTR with a change in hot gas inlet temperature to get a higher thermal efficiency.

Table 6.3 shows that the working fluid generated in the generator is 1.72 kg/s, and 0.27 kg/s of vapor is liberated for the expansion. The balance fluid, that is, 1.45 kg/s, is the weak solution used for heat recovery and also to dilute the vapor mixture at the exit of turbine from 0.85 to 0.4 concentration. Since the SH is not used, the temperature difference between hot gas temperature (175°C) and turbine inlet temperature (160°C) is only TTD.

TABLE 6.3 Material Balance Results of KCS with LTR.

State	P (bar)	T (°C)	x	m (kg/s)	h (kJ/kg)	s (kJ/kg K)
1	37.67	160.00	0.85	0.27	1707.04	4.89
2	2.58	64.37	0.85	0.27	1386.13	5.20
3	2.58	76.09	0.40	1.72	663.47	2.63
4	2.58	63.03	0.40	1.72	428.58	1.95
5	2.58	35.00	0.40	1.72	−72.57	0.39
6	37.67	35.82	0.40	1.72	−65.92	0.40
7	37.67	88.57	0.40	1.72	169.55	1.10
8	37.67	141.31	0.40	1.72	424.79	1.76
9	37.67	160.00	0.40	1.72	737.82	2.50
10	37.67	160.00	0.32	1.45	530.71	1.99
11	2.58	78.90	0.32	1.45	533.15	2.16
12	1.01	30.00	0.00	14.99	1.39	0.00
13	1.01	38.00	0.00	14.99	3.63	0.01
14	1.01	175.00	0.00	36.89	157.21	0.43
15	1.01	161.31	0.00	36.89	142.61	0.39
16	1.01	150.13	0.00	36.89	130.70	0.37

6.5 KCS WITH HIGH-TEMPERATURE REGENERATOR

Similar to LTR, HTR also plays a same role on efficiency boosting. In this section, HTR alone is studied on performance variations. Figure 6.10 shows the schematic flow lines of plant arrangement of KCS with incorporation of HTR. The hot line of weak solution and cold line of strong solution are combined in HTR to transfer the heat from hot weak solution to cold strong solution. The weak solution needs to be cooled in the absorber and the strong solution needs to be heated in the boiler. Therefore, the addition of HTR saves the heat load in absorber and boiler and decreases its capacity and size. The strong solution exit temperature, T_6, plays an important role on heat recovery in HTR. The BPT is the maximum possible liquid solution temperature in the HTR. So, the maximum temperature rise is $(T_7 - T_5)$. The degree of heat recovery from the weak solution depends on the wetness fraction in the boiler. A more wetness fraction results in more amount of weak solution and more scope for the heat recovery. Therefore, the actual temperature raise is determined with the proportionality of wetness fraction and the maximum possible temperature difference as follows:

$$T_6 = T_5 + (1 - VF_{separator})(T_7 - T_5) \qquad (6.17)$$

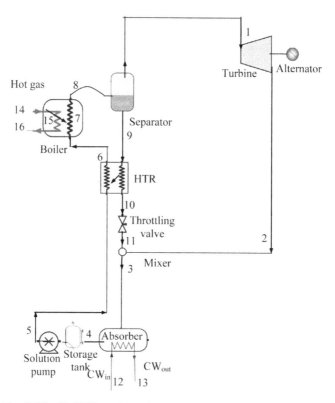

FIGURE 6.10   KCS with HTR—schematic material flow diagram.

Figure 6.11 shows the (a) enthalpy–concentration and (b) temperature–entropy charts. The properties of states shown in the schematic are referred in Figure 6.11 to understand the natures of processes. The dryness fraction in the generator is 15% only at the stated assumptions. It indicates that 85% of mass is available as weak solution for heat recovery. With reference to Table 6.4 data, the weak solution temperature is decreased from 160.00°C to 55.76°C with a rise in strong solution temperature from 35.82°C to 124.90°C. The resulted strong solution is 2.99 kg/s with the hot gas flow of

36.89 kg/s. The circulating water demand in the absorber is 9.81 kg/s. Since the heat recovery in HTR is more than the LTR, the generated working fluid in the boiler is more.

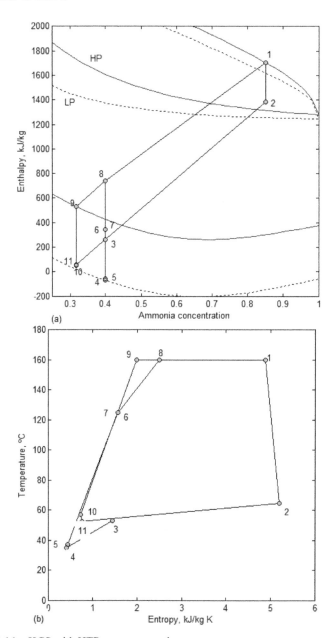

(a)

(b)

FIGURE 6.11 KCS with HTR on property chart.

TABLE 6.4   Material Balance Results of KCS with HTR.

State	P (bar)	T (°C)	x	m (kg/s)	h (kJ/kg)	s (kJ/kg K)
1	37.67	160.00	0.85	0.46	1707.04	4.89
2	2.58	64.37	0.85	0.46	1386.13	5.20
3	2.58	52.74	0.40	2.99	255.34	1.43
4	2.58	35.00	0.40	2.99	−72.57	0.39
5	37.67	35.82	0.40	2.99	−65.92	0.40
6	37.67	124.90	0.40	2.99	342.37	1.56
7	37.67	124.90	0.40	2.99	342.37	1.56
8	37.67	160.00	0.40	2.99	737.82	2.50
9	37.67	160.00	0.32	2.52	530.71	1.99
10	37.67	55.76	0.32	2.52	46.73	0.71
11	2.58	51.94	0.32	2.52	49.17	0.73
12	1.01	30.00	0.00	9.81	2.13	0.01
13	1.01	38.00	0.00	9.81	5.54	0.02
14	1.01	175.00	0.00	36.89	157.21	0.43
15	1.01	144.90	0.00	36.89	125.17	0.35
16	1.01	144.93	0.00	36.89	125.17	0.35

6.6   KCS WITH DEPHLEGMATOR (HEAT REJECTION TO SURROUNDINGS)

The throttling process in the vapor absorption refrigeration is a key process to result the low temperature in evaporator in case of cooling generation or refrigeration. In KCS system, there is no much importance of dephlegmator on performance of plant. In fact, it drops the power output in the turbine as the source temperature and fluid flow drops due to heat rejection in the dephlegmator. The main function of dephlegmator is the distillation of fluid by removing the water by condensation. In this case, the heat is rejected to the surroundings. To condense the water from the mixture, vapor from the generator needs to be cooled. Since the dephlegmator is increasing the concentration of fluid at the turbine, it is possible to operate the plant's strong solution line at the lower concentration side.

Figure 6.12 shows the use of dephlegmator in a KCS to increase the concentration of working fluid in turbine and at lower pressures. The generator concentration results in high pressure. The pressure increases with increase in the generator concentration, that is, strong solution concentration.

For small capacity KCS plant, the operation of the plant at low pressures is recommended. The operation of KCS without dephlegmator demands high pressure and results in more power. For the same turbine concentration, the dephlegmator results in lower concentration in the generator. Figure 6.13 shows the (a) enthalpy–concentration chart and (b) temperature–entropy chart prepared on scale with reference to Figure 6.12. The area under temperature–entropy diagram is the heat transfer. It is clear from the diagram that the heat load in the economizer is more than the vapor generator (evaporator). The temperature rise in economizer is more than the temperature rise in the evaporator.

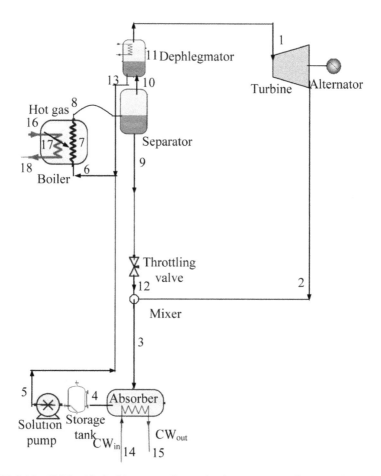

FIGURE 6.12   KCS with dephlegmator (heat rejection to surroundings)—schematic flow diagram.

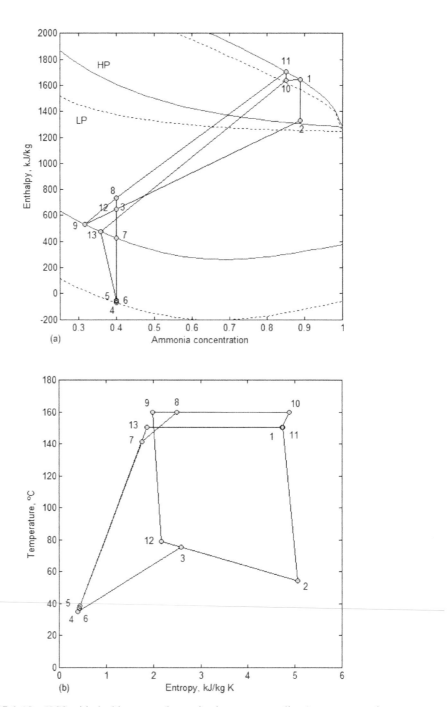

FIGURE 6.13   KCS with dephlegmator (heat rejection to surroundings) on property chart.

From the tabulated plant results shown in Table 6.5, the use of dephleg-mator in KCS is resulted as 1.7 kg/s of working fluid which is low compared to KCS without dephlegmator use. The circulating water demand in condenser is 21.51 kg/s with the hot gas flow rate of 36.89 kg/s. The entropy is increased from 4.75 to 5.06 kJ/kg K in the turbine. The degree of dephleg-mator is selected from the effectiveness of dephlegmator. The definition and details of effectiveness of dephlegmator are outlined in the earlier chapter. In this section, the assumed degree of dephlegmator is 25%.

TABLE 6.5 Material Balance Results of KCS with Dephlegmator (Heat Rejection to Surroundings).

State	P (bar)	T (°C)	x	m (kg/s)	h (kJ/kg)	s (kJ/kg K)
1	37.67	150.32	0.89	0.25	1642.72	4.75
2	2.58	54.39	0.89	0.25	1331.10	5.06
3	2.58	75.23	0.40	1.70	646.73	2.59
4	2.58	35.00	0.40	1.70	−72.57	0.39
5	37.67	35.82	0.40	1.70	−65.92	0.40
6	37.67	37.18	0.40	1.72	−59.84	0.42
7	37.67	141.41	0.40	1.72	425.31	1.76
8	37.67	160.00	0.40	1.72	736.70	2.49
9	37.67	160.00	0.32	1.45	530.71	1.99
10	37.67	160.00	0.85	0.27	1707.04	4.89
11	37.67	150.32	0.85	0.27	1638.38	4.73
12	2.58	78.90	0.32	1.45	533.15	2.16
13	37.67	150.32	0.36	0.02	474.54	1.87
14	1.01	30.00	0.00	21.51	0.97	0.00
15	1.01	38.00	0.00	21.51	2.53	0.01
16	1.01	175.00	0.00	36.89	157.21	0.43
17	1.01	161.41	0.00	36.89	142.71	0.39
18	1.01	140.16	0.00	36.89	120.11	0.34

6.7 KCS WITH DEPHLEGMATOR (INTERNAL HEAT RECOVERY)

In the earlier model, heat from the dephlegmator is rejected to the surround-ings. It demands additional heat in the boiler and so lower efficiency. If this heat is recovered in the cycle, the efficiency will increase. Therefore, in this model, instead of rejecting the heat to the surroundings, the heat from the dephlegmator is transferred to the strong solution at the exit of the pump. Figure 6.14 shows the schematic view of KCS with dephlegmator where the heat is transferred to the strong solution. As per the required heat load, strong solution is separated to dephlegmator. After the heat transfer to the solution, the heated solution is mixed to strong solution at the inlet of boiler. So, the heat load in the boiler and vapor at the exit of dephlegmator decreases. The drop in power due to decreased vapor is more significant than a drop in heat load in the boiler. Finally, the power and efficiency will decrease with the adoption of dephlegmator but is operated at a lower boiler pressure as per the discussion in the earlier section. But the drop in efficiency is not much compared to the earlier model, that is, dephlegmator heat loss to surroundings.

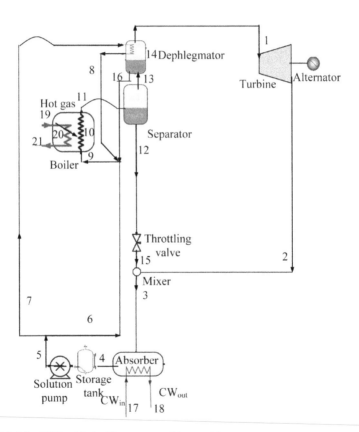

FIGURE 6.14   KCS with dephlegmator and internal heat recovery.

Figure 6.15 outlines the KCS on (a) enthalpy–concentration chart and (b) temperature–entropy chart. It clearly shows that the concentration of turbine fluid is increased from state 13 to state 1 with the use of dephlegmator. A small rise in fluid temperature at the inlet of boiler is observed due to mixing of return fluid from the dephlegmator. So, it saves the heat load in the generator.

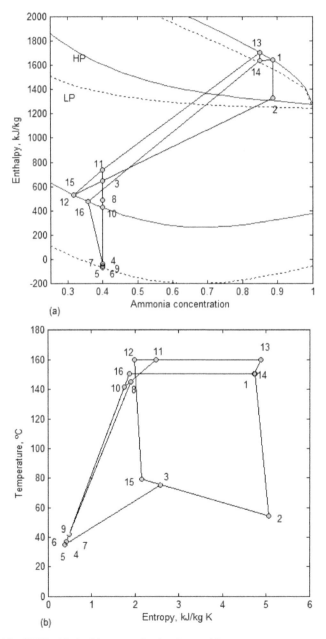

(a)

(b)

FIGURE 6.15 KCS with dephlegmator having internal heat recovery on property chart.

The cycle properties are also tabulated in Table 6.6. The high pressure and the low pressure in the cycle are, respectively, 37.67 and 2.58 bar. The temperature drop is nearly 100°C in the turbine with an entropy rise from 4.75 to 5.06 kJ/kg K. The vapor concentration is increased from 0.85 to 0.89 with the use of dephlegmator. The considered degree of dephlegmator is low (25%) due to power plant application.

TABLE 6.6 Material Balance Results of KCS with Dephlegmator (Internal Heat Recovery).

State	P (bar)	T (°C)	x	m (kg/s)	h (kJ/kg)	s (kJ/kg K)
1	37.67	150.32	0.89	0.25	1642.72	4.75
2	2.58	54.39	0.89	0.25	1331.10	5.06
3	2.58	75.23	0.40	1.70	646.73	2.59
4	2.58	35.00	0.40	1.70	−72.57	0.39
5	37.67	35.82	0.40	1.70	−65.92	0.40
6	37.67	35.82	0.40	1.67	−65.92	0.40
7	37.67	35.82	0.40	0.03	−65.92	0.40
8	37.67	145.00	0.40	0.03	487.83	1.91
9	37.67	40.82	0.40	1.72	−43.88	0.47
10	37.67	141.41	0.40	1.72	425.31	1.76
11	37.67	160.00	0.40	1.72	736.70	2.49
12	37.67	160.00	0.32	1.45	530.71	1.99
13	37.67	160.00	0.85	0.27	1707.04	4.89
14	37.67	150.32	0.85	0.27	1638.38	4.73
15	2.58	78.90	0.32	1.45	533.15	2.16
16	37.67	150.32	0.36	0.02	474.54	1.87
17	1.01	30.00	0.00	21.51	0.97	0.00
18	1.01	38.00	0.00	21.51	2.53	0.01
19	1.01	175.00	0.00	36.89	157.21	0.43
20	1.01	161.41	0.00	36.89	142.71	0.39
21	1.01	161.43	0.00	36.89	142.71	0.39

Figure 6.16 shows the role of dephlegmator and its effectiveness of performance of KCS on (a) power and (b) cycle efficiency. The zero effectiveness of dephlegmator indicates the operation of the KCS plant without

the use of dephlegmator. The power and efficiency of KCS are high at the lower range of generator vapor concentration. But decreasing the vapor concentration also decreases the strong solution concentration. It leads to the dilution of solution and increase of suitable range of low-temperature source. Therefore to design a KCS at lower source temperature, the strong solution concentration should be high. But the performance trends are recommending lower side of cycle concentrations. Therefore to gain the two benefits, that is, operation of cycle at lower temperature and gaining higher performance, a small degree of dephlegmator is required to increase the vapor concentration. The incorporation of dephlegmator also shows that the optimum generator vapor concentration increases with increase in effectiveness of dephlegmator to gain the higher efficiency. The concentration levels are increased in the cycle with the help of dephlegmator to gain the advantage of low-temperature source operation but with a penalty in output and efficiency. The high pressure is related to generator vapor concentration. Therefore, the pressure is increased with increase in vapor concentration but not with effectiveness of dephlegmator. Without compromising the turbine concentration, cycle is operated at lower pressure for small-scale applications using dephlegmator.

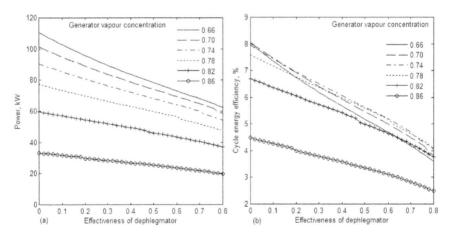

FIGURE 6.16   Influence of dephlegmator effectiveness on (a) power output and (b) cycle efficiency.

Figure 6.17 shows the relation of dephlegmator's effectiveness with strong solution concentration on (a) plant power generation capacity and (b) cycle thermal efficiency. As usual, the plant power and efficiency are decreasing with increase in dephlegmator effectiveness. The considered configuration, that is, internal heat recovered dephlegmator, is demanding high strong solution concentration for more power and efficiency. The addition of other accessories may change the demand of high strong solution concentration. At the end of outlining all the KCS configurations, all the configurations are compared under common agenda to highlight the relative merits and demerits of use of components in KCS.

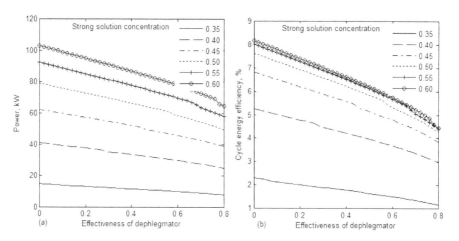

FIGURE 6.17   Influence of dephlegmator effectiveness with strong solution concentration on (a) power and (b) cycle efficiency.

6.8   KCS WITHOUT DEPHLEGMATOR

With reference to the earlier discussion, KCS can be constructed with dephlegmator or without dephlegmator depending on the source temperature, pressure, and application. Figure 6.18 outlines the schematic view of the KCS without dephlegmator and with the accessories. In this model except dephlegmator, all the components, that is, SH, LTR, and HTR, are included. The property charts, that is, (a) enthalpy–concentration and (b) temperature–entropy, are shown in Figure 6.19. These values are tabulated in Table 6.7. In this configuration, HTR and boiler are connected in series. There is another possibility of connecting HTR with the boiler. This is the parallel arrangement of HTR with the boiler. The influence of dephlegmator is compared with this plant to show the role of dephlegmator. After analyzing the role of each component on performance, the three layouts of KCS are compared. The three plants are (1) KCS without dephlegmator, (2) KCS plant with HTR and boiler connected in series, and (3) KCS plant with HTR and boiler connected in parallel lines. Therefore, KCS without dephlegmator configuration is required to compare with the other plants (Fig. 6.18).

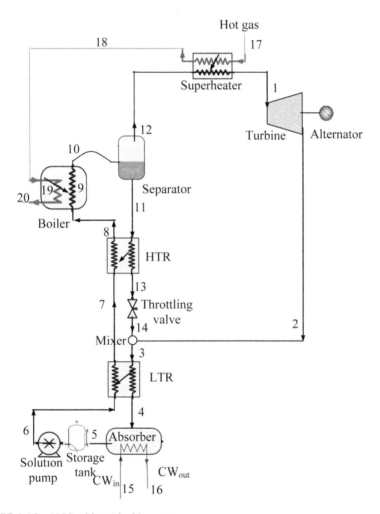

FIGURE 6.18 KCS without dephlegmator.

The vapor in turbine is 0.61 kg/s separated from 3.14 kg/s of strong solution in boiler. The hot gas flow rate is 36.89 kg/s at 175°C. The weak solution flow rate is 2.52 kg/s with 0.29 concentration. The specific entropy is increased from 5.21 to 5.48 kJ/kg K in turbine. The use of SH in this cycle lowers the boiler pressure and so more heat recovery and working fluid.

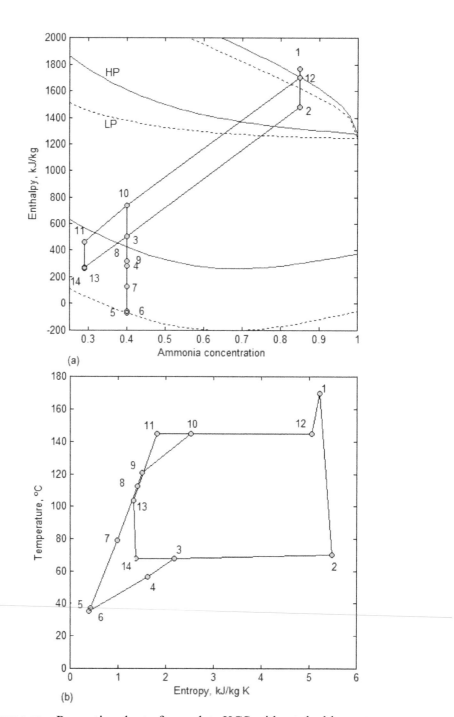

FIGURE 6.19 Properties chart of complete KCS without dephlegmator.

TABLE 6.7 Material Balance Results of KCS Without Dephlegmator.

State	P (bar)	T (°C)	x	m (kg/s)	h (kJ/kg)	s (kJ/kg K)
1	25.04	170.00	0.85	0.61	1772.18	5.21
2	2.58	70.09	0.85	0.61	1482.52	5.48
3	2.58	67.32	0.40	3.14	502.22	2.17
4	2.58	56.16	0.40	3.14	313.15	1.60
5	2.58	35.00	0.40	3.14	−72.57	0.39
6	25.04	35.64	0.40	3.14	−67.81	0.40
7	25.04	78.15	0.40	3.14	121.27	0.97
8	25.04	112.34	0.40	3.14	280.83	1.41
9	25.04	120.67	0.40	3.14	321.22	1.51
10	25.04	145.00	0.40	3.14	736.83	2.53
11	25.04	145.00	0.29	2.52	462.14	1.82
12	25.04	145.00	0.85	0.61	1701.95	5.05
13	25.04	102.62	0.29	2.52	263.24	1.32
14	2.58	67.59	0.29	2.52	265.68	1.36
15	1.01	30.00	0.00	11.53	1.81	0.01
16	1.01	38.00	0.00	11.53	4.71	0.02
17	1.01	175.00	0.00	36.89	157.21	0.43
18	1.01	173.93	0.00	36.89	156.04	0.42
19	1.01	140.67	0.00	36.89	120.68	0.34
20	1.01	137.44	0.00	36.89	117.24	0.33

6.9 FULL KCS WITH HTR AND VAPOR GENERATOR IN SERIES

Full KCS means all the accessories including dephlegmator. In this configuration, the HTR and vapor generator are connected in series usually. The schematic flow diagram of the KCS plant with hot gas as heat source is shown in Figure 6.20. The hot gas (27) from the process industry is recovered in the SH and boiler (HRVG). In separator, the working fluid is separated into vapor (16) and weak liquid mixture (15). The ammonia concentration is increased in dephlegmator (16–21) by rejecting heat with a recovery. The vapor's temperature is increased in SH at inlet of the turbine (1). The vapor (1) is expanded in mixture turbine to generate power and it is diluted by mixing with a weak solution (19). The liquid weak solution from separator has been throttled (18–19) after rejecting heat (15–18) at HTR and mixed with turbine exit fluid (2). The mixture (3) again rejects heat (3–4) at LTR and condenses into a saturated liquid (5). The condensate is pumped to separator pressure (6) and preheated at LTR (7–10) and HTR (10–11). Some amount of liquid mixture is used to absorb the heat (8–9) from the dephlegmator which allows internal heat recovery without wasting the heat to surroundings. The three liquid streams (9, 20, and 11) are mixed (12) at the HRVG inlet. The preheated liquid mixture (12) is converted into liquid vapor mixture (14) in economizer and evaporator sections of HRVG. The saturated vapor (16) is cooled in dephlegmator and heated in a SH before entering into the turbine. This cycle repeats for continuous power supply.

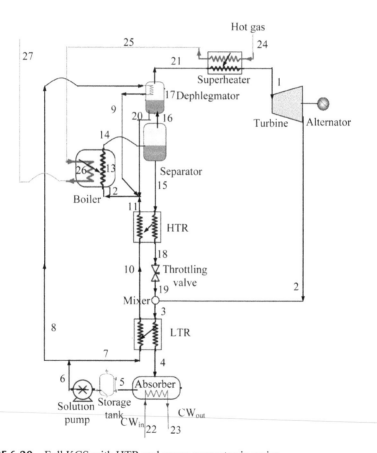

FIGURE 6.20   Full KCS with HTR and vapor generator in series.

Figure 6.21 shows (a) enthalpy–concentration and (b) temperature–entropy diagram of KCS. In KCS diagram, liquid–vapor mixture (14) in the generator is separated into liquid (15) and vapor (16). Similarly after cooling at dephlegmator, the liquid–vapor mixture (17) in dephlegmator is separated into liquid (20) and vapor (21). Due to the separation of working fluid before the turbine, the fluid in the turbine decreases.

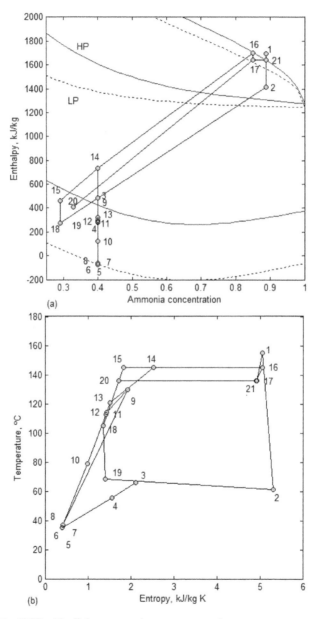

FIGURE 6.21 KCS with all the accessories on property chart.

The solution of KCS involves the handling of assumptions and development of mass and energy balance equations. Assumptions used in KCS analysis are as follows. TTD at HRVG inlet with hot gas is taken at 15°C. DSH is 10°C. PP in boiler is 20°C. AP in the boiler is 2°C. TTD in dephlegmator is 25°C. Vapor concentration at exit of separator, that is, generator, is 0.85. Strong solution concentration is 0.4. Effectiveness of dephlegmator is 25%. The isentropic efficiency of vapor turbine and solution pump is considered as 80% and 75%, respectively. The mechanical efficiency of the solution pump $(\eta_{m,p})$ and mixture turbine $(\eta_{m,t})$ is taken at 96%. Electrical generator efficiency (η_{ge}) is taken as 98%. The condensate leaving the condenser is assumed as saturated liquid. Pressure drop and heat loss in pipelines are neglected.

The turbine inlet temperature is

$$T_1 = T_{24} - \text{TTD} \tag{6.18}$$

The separator temperature is resulted from the turbine inlet temperature and DSH. In this case, DSH is the temperature difference between turbine inlet and separator.

Separator temperature is

$$T_{16} = T_{24} - \text{DSH} \tag{6.19}$$

From the effectiveness of dephlegmator, the turbine concentration is

$$x_{21} = x_{16} + \varepsilon_{\text{dephlegmator}} (1 - x_{16}) \tag{6.20}$$

The high pressure (P_{16}) is determined from the separator temperature and vapor concentration (T_{16}, x_{16}), since it is function of temperature and concentration at the saturated vapor state. The separator temperature, T_{15} at the liquid line, is the BPT at the high pressure and concentration, x_{15}. From this relation, the liquid portion concentration, that is, weak solution concentration x_{12}, is determined through the iteration. The low pressure is determined from mixture ammonia concentration (x_5) and temperature (T_5) at condenser outlet.

At inlet of generator, three streams are mixing; two from dephlegmator and one from HTR. The condition of mixed stream is calculated from the iteration of loops in the cycle.

The economizer's exit temperature, T_{13}, is maintained below the BPT to avoid the sharp transition at the evaporator inlet. This temperature difference is equal to AP.

The temperature of strong solution at the evaporator inlet is

$$T_{13} = T_{bp} - AP \tag{6.21}$$

where T_{bp} is the BPT at boiler pressure and concentration.

The hot fluid temperature at the evaporator section of boiler is

$$T_{29} = T_{bp} + PP \tag{6.22}$$

For specific results, KCS is solved at unit mass of strong solution at separator inlet (14). In separator, out of one unit mass of mixture, VF is the vapor portion and $(1 - VF)$ is the liquid portion to be separated. After applying lever rule for separation process,

$$VF_{separator} = \frac{x_{14} - x_{15}}{x_{16} - x_{15}} \tag{6.23}$$

Similarly, the dryness fraction at dephlegmator is evaluated as follows:

$$DF_{dephlegmator} = \frac{x_{16} - x_{20}}{x_{21} - x_{20}} \tag{6.24}$$

The simplifications of mass balance equations result in the unknown mass flow rate in the cycles.

$$\begin{aligned}
m_{21} &= DF_{dephlegmator} m_{16} \\
&= DF_{dephlegmator} DF_{separator} m_{14} \\
&= DF_{dephlegmator} DF_{separator} (m_6 + m_{20}) \\
&= DF_{dephlegmator} DF_{separator} (m_6 + (1 - DF_{dephlegmator}) m_{16}) \\
&= DF_{dephlegmator} DF_{separator} \left(m_6 + (1 - DF_{dephlegmator}) \frac{m_{21}}{DF_{dephlegmator}} \right) \\
&= DF_{dephlegmator} DF_{separator} m_6 + DF_{separator} (1 - DF_{dephlegmator}) m_{21}
\end{aligned}$$

$$m_{21} - DF_{separator} (1 - DF_{dephlegmator}) m_{21} = DF_{dephlegmator} DF_{separator} m_6$$

$$m_{21} = \frac{DF_{dephlegmator} DF_{separator} m_6}{1 - DF_{separator} (1 - DF_{dephlegmator})}$$

$$= \frac{DF_{dephlegmator} DF_{separator} m_6}{1 - DF_{separator} + DF_{separator} DF_{dephlegmator}}$$

Therefore, the vapor generation from the dephlegmator is

$$m_{21} = \frac{DF_{dephlegmator}DF_{separator}}{1 + DF_{separator}DF_{dephlegmator} - DF_{separator}}m_4 \qquad (6.25)$$

After mixing the streams, the fluid entering into the separator is

$$m_{14} = \frac{m_{16}}{DF_{separator}}$$

$$m_{14} = \frac{m_{21}}{DF_{separator}DF_{dephlegmator}} \qquad (6.26)$$

The liquid fluid from the separator is

$$m_{15} = (1 - DF_{separator})m_{14}$$

Substituting m_{14} from the above equation,

$$m_{15} = \frac{\left(1 - DF_{separator}\right)}{DF_{separator}DF_{dephlegmator}}m_{21} \qquad (6.27)$$

The fluid at the inlet of dephlegmator is

$$m_{16} = \frac{m_{21}}{DF_{dephlegmator}} \qquad (6.28)$$

$$m_{20} = (1 - D_{dephlegmator})m_{16}$$

With reference to Figure 6.22, it is observed that the mass flow rates in all the lines are determined from the above simplified equations, that is, m_{21}, m_{14}, m_{15}, m_{16}, and m_{20} at unit mass of m_4.

The heat load in the dephlegmator results in the fluid flow

$$m_8 = \frac{h_{16} - h_{17}}{h_9 - h_8} \times m_{16} \qquad (6.29)$$

The turbine exit temperature is iterated by entropy matching in isentropic expansion and the actual temperature from the isentropic efficiency. The same procedure is adopted for solution pump.

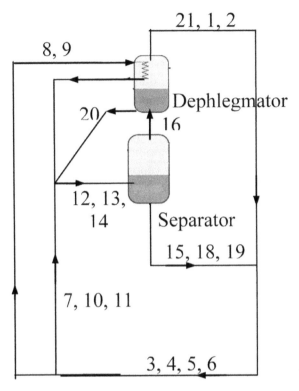

FIGURE 6.22   Mass balance.

The energy interactions in the plant components are as follows:

Turbine output,

$$W_t = m_1(h_1 - h_2)\eta_{m,t}\eta_{ge} \qquad (6.30)$$

Work input to pump,

$$W_p = \frac{m_5(h_6 - h_5)}{\eta_{m,p}} \qquad (6.31)$$

Net output from plant,

$$W_{net} = (W_t - W_p) \qquad (6.32)$$

The methodology adopted for KCS at LTHR is outlined in the flow chart shown in Figure 6.23.

Table 6.8 shows the plant (Fig. 6.21) properties (pressure, temperature, concentration, flow rate, specific enthalpy, and specific entropy) at the hot gas temperature of 175°C and 36.89 kg/s (100,000 N m³/h). The generated working fluid in the generator is 3.18 kg/s. After separation and distillation in dephlegmator, the available fluid in turbine is only 0.58 kg/s. The weak and strong solution flow rates are 2.56 and 3.14 kg/s, respectively. The high and low pressures in the cycle are 25.04 and 2.58 bar at the source and sink fluid inlet temperature of 175°C and 30°C, respectively. The cycle concentrations are 0.29 for weak solution, 0.33 at dephlegmator return, 0.40 of strong solution, 0.85 at separator vapor, and 0.89 turbine fluid.

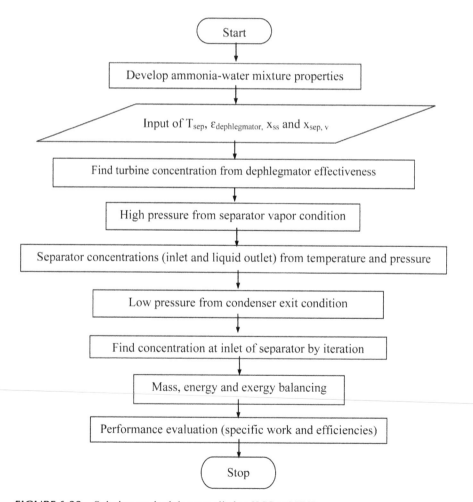

FIGURE 6.23   Solution methodology applied to KCS at LTHR.

TABLE 6.8 Material Balance Results of KCS with HTR and Vapor Generator in Series.

State	P (bar)	T (°C)	x	m (kg/s)	h (kJ/kg)	s (kJ/kg K)
1	25.04	155.00	0.89	0.58	1693.78	5.04
2	2.58	61.11	0.89	0.58	1413.38	5.31
3	2.58	65.98	0.40	3.14	479.01	2.10
4	2.58	55.03	0.40	3.14	294.05	1.55
5	2.58	35.00	0.40	3.14	−72.57	0.39
6	25.04	35.64	0.40	3.14	−67.81	0.40
7	25.04	35.64	0.40	3.07	−67.81	0.40
8	25.04	35.64	0.40	0.07	−67.81	0.40
9	25.04	130.00	0.40	0.07	482.92	1.91
10	25.04	78.25	0.40	3.07	121.70	0.97
11	25.04	112.59	0.40	3.07	282.00	1.41
12	25.04	113.80	0.40	3.18	287.95	1.43
13	25.04	120.86	0.40	3.18	322.25	1.51
14	25.04	145.00	0.40	3.18	734.53	2.52
15	25.04	145.00	0.29	2.56	462.14	1.82
16	25.04	145.00	0.85	0.62	1701.95	5.05
17	25.04	136.17	0.85	0.62	1643.36	4.91
18	25.04	103.98	0.29	2.56	269.43	1.33
19	2.58	67.94	0.29	2.56	271.88	1.38
20	25.04	136.17	0.33	0.04	409.01	1.71
21	25.04	136.17	0.89	0.58	1640.41	4.92
22	1.01	30.00	0.00	10.96	1.91	0.01
23	1.01	38.00	0.00	10.96	4.96	0.02
24	1.01	175.00	0.00	36.89	157.21	0.43
25	1.01	174.24	0.00	36.89	156.38	0.42
26	1.01	140.86	0.00	36.89	120.88	0.34
27	1.01	138.09	0.00	36.89	117.93	0.34

6.10 PERFORMANCE ANALYSIS OF KCS MODELS WITH ACCESSORIES

Incorporation of components in KCS changes the plant and cycle behavior. In this comparative study, eight KCS configurations are considered and they are basic KCS, SH, LTR, HTR, dephlegmator with heat rejection to surroundings, dephlegmator with internal heat recovery, full KCS without dephlegmator, full KCS with heaters in series, and full KCS with heaters in parallel lines. Figure 6.24 compares the KCS plants with reference to a change in source temperature on (a) power and (b) cycle efficiency. Addition of LTR and HTR in KCS effects on heat recovery and efficiency without disturbing the power output. The optimum source temperature is different with different models of KCS. For basic KCS, the optimum source temperature is 150°C. Addition of LTR and HTR to basic KCS increases the optimum source temperature with reference to both power and efficiency. The highest power is resulted in full KCS without dephlegmator. The power and efficiency are maximum with the full KCS without dephlegmator.

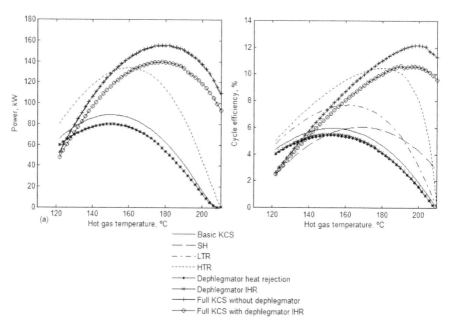

FIGURE 6.24   Comparative study of KCS configurations with hot gas inlet temperature.

Figure 6.25 shows the influence of strong solution concentration on performance of the various KCS configurations. The power is increasing with increase in strong solution concentration. The efficiency of cycle is maximizing at a particular strong solution concentration. All KCS cycles are resulting in high power at high strong solution concentration. But to result in high thermal efficiency, the strong solution concentration needs to be limited at the optimum concentration.

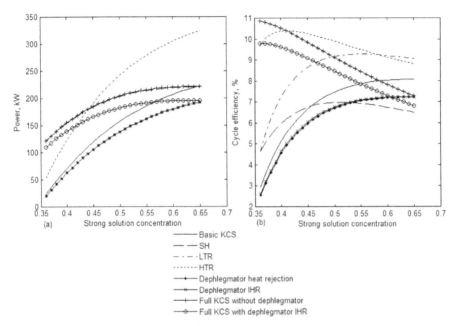

FIGURE 6.25   Comparative study of KCS configurations with strong solution concentration.

Figure 6.26 compares the KCS configurations with a change in generator vapor concentration on (a) power output and (b) cycle efficiency. At lower generator vapor concentration, SH configuration and HTR configuration are resulting in higher output. At high generator vapor concentration, the cycle efficiency is maximized with full KCS without dephlegmator and followed by a full KCS with dephlegmator and internal heat recovery.

Figure 6.27 demonstrates the influence of dephlegmator effectiveness on (1) power and (2) cycle efficiency of KCS plants. Without dephlegmator, the role of effectiveness is not applicable. The three layouts with dephlegmator are dephlegmator with heat rejection to surroundings, dephlegmator with internal heat recovery, and dephlegmator with internal heat recovery and all components. Out of these three, dephlegmator with internal heat recovery results in more power and efficiency. The power and efficiency decrease with increase in dephlegmator effectiveness.

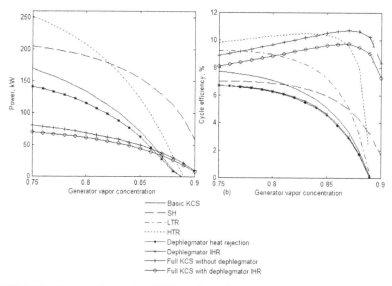

FIGURE 6.26   Comparative study of KCS configurations with generator's vapor concentration.

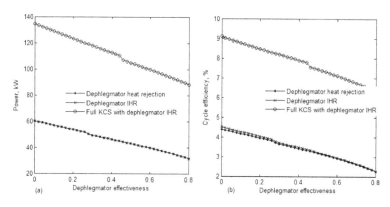

FIGURE 6.27   Comparative study of KCS configurations with dephlegmator's effectiveness.

6.11   FULL KCS WITH HTR AND VAPOR GENERATOR IN PARALLEL

The unique feature of KCS compared to steam Rankine is its internal heat recovery. The HTR, economizer, and evaporators are serially connected in a regular low-temperature KCS plant. Obviously, a more amount of vapor from generator and followed by in separator gives high power output by a turbine. But at this condition, low heat recovery in HTR results in low thermal efficiency with the serially connected heat exchangers. It is solved by sharing of heat load between HTR and boiler (economizer + partial evaporator) based on available heat with parallel arrangement in place of serial. It also facilitates the flexible operation of heaters with the heat source conditions.

The process flow diagram of KCS with waste heat recovery is shown in Figure 6.28. In this configuration, the HTR and boiler are arranged in parallel. Figure 6.29a and b shows the enthalpy–concentration and tempera-ture–entropy diagram, respectively, for the process flow diagram of Figure 6.29. The superheated vapor (1) expands in the turbine (1–2) and then mixes with the throttled fluid (22). To run the thermodynamic cycle, the mixture (3) needs to be cooled at condenser whereas the pumped fluid (6) needs to be heated at the boiler. The LTR reduces the condenser and boiler load. Similarly, HTR also reduces the boiler load. These two components internally recover the heat and increase the cycle thermal efficiency. In the

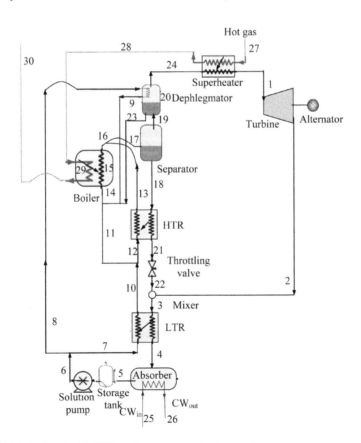

FIGURE 6.28 KCS with HTR and vapor generator in parallel.

earlier configuration, HTR and boiler are connected in serial line. Now in the current configurations, the parallel arrangement of these two heaters allows a greater flexibility in heat sharing between source and heat recovery. By splitting the working fluid at the entry of boiler shares the source heat load. It may increase the efficiency due to share in the heat supply.

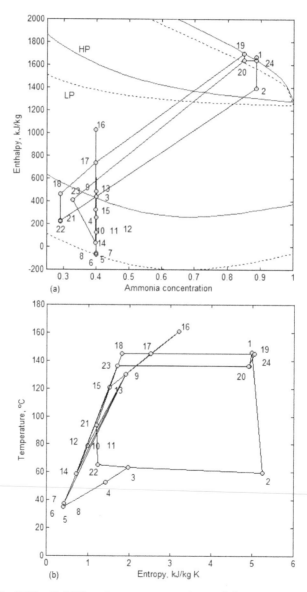

FIGURE 6.29   KCS with HTR and vapor generator in parallel on property chart.

Table 6.9 shows the mass and energy balance results of KCS with HTR and boiler connected in parallel. The high pressure and low pressure in the cycle are 25.04 and 2.58 bar, respectively. In the turbine, the entropy is increased from 4.98 to 5.25 kJ/kg K. With the hot fluid flow rate of 36.89 kg/s, the working fluid flow rate in the turbine is 0.73 kg/s. The total fluid generated in boiler and HTR is 4.03 kg/s.

TABLE 6.9 Material Balance Results of KCS with HTR and Vapor Generator in Parallel.

State	P (bar)	T (°C)	x	m (kg/s)	h (kJ/kg)	s (kJ/kg K)
1	25.04	146.17	0.89	0.73	1668.96	4.98
2	2.58	59.56	0.89	0.73	1393.01	5.25
3	2.58	63.45	0.40	3.98	435.82	1.97
4	2.58	52.48	0.40	3.98	250.86	1.41
5	2.58	35.00	0.40	3.98	−72.57	0.39
6	25.04	35.64	0.40	3.98	−67.81	0.40
7	25.04	35.64	0.40	3.89	−67.81	0.40
8	25.04	35.64	0.40	0.08	−67.81	0.40
9	25.04	130.00	0.40	0.08	482.92	1.91
10	25.04	78.25	0.40	3.89	121.70	0.97
11	25.04	78.25	0.40	1.73	121.70	0.97
12	25.04	78.25	0.40	2.17	121.70	0.97
13	25.04	130.00	0.40	2.17	482.92	1.91
14	25.04	58.45	0.40	1.86	33.24	0.72
15	25.04	121.08	0.40	1.86	323.46	1.52
16	25.04	160.88	0.40	1.86	1027.07	3.20
17	25.04	145.00	0.40	4.03	734.53	2.52
18	25.04	145.00	0.29	3.25	462.14	1.82
19	25.04	145.00	0.85	0.78	1701.95	5.05
20	25.04	136.17	0.85	0.78	1643.36	4.91
21	25.04	93.25	0.29	3.25	221.11	1.20
22	2.58	65.19	0.29	3.25	223.53	1.24
23	25.04	136.17	0.33	0.05	409.01	1.71
24	25.04	136.17	0.89	0.73	1640.41	4.92
25	0.00	30.00	0.00	9.67	2.16	0.01
26	0.00	38.00	0.00	9.67	5.62	0.02
27	1.01	175.00	0.00	36.89	157.21	0.43
28	1.01	174.49	0.00	36.89	156.65	0.43
29	1.01	141.08	0.00	36.89	121.12	0.34
30	1.01	127.25	0.00	36.89	106.46	0.31

6.12   COMPARATIVE ANALYSIS OF KCS PLANTS WITH ARRANGEMENTS OF HTR AND BOILER

Figure 6.30 compares the performance of three KCS plants with a change in source temperature. The studied results are (a) power and (b) cycle efficiency of (1) KCS without dephlegmator, (2) KCS with HTR and boiler are arranged in series, and (3) KCS with HTR and boiler are arranged in parallel. For maximum power point of view, KCS with heaters in parallel connection is recommended. The maximum efficiency condition recommends a KCS without dephlegmator. The efficiency of serial and parallel-arranged heaters in KCS is nearly the same efficiency. The drop in power generation with dephlegmator in the series model is compensated with the parallel option.

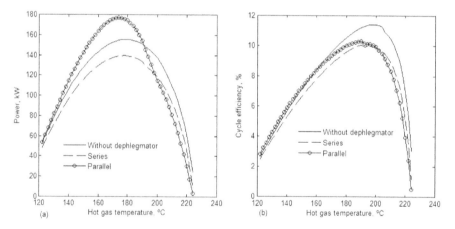

FIGURE 6.30   Comparative study of KCS plants without dephlegmator, series-connected heaters, and parallel-connected heaters with change in source temperature.

Figure 6.31 compares the three KCS plants considered with a change in strong solution concentration. The strong solution concentration is optimized for parallel model for the maximum power generation. The optimized strong solution concentration in this plant is 0.5. As stated in the earlier section, parallel heaters plant results in higher power generation compared to the other two plants. In view of efficiency point of view, KCS without dephlegmator has higher efficiency. The efficiency of these three plants is decreasing with increase in strong solution concentration.

Figure 6.32 compares the (a) power and (b) cycle efficiency of three KCS plants and the consideration in the light of generator vapor concentration.

The generation vapor concentration is an important parameter to find the boiler pressure, that is, high pressure in the cycle. The increase in the concentration indicates the increased high pressure. Nearly up to 85% vapor concentration, the parallel configured plant, offers higher power generation and after that KCS without dephlegmator offers higher power. As usual, the use of dephlegmator in series heaters plant drops output compared to the output of KCS without dephlegmator. The cycle efficiency is maximized after 85% vapor concentration but below 90% for all the three plants.

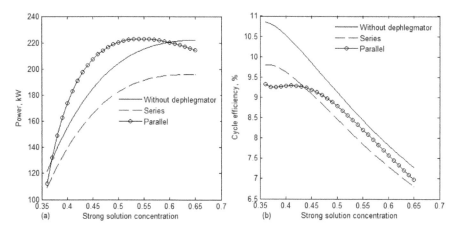

FIGURE 6.31 Comparative study of KCS plants without dephlegmator, series-connected heaters, and parallel-connected heaters with change in strong solution concentration.

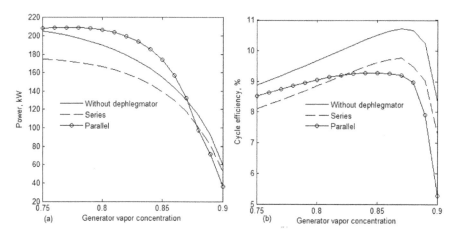

FIGURE 6.32 Comparative study of KCS plants without dephlegmator, series-connected heaters, and parallel-connected heaters with change in generator vapor concentration.

Figure 6.33 is also a similar comparison with reference to dephlegmator effectiveness on the performance of three plants. The KCS without dephlegmator has no significance on dephlegmator effectiveness. The power and efficiencies of KCS with series configuration and KCS with parallel configuration are decreasing with increase in dephlegmator effectiveness. Therefore, in case dephlegmator is incorporated in KCS, a small degree of dephlegmation is enough. The specific power of (a) KCS without dephlegmator, (b) KCS with series heaters, and (c) KCS with parallel heaters are compared with each other. The highest power generation is through the KCS with parallel configuration plant. The lowest is the KCS with series configuration.

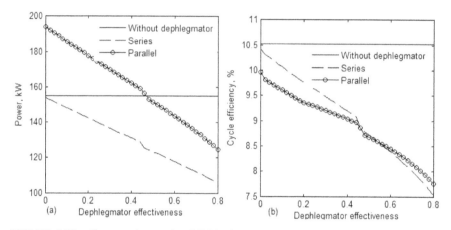

FIGURE 6.33 Comparative study of KCS plants without dephlegmator, series-connected heaters, and parallel-connected heaters with change in dephlegmator effectiveness.

Figure 6.34 shows the specific power of the three plants with change in generator temperature and generator vapor concentration. The optimum generator temperature is changing for the maximum power with change in vapor concentration. With increase in generator vapor concentration, the power is decreasing. The optimum source temperature is also decreasing with increase in vapor concentration.

Figure 6.35 compares the cycle efficiencies of KCS without dephlegmator, KCS with series configuration in heaters, and KCS with parallel arrangement. The optimum source temperature is decreasing with increase in generator vapor concentration for the KCS without dephlegmator and KCS with series configuration. The parallel configuration demands higher optimum source temperature compared to the other two plants. In parallel configuration, the efficiency variations are minor with change in vapor concentration compared to other two plants.

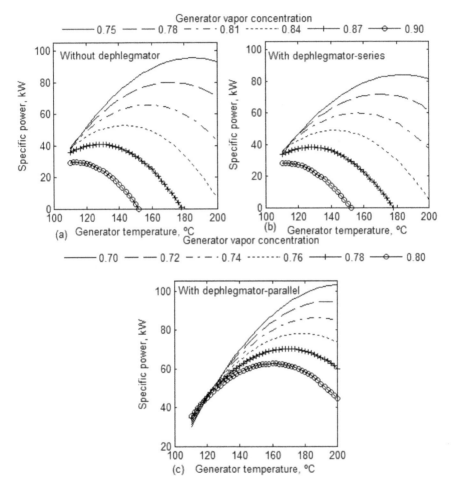

FIGURE 6.34 Variations in specific power of KCS configurations (a) without dephlegmator, (b) series-connected heaters, and (c) parallel-connected heaters with generator temperature and vapor concentration.

Figure 6.36 is plotted to study the generator pressure with a change in generator temperature and generator vapor concentration for the three plants considered. The pressure is increasing with increase in generator vapor concentration and also with generator temperature as the pressure is the function of temperature and concentration.

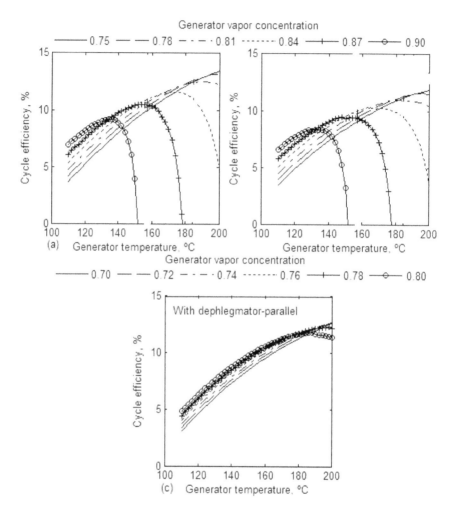

FIGURE 6.35 Cycle efficiency variations in KCS configurations (a) without dephlegmator, (b) series-connected heaters, and (c) parallel-connected heaters with generator temperature and vapor concentration.

Relatively lower boiler pressure is maintained with parallel heaters plant compared to the other two plants.

Table 6.10 concludes the performance details of KCS configurations and compares with the basic KCS. The degree of improvement in first law efficiency and second law efficiency is given with reference to the reference plant, that is, basic KCS. Since the change in power is directly related to second law efficiency, one result only tabulated instead of repetition. The dryness fraction of fluid at the exit of turbine is above 90% to all the plants.

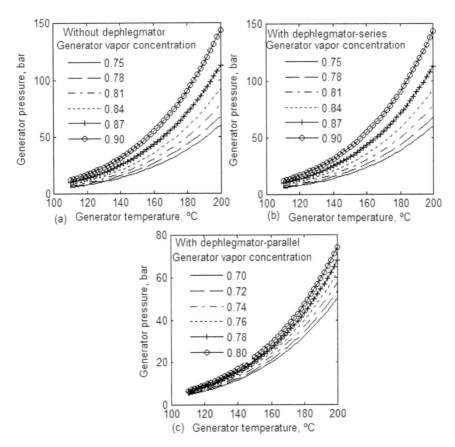

FIGURE 6.36 Changes in generator pressure with KCS plant's configurations (a) without dephlegmator, (b) series-connected heaters, and (c) parallel-connected heaters with generator temperature and vapor concentration.

Therefore, there is no issue of wetness or moisture in the turbine during the expansion. The maximum dryness fraction is resulted to full KCS without dephlegmator among the plant under consideration. The working fluid available in turbine is decreased in LTR, dephlegmator with heat rejection to surroundings, and dephlegmator with internal heat recovery compared to the basic KCS. In other plants, such as HTR and full KCS (three models), the working fluid flow rate is increased. The turbine fluid in SH plant and basic KCS is the same. The maximum limit of source temperature is around 195°C. The separator temperature and turbine inlet temperature are the same, at 160°C without use of SH. In the SH model, the turbine

TABLE 6.10 Comparison of Performance of Various KCS Configurations at the Hot Gas Temperature of 175°C and 100,000 N m³/kg.

Details	Basic KCS	SH	LTR	HTR	Dephleg-mator heat rejection	Dephleg-mator IHR	Full KCS without dephlegmator	Full KCS serial heaters	Full KCS with parallel heaters
1. Turbine exit dryness fraction (%)	90.74	94.00	90.74	90.74	91.08	91.08	94.56	94.20	93.42
2. Turbine fluid (kg/s)	0.27	0.27	0.25	0.46	0.25	0.25	0.61	0.58	0.73
3. Maximum limit in generator temperature (°C)	195.00	195.00	195.00	195.00	195.00	195.00	195.00	195.00	195.00
4. Generator temperature (°C)	160.00	145.00	160.00	160.00	160.00	160.00	145.00	145.00	145.00
5. Power (kW)	70.70	154.27	70.70	122.78	62.39	62.39	155.15	139.44	173.85
6. Thermal efficiency (%)	5.11	5.97	7.23	10.39	4.56	4.65	10.52	9.62	9.28
7. % Change in thermal efficiency	–	16.80	41.44	103.24	–10.79	–8.79	105.88	88.27	81.70
8. Second law efficiency (%)	6.32	5.96	6.38	11.08	5.63	5.63	14.00	12.58	15.69
9. % Change in second law efficiency	–	118.21	0	73.66	–11.75	–11.75	119.45	97.22	145.90

inlet temperature is 160°C and the separator temperature is 145°C. In the plant configurations, namely, LTR, HTR, dephlegmator with heat rejection to surroundings, and dephlegmator with internal heat recovery, SH is not integrated. Therefore, the separator temperature is 160°C. The full KCS with SH configuration results in the separator temperature as 145°C. In the basic KCS plant, the power generation with the specified hot gas conditions is 70.70 kW. In LTR, there is no change in the boiler section and so same working fluid is generated as basic KCS. The power generation in basic KCS and LTR is the same.

Table 6.11 shows the capacity and specifications of various KCS plants under this study. This comparison can assist the designer to visualize the relative capacity and size of various heat exchangers and machines in the plant and further design of systems. The capacity of LTR is high in full KCS without dephlegmator and low in LTR configuration. But the HTR load is highest in HTR configuration only and least in full KCS without dephlegmator. As per the assumptions in full KCS plants, more or less LTR and HTR loads are equal. The dephlegmator load is increased from 88.87 kW in series-connected plant to 112.75 kW in parallel-connected plant. In HTR plant, the strong solution temperature crossed the BPT; so, the heat duty in economizer is zero. The adoption of SH in plant decreases the boiler pressure and so more heat recovery. The highest generator load is in SH configuration and the lowest at LTR. The economizer role is increased in SH and also full KCS plants. The superheat load is increased from 26.42 to 43.15 kW from SH model to full KCS without dephlegmator. Because of SH role, the maximum heat rejection in absorber is 2319.60 kW in SH plant and decreased to minimum of 862.56 kW in LTR plant. The maximum turbine gross power (189.44) is from full KCS with parallel heaters. The minimum pump input (9.99 kW) is at dephlegmator configuration having heat rejection to surroundings and also internal heat recovery. Finally, full KCS with parallel heaters resulted in higher power output.

6.13 SUMMARY

This chapter is focused on KCS with low-temperature heat recovery. The role of each and every component on performance of KCS has been detailed. The studied models are basic KCS, SH, LTR, HTR, dephlegmator with heat rejection to surroundings, dephlegmator with internal heat recovery, full KCS without dephlegmator, full KCS with dephlegmator and heaters in series, and full KCS with dephlegmator and heaters in parallel. The

TABLE 6.11 Comparison of KCS Plants Specifications at the Hot Gas Temperature of 175°C and 100,000 N m³/kg.

Details	Basic KCS	SH	LTR	HTR	Dephleg-mator heat rejection	Dephleg-mator IHR	Full KCS without dephlegmator	Full KCS serial heaters	Full KCS with parallel heaters
1. LTR (kW)	–	–	404.29	–	–	–	593.50	579.85	735.62
2. HTR (kW)	–	–	–	1221.70	–	–	502.10	493.30	782.74
3. Dephlegmator (kW)	–	–	–	–	39.12	39.12	–	88.87	112.75
4. Economizer (kW)	844.62	1236.80	439.31	0.00	833.54	806.12	126.79	108.95	540.60
5. Evaporator	538.77	1321.30	538.77	1182.10	535.01	535.01	1304.60	1309.50	1310.70
6. Generator (kW)	1383.39	2558.10	978.09	1182.10	1368.60	1341.10	1431.40	1418.50	1851.30
7. Superheater (kW)	–	26.41	–	–	–	–	43.15	30.69	20.83
8. Absorber (kW)	1266.90	2319.60	862.56	980.17	1222.30	1222.30	1210.70	1418.50	1286.40
9. Turbine (kW)	80.82	166.73	80.82	140.36	72.39	72.39	167.45	151.72	189.44
10. Pump (kW)	10.12	12.46	10.12	17.58	9.99	9.99	12.30	12.28	15.59
11. Net power (kW)	70.70	154.27	70.70	122.78	62.39	62.39	155.14	139.44	173.85

highest increment in power and second law efficiency is observed in full KCS with parallel heaters. The highest increment in cycle efficiency is resulted in full KCS without dephlegmator. Dephlegmator is recommended only for small capacity only to operate the plant with lower strong solution concentration. The role of HTR is more compared to LTR on performance improvement. Only dephlegmator addition to basic KCS leads to drop in power and efficiency. The addition of other elements such as SH, LTR, and HTR compensates this loss. The parallel configuration KCS showed many benefits over the other KCS configurations. Finally, KCS with HTR and boiler in parallel and all other components give the higher performance over the others.

KEYWORDS

- **Kalina cycle system**
- **direct fuel burning**
- **solar thermal collectors**
- **water heat recovery**
- **low-temperature heat recovery**

CHAPTER 7

KALINA CYCLE SYSTEM WITH INTERMEDIATE-TEMPERATURE HEAT RECOVERY

ABSTRACT

Majority of waste heat falls under the category of intermediate temperature range. The Kalina cycle designed at the temperature operated
between two pressures but without throttling the fluid. The plant layout
is complex but it offers high performance compared to other KCS plants.
A low-capacity steam power plant also can operate at this temperature
range. Therefore, KCS at this temperature has been compared with the
steam power plant to find the highlights of the Kalina plant. This chapter
presents modeling, simulation, analysis, and system specifications at the
recommended conditions.

7.1 INTRODUCTION

In this chapter, the intermediate temperature range is considered from 250°C
to 300°C. The Kalina cycle system (KCS) configuration permits this narrow
range of temperature due to many constraints in the heat exchangers and
heat recoveries. To meet these constrains, the source temperature is varied in
this range only. Many industries reject heat at this temperature range such as
cement, steel, ferroalloys, etc. The configuration studied in this temperature
range has many specialties such as extra fluid in turbine, absence of throttling, etc. This chapter presents the description, modeling, simulation and
analysis, and parametric optimization of the KCS suitable at the intermediate-
temperature heat recovery.

7.2   DESCRIPTION OF CYCLE

The schematic flow diagram of KCS operating at moderate temperature heat has been depicted in Figure 7.1 and the corresponding enthalpy–concentration and temperature–entropy diagrams in Figure 7.2a and b, respectively. The mixture (3) is separated into liquid (4) and vapor (5) at the exit of evaporator 1. In this cycle, the condensation (8–9) and boiling processes (16–19) happens at variable temperature with the use of binary mixture. The fully condensed (9) liquid is pumped to high-pressure stream (10). The fluid moves to economizer, where it is heated by a counter-flow stream (7) from

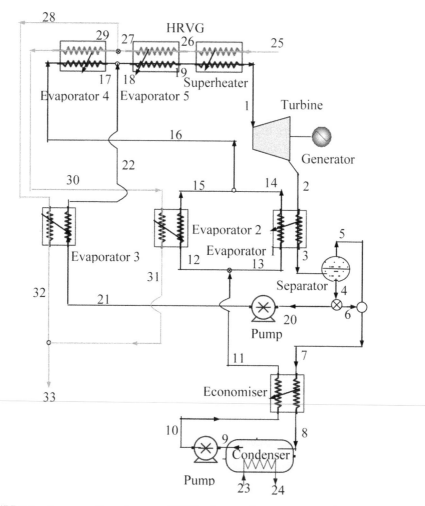

FIGURE 7.1   Processes flow diagram of KCS at intermediate-temperature heat recovery.

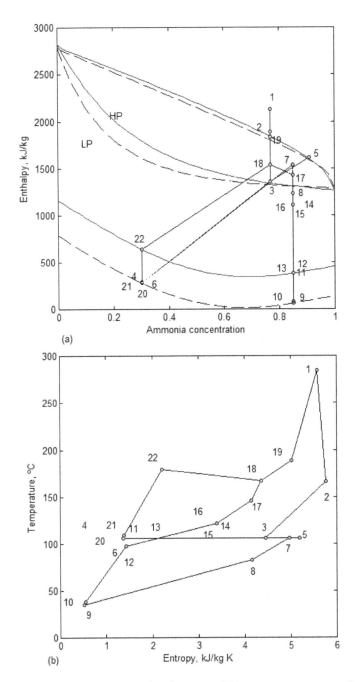

FIGURE 7.2 (a) Enthalpy–concentration diagram and (b) temperature–entropy diagram for KCS with reference to the process flow diagram shown in Figure 7.1.

the separator. It is the first heat recovery during which a preheated solution (11) and condensed inlet fluid (8) are produced. The preheated fluid (11) is at a state of saturated or slightly subcooled state. The saturated liquid is separated into two substreams. One of the two substreams (13) is passed through an economizer 1 where it is heated by a spent working fluid (2) from the turbine, resulting as partially vaporized (14) and partially condensed (3) vapor–liquid mixture. The other of the two substreams (12) is passed though evaporator 2 where it is heated by a counter-flow cooled heat-source stream (22) and resulted as partially vaporized fluid (15). The partially vaporized fluids from evaporator 1 and evaporator 2 are mixed into liquid–vapor mixture (16). The combined fluid is then sent through evaporator 4, where it is further heated and vaporized by a counter-flow cooled heat-source stream (29) resulting as saturated vapor (17) (complete vaporization) with a small amount of wetness and cooled heat source fluid stream (30).

The saturated vapor (17) is mixed with a recirculating solution (22) in a state of saturated liquid or liquid–vapor mixture (18). The working fluid then passes through evaporator 5, heated and vaporized in counter-flow with the heat-source fluid (26) resulting saturated vapor (19) and a cooled heat source working fluid (27). The saturated vapor is superheated in a superheater by a counter-flow source fluid (25) producing a superheated vapor (1) and a first cooled hot gas (24). The superheated vapor is expanded in a turbine producing power and expanded vapor (2). The spent vapor is passed to evaporator 2, releases heat during condensation. The vapor–liquid mixture (3) is separated into a saturated vapor (5) and saturated liquid (4). The saturated liquid is divided into two streams, namely, fluid at state 6 and fluid at state 20. The fluid (6) is combined with saturated vapor (5) forming a working solution (7). The fluid (7) is transferred heat in economizer and cooled to mixture (8), which is then passed through a condenser, fully condensed (9) by counter-flow water-cooled condenser. The other one of the two separated streams (20) is pumped to evaporator 3 (21).

The subcooled liquid is allowed to evaporator 3, where it is heated by a counter-flow working fluid (22) producing a saturated or slightly subcooled liquid (22) and a cooled heat source stream (32). The cooled hot gases (32) and (31) are combined into a spent fluid (33) which is then exhausted to stack. In the cycle, the mixing at the inlet of evaporator 5 increases the flow rate of the working solution passing through the turbine and thus augments the power output.

Table 7.1 shows the thermodynamic properties of KCS developed from material balance equations. The pressure, temperature, concentration, flow rate, energy, and entropy values are shown at all the states shown in

Figure 7.1. At hot gas supply of 100,000 m³/h (36.89 kg/s), the resulted working fluid at inlet of turbine is 8.33 kg/s. The cooling water requirement in the condenser is 258.6 kg/s. Since the separator is located at low-pressure (LP) side, there is no need of throttling device in this layout. Nearly 30% of extra working fluid has been found in the turbine for expansion against the decreased amount in other KCS configurations.

TABLE 7.1 Material Balance Results of KCS at Intermediate Temperature Heat Recovery.

State	P (bar)	T (°C)	x	m (kg/s)	h (kJ/kg)	s (kJ/kg K)
1	50.00	285.00	0.77	7.35	2126.31	5.59
2	11.31	166.15	0.77	7.35	1885.47	5.78
3	11.31	106.09	0.77	7.35	1358.22	4.46
4	11.31	106.09	0.30	1.73	279.06	1.37
5	11.31	106.09	0.91	5.61	1583.44	5.13
6	11.31	106.09	0.30	0.63	279.06	1.37
7	11.31	106.09	0.85	6.24	1537.42	4.98
8	11.31	82.42	0.85	6.24	1231.80	4.16
9	11.31	35.00	0.85	6.24	57.00	0.51
10	50.00	37.90	0.85	6.24	72.64	0.54
11	50.00	97.47	0.85	6.24	376.25	1.43
12	50.00	97.47	0.85	0.94	376.25	1.43
13	50.00	97.47	0.85	5.31	376.25	1.43
14	50.00	121.78	0.85	5.31	1104.55	3.40
15	50.00	121.78	0.85	0.94	1104.55	3.40
16	50.00	121.78	0.85	6.24	1104.55	3.40
17	50.00	146.09	0.85	6.24	1424.10	4.15
18	50.00	166.72	0.77	7.35	1534.80	4.37
19	50.00	188.40	0.77	7.35	1835.71	5.02
20	11.31	106.09	0.30	1.10	279.06	1.37
21	50.00	108.71	0.30	1.10	287.44	1.38
22	50.00	179.23	0.30	1.10	633.11	2.21
23	1.01	25.00	0.00	219.38	0.00	0.00
24	1.01	33.00	0.00	219.38	33.44	0.11
25	1.01	300.00	0.00	36.89	292.91	0.69
26	1.01	247.20	0.00	36.89	235.04	0.59
27	1.01	191.72	0.00	36.89	175.12	0.47
28	1.01	191.72	0.00	5.26	1227.24	3.26
29	1.01	191.72	0.00	31.63	204.26	0.54
30	1.01	132.51	0.00	31.63	130.67	0.37
31	1.01	112.05	0.00	31.63	105.51	0.31
32	1.01	123.71	0.00	5.26	720.10	2.08
33	1.01	113.72	0.00	36.89	92.21	0.27

7.3 MATHEMATICAL MODELING AND FORMULATION

The Kalina cycle is solved for 1 kg/s of strong solution at the condensate feed pump (9) to get the specific power output and other results. Later, the mass flow rates are normalized to hot gas flow rate. In this KCS, five variables are identified as key parameters for thermodynamic analysis. They are strong solution concentration, hot fluid inlet temperature, turbine inlet pressure, split ratio 1 (SR_1), and split ratio 2 (SR_2). LP is determined from condenser condition. Thus, P_9 is determined from T_9 (bubble-point temperature) and x_9.

Figure 7.3 shows the generation of fluid loops to identify the fluid flows and mass balance formulations. It is useful to find the unknown masses in the cycle loops. As per the assumptions used in the thermodynamic analysis, atmospheric condition is taken as 1.01325 bar and 25°C. The fluid concentration in condenser is 0.85. The vaporization process completes in evaporator 4 and evaporator 5. Therefore, T_{17} and T_{19} are the dew-point temperatures. Turbine inlet pressure is 50 bar. The working fluid in the turbine has been separated into two lines after the expansion; one toward the condenser (unit mass considered) and another to evaporator 3. SR_1 is defined as the mass ratio fluid in 3rd evaporator to turbine fluid. It is assumed as 0.10. Similarly at the exit of pump and economizer, the fluid is divided between first evaporator

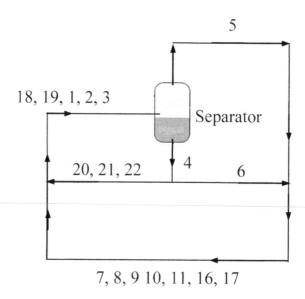

FIGURE 7.3 Looping of KCS fluids for mass balance simplification.

and second evaporator based on the heat available for recovery. SR_2 is defined as the mass ratio of fluid in first evaporator to pumped fluid and it is taken at 0.8. The isentropic efficiency of both solution pump and mixture turbine is 75%. The mechanical efficiency of the solution pump (η_{mp}) and mixture turbine (η_{mt}) is taken at 96%. Electrical generator efficiency (η_{ge}) is taken as 98%. The condensate leaving the condenser is assumed as saturated liquid. The cooling water inlet temperature has been considered as 25°C. Pressure drop and heat loss in pipelines are neglected. The LP is determined from condenser exit condition. Thus, P_9 is determined from T_9 (bubble-point temperature) and x_9.

Let the strong solution flow rate, $m_7 = 1$ kg/s. An equal temperature difference is considered in evaporator 1 and evaporator 2.

$$T_{14} = T_{13} + \frac{(T_{17} - T_{13})}{2} \tag{7.1}$$

According to the SR_1, the turbine mass flow,

$$m_{20} = \left(\frac{SR_1}{1 - SR_1}\right) m_7 \tag{7.2}$$

The turbine fluid,

$$m_1 = m_7 + m_{20} \tag{7.3}$$

The SR_2 results m_{13},

$$m_{13} = SR_2 m_{11} \tag{7.4}$$

The separator temperature is solved through iteration procedure. The separator condition, that is, temperature (T_3) and pressure (P_3) gives the liquid concentration (x_4) and vapor concentration (x_5). It will help to develop mass and energy balance equations. T_{22} is the bubble-point temperature. At inlet of evaporator, the concentration balance equation results x_{18}. T_{19} is dew-point temperature of the fluid at the new concentration. The pump and turbine exit temperatures is predicted by thermodynamic evaluation of components with isentropic efficiencies.

From the energy balance at evaporator 5 and superheater,

$$m_1 = \frac{m_{25}(h_{25} - h_{27})}{(h_1 - h_{18})} \tag{7.5}$$

Similarly from evaporator 4,

$$m_{28} = \frac{m_{21}(h_{22} - h_{21})}{(h_{28} - h_{32})} \tag{7.6}$$

The unknown hot fluid temperatures are found from the energy balance equations applied to superheater, evaporator 5, and evaporator 4.

The energy interactions in the plant components are

Mixture turbine output,

$$W_t = m_1(h_1 - h_2)\eta_{mt}\eta_{ge} \tag{7.7}$$

Work input to pump,

$$W_p = \frac{m_{10}(h_{10} - h_9)}{\eta_{mp}} + \frac{m_{20}(h_{21} - h_{20})}{\eta_{mp}} \tag{7.8}$$

Net output from Kalina cycle,

$$W_{net} = (W_t - W_p) \tag{7.9}$$

The total heat supply to heat recovery vapor generator (HRVG),

$$Q_{supply} = m_{18}(h_1 - h_{18}) + m_{16}(h_{17} - h_{16}) + m_{21}(h_{22} - h_{21}) + m_{15}(h_{15} - h_{12}) \tag{7.10}$$

Kalina cycle energy efficiency,

$$\eta_{I,KCS} = \frac{W_{net}}{Q_{supply}} \times 100 \tag{7.11}$$

The methodology adopted for KCS at intermediate-temperature heat recovery (ITHR) is outlined in Figure 7.4.

The performance of the KCS at ITHR is compared with steam Rankine cycle (SRC) under the same heat supply conditions. Figure 7.5 outlines the flow lines of the steam power plant with a deaerator. Apart from the deaerator, the plant consists of basic components such as steam turbine, condenser, pumps, and heat recovery steam generator. Compared to KCS, the boiling and condensation happens at a fixed saturated temperature. But in KCS, the boiling starts at bubble point and ends at the dew point. In KCS condenser, condensation starts at dew point and ends at bubble point.

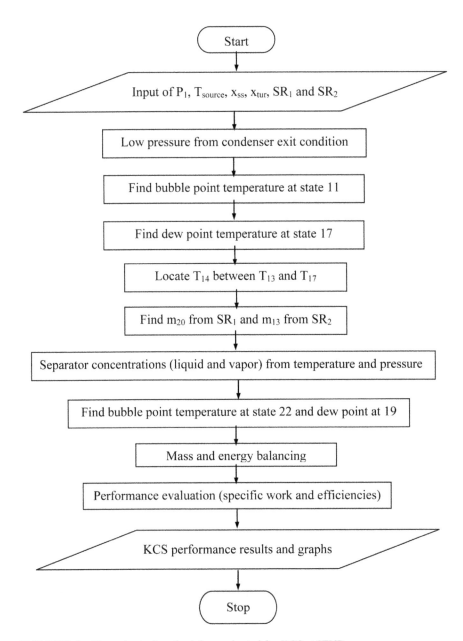

FIGURE 7.4   Flow chart of methodology adopted for KCS at ITHR.

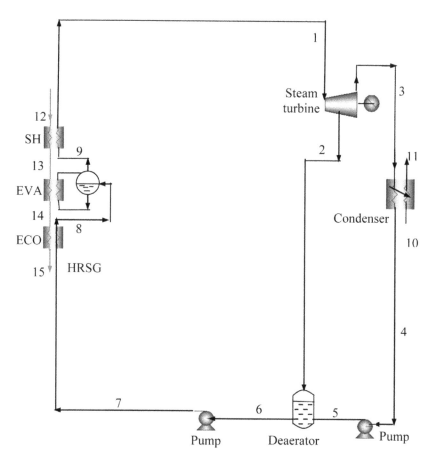

FIGURE 7.5   Schematic layout of thermal power plant with steam as working fluid.

7.4   ANALYTICAL RESULTS AND INFERENCE

This section is focused on power production and thermal efficiency of the KCS cycle with the hot gas flow rate of 100,000 N m³/h and temperature of 300°C. The identified variable operating conditions are strong solution concentration, hot gas supply temperature in place of 300°C, turbine inlet pressure, and fluid mass split ratio at both the splitters.

Figure 7.6 shows (a) power and (b) thermal efficiency, that is, cycle energy efficiency. The power and efficiency variations are observed by changing the source temperature with the association of strong solution concentration. The power is increasing with increase in heat source temperature. The increase in strong solution concentration is also assisting in power and efficiency augmentation. But there is a drop in efficiency with increase in hot gas inlet temperature.

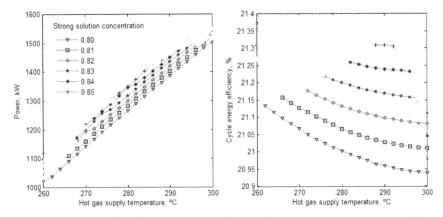

FIGURE 7.6   (See color insert.) Effect of hot gas inlet temperature with strong solution concentration on (a) plant's power generation and (b) cycle thermal efficiency.

Figure 7.7 shows (a) power and (b) thermal efficiency, that is, cycle energy efficiency with a change in turbine inlet pressure with the strong solution concentration changes. The power and efficiency increase with the increase in turbine inlet pressure. The role of concentration is minor on power and efficiency compared to influence of turbine pressure.

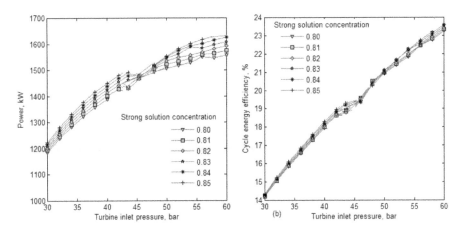

FIGURE 7.7   (See color insert.) Effect of turbine inlet pressure with strong solution concentration on (a) plant's power generation and (b) cycle thermal efficiency.

Figure 7.8 plots the role of mass separation ratio at splitter 1 and splitter 2. Split ratio at 1 is the ratio of mass of fluid pumped at the exit of separator to the turbine fluid (m_{20}/m_3). Similarly, SR_2 is also defined as the ratio of mass shifted to first evaporator from the pumped fluid (m_{13}/m_{11}). The increase in mass ratio in splitter 1 decreases the power but increases the efficiency. For heat recovery applications, a lower limit of split ratio at splitter 1 is recommended. The splitting of mass at second splitter has no significant role on power generation but it influences the efficiency in a considerable amount. The higher split ratio at splitter 2 is recommendable to gain higher thermal efficiency. Finally, a lower SR_1 with no restriction in SR_2 is recommended for power and higher SR_1 and SR_2 are suggested for efficiency enhancement.

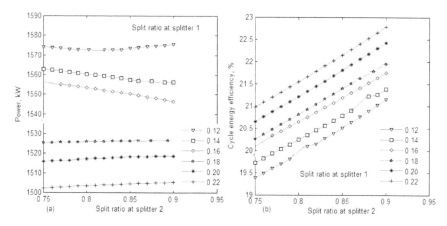

FIGURE 7.8 **(See color insert.)** Effect of separations in the cycle on (a) plant's power generation and (b) cycle thermal efficiency.

Figure 7.9 compares the KCS with the steam power plant under the same heat source and sink conditions. Figure 7.9a compares the power and Figure 7.9b plots the thermal efficiency. The KCS is proved over the SRC by generating more power over the SRC. But it failed to show higher efficiency over the SRC. Therefore, for heat-recovery-based plant configuration, KCS is better than the SRC at this temperature range. But if the target is the efficiency improvement, steam power plant is better than KCS.

Table 7.2 summarizes the specifications of KCS at ITHR. The heat supplied to the cycle is equal to the heat load in HRVG and it is 7404.10 kW. Out of this input to the cycle, the net power output is 1553.20 kW. That means the cycle thermal efficiency is around 21%. The table presents heat load in all the heat exchangers, that is, economizer, all the evaporators, superheater, and condenser. The vapor generated at the exit of turbine after the separation is considerable and is 76%.

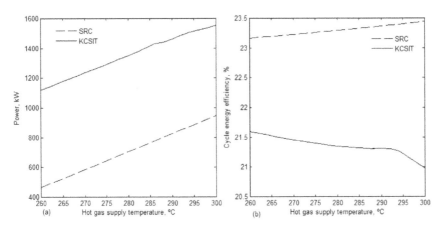

FIGURE 7.9 Comparison of KCS at ITHR with SRC under the same heat supply conditions.

TABLE 7.2 Specifications of the KCS at ITHR.

Description	Result
Heat load in HRVG (kW) (total)	7404.10
Heat load in economizer (kW)	1895.90
Heat load in evaporator 1 (kW)	3865.70
Heat load in evaporator 2 (kW)	682.17
Heat load in evaporator 3 (kW)	380.92
Heat load in evaporator 4 (kW)	1995.50
Heat load in evaporator 5 (kW)	2210.60
Heat load in superheater (kW)	2134.90
Heat load in condenser (kW)	7336.00
Separator dryness fraction (%)	76.00
Work output of vapor turbine (kW)	1664.60
Work input to solution pump (kW)	111.34
Net electricity output (kW)	1553.20
Kalina cycle energy efficiency (%)	20.98
Hot water requirement (kg/s)	258.60

7.5   SUMMARY

KCS at ITHR is modeled, simulated, and analyzed to draw the potential of the configuration. It allows an operation in a narrow range of the heat-source temperature. The configuration is demanding higher strong solution concentration for power and efficiency improvement. A lower SR_1 with no restriction on SR_2 is recommended for power boosting. A higher SR_1 and SR_2 are required to enhance the cycle efficiency. KCS is recommended over steam power plant for higher power generation under the same heat source conditions but not to expect the cycle efficiency over the steam power cycle.

KEYWORDS

- intermediate temperature range
- Kalina cycle system
- source temperature
- throttling
- enthalpy

CHAPTER 8

KALINA CYCLE SYSTEM WITH HIGH-TEMPERATURE HEAT RECOVERY

ABSTRACT

The Kalina cycle system (KCS) at high-temperature source resembles a Rankine cycle. Compared to the Rankine, this KCS has internal heat recovery at the exit of turbine and fluid dilution for easy condensation. It has full-pledged economizer, evaporator, and superheater. It is a three-pressure level configuration with two condensers. Similar to low- and intermediate-temperature KCS plants, the system modeling, simulation, and analysis are conducted to recommend the suitable best operating conditions.

8.1 INTRODUCTION

In the earlier chapters, Kalina cycle system (KCS) at low- and intermediate-temperature heat recoveries is studied. Similar to low-temperature and intermediate-temperature heat recoveries, KCS can be operated at a temperature range of 400–600°C which is termed as high-temperature heat recovery (HTHR) in this section. In this configuration, vapor is superheated directly from the heat source. In the same temperature range, steam power plant also can be operated. The difference between these two is the heat addition and heat rejection. In KCS, the heat source and sink operates at the variable temperature where it is constant temperature of working fluid in Rankine cycle.

8.2 CYCLE DESCRIPTION

The KCS matched to HTHR is depicted in Figure 8.1. Heat-recovery vapor generator (HRVG) is the source of KCS where the working fluid is generated from liquid. Later, it is superheated (13) and followed by an expansion in the turbine (13–14). In steam Rankine cycle (SRC), steam expands till

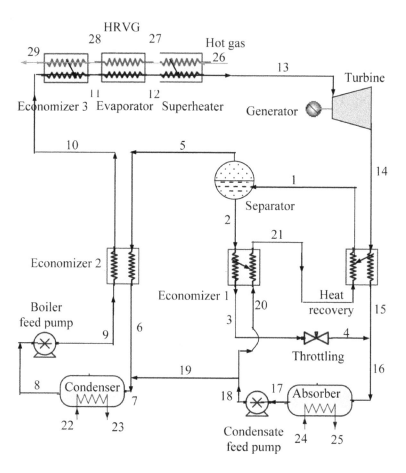

FIGURE 8.1   Schematic flow diagram of the Kalina cycle suitable to high-temperature heat recovery.

the condenser pressure and temperature. There is no scope to recover the internal heat in a steam power plant. In KCS, the fluid temperature at the exit of turbine is enough high and so there is a chance for internal heat recovery. The heat from the expanded fluid is tapped in the heat recovery placed at the exit of the turbine. To overcome the difficulty in condenser, the fluid (15) is diluted by mixing with the weak solution (4) at the low pressure (LP). Otherwise without dilution of turbine fluid, the condensation process in the absorber demands very low temperature which needs energy for production of chilled water. The turbine fluid concentration decreases and the weak solution concentration rise in the mixing process. The condensed fluid is called strong solution. The strong solution is pumped (17–18) from the LP to intermediate pressure (IP) and followed by a splitting of the fluid to condenser (19) and economizer (20). The splitted fluid is heated in the economizer 1 and heat recovery results liquid–vapor mixture. The liquid–vapor mixture is separated into weak solution (2) and vapor (5) in the separator. KCS operating with low-temperature heat recovery (LTHR) generated fluid at the state of liquid–vapor mixture before the turbine. In the HTHR-based KCS, the fluid is superheated at the inlet of the turbine, that is, in the boiler, the vaporization process is complete, whereas in the LTHR-based KCS, the vaporization is incomplete.

Another fluid (19) from the splitter is mixed with the vapor (6) and restores the original concentration. After mixing, the fluid moves to condenser. Since the pressure is maintained at the intermediate level, the high concentration is not the issue in the condenser. But it is the main problem (without dilution) in the low-pressure absorber for condensation. The weak mixture rejects the heat in the economizer 1 and throttles to LP (4). Now, this weak solution is used for dilution of the turbine mixture. The boiler-feed pump increases the pressure from IP to high pressure (HP) (8–9). The pumped fluid is supplied to the economizer 2 where the internal heat is recovered from the vapor. On overall basis, the plant works on three pressures, namely, LP, IP, and HP. After economizer 2, the fluid flow to economizer 3, evaporator, and superheater. The economizer 3, evaporator, and superheater gain heat from the external hot fluid. The superheated fluid expands in turbine and the cycle repeats.

The perfect thermal design of thermodynamic cycle involves the minimization of the losses and improvement of the efficiency. To solve the specific performance of the KCS, the fluid inlet flow rate at the turbine inlet is 1 kg/s. The properties at every state are evaluated by simplifying the mass balance and energy balance equations. Lever rule is applied in the mixing and separation process solutions. The properties and the mass flow rates around the loop were calculated. The liquid and vapor mixtures are separated in the separator. At IP, T_{17} is the saturated liquid temperature determined at the sink temperature.

As a preset of thermodynamic assumptions, the turbine inlet pressure, temperature, and concentration are, respectively, 100 bar, 400°C, and 0.75. The separator temperature is 80°C. For the schematic sketch shown in Figure 8.1, enthalpy–concentration (h–x) (Fig. 8.2a) and temperature–entropy (t–s) (Fig. 8.2b) are prepared. Totally, four concentrations are maintained in the fluid flow lines and they are 0.4, 0.59, 0.75, and 0.97. The mixing processes, separation, expansion, and heat transfer processes can be studied in these two diagrams to understand the nature of the process. The heat addition in the source and heat rejection in the sinks, that is, absorber and condenser are depicted as a variable temperature heat transfer process.

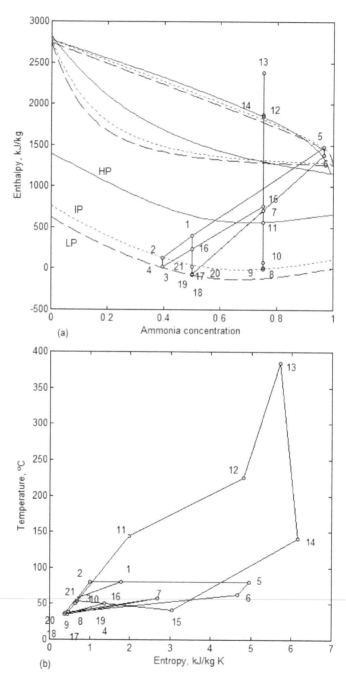

FIGURE 8.2 (a) Enthalpy–concentration diagram and (b) temperature–entropy diagram for KCS at high-temperature heat recovery for the drawing shown in Figure 8.1.

Table 8.1 results the mass balance and energy balance results of the KCS work. At the assumed listed, the resulted working fluid is 5.27 kg/s at the turbine inlet.

TABLE 8.1   Process Flow Conditions and Its Properties for the Drawing Shown in Figure 8.1.

State	Pressure (bar)	Tempera-ture (°C)	Ammonia concentration	Flow rate (kg/s)	Specific enthalpy (kJ/kg)	Specific entropy (kJ/kg K)
1	9.67	80.00	0.50	15.59	393.90	1.77
2	9.67	80.00	0.40	12.77	129.33	1.00
3	9.67	54.79	0.40	12.77	16.10	0.67
4	4.41	52.83	0.40	12.77	18.53	0.68
5	9.67	80.00	0.97	2.82	1465.85	4.91
6	9.67	60.06	0.97	2.82	1360.55	4.61
7	9.67	56.95	0.75	5.27	693.30	2.65
8	9.67	35.00	0.75	5.27	−6.22	0.44
9	100.00	38.00	0.75	5.27	13.24	0.46
10	100.00	50.00	0.75	5.27	69.37	0.63
11	100.00	143.36	0.75	5.27	563.64	1.97
12	100.00	224.84	0.75	5.27	1849.68	4.80
13	100.00	385.00	0.75	5.27	2383.64	5.70
14	4.41	140.38	0.75	5.27	1866.26	6.16
15	4.41	40.76	0.75	5.27	751.28	3.02
16	4.41	49.81	0.50	18.04	229.75	1.36
17	4.41	35.00	0.50	18.04	−82.98	0.37
18	9.67	36.68	0.50	18.04	−74.97	0.39
19	9.67	36.68	0.50	2.45	−74.97	0.39
20	9.67	36.68	0.50	15.59	−74.97	0.39
21	9.67	56.94	0.50	15.59	16.72	0.68
22	0.00	30.00	0.00	51.99	20.90	0.07
23	0.00	46.95	0.00	51.99	91.76	0.30
24	0.00	30.00	0.00	137.58	20.90	0.07
25	0.00	39.81	0.00	137.58	61.90	0.20
26	1.01	400.00	0.00	36.89	404.64	0.87
27	1.01	332.07	0.00	36.89	328.41	0.75
28	1.01	163.36	0.00	36.89	144.79	0.40
29	1.01	96.59	0.00	36.89	74.22	0.22

8.3 MATHEMATICAL FORMULATION AND SYSTEM MODELING

The thermodynamic mathematical modeling is started with the assumption involved for the plant simulation. At the exit of the turbine, the vapor to be diluted for condensation at the normal sink temperature and the diluted solution is called strong solution as the concentration of this solution is more than the weak solution which is separated from the in the vapor separator. The strong solution concentration considered is 0.5. Turbine concentration is 75%. The separator temperature is 80°C. The HRVG pressure is 100 bar. Terminal temperature difference (TTD) in condenser is 5 K.

The unknown cycle properties have been obtained using mass and energy balance equations (eqs 8.1–8.16).

The turbine inlet mass flow rate is considered as $m_{13} = 1$ kg/s.

The temperature between the evaporator and superheater (T_{12}) is determined from the pressure and concentration as a dew-point temperature. From temperature and concentration at state 8, the IP is evaluated. The liquid concentration (x_2) and vapor concentration (x_5) is found at separator pressure (IP) and its temperature. The vapor fraction or dryness fraction (mass ratio vapor to total mixture) are resulted from the three concentrations. In separation process, after applying the lever rule,

$$DF = \frac{m_5}{m_1} = \frac{x_1 - x_2}{x_5 - x_2} \tag{8.1}$$

$$m_6 = m_5 = DFm_1 \tag{8.2}$$

$$m_4 = m_2 = (1 - DF)m_1 \tag{8.3}$$

At the mixing of fluids after the turbine and before absorber,

$$m_4 + m_{15} = m_{16} \tag{8.4}$$

$$m_4 x_4 + m_{15} x_{15} = m_{16} x_{16} \tag{8.5}$$

With simple loops, the complex configuration of Kalina cycle is simplified as shown in Figure 8.3 to solve the unknown mass at unit mass of turbine flow rate. The properties at the points (2, 5, 17, and 19) are calculated by fixing the concentration at 1 and 13.

Combining eq 8.4 and eq 8.5,

$$\frac{m_4}{m_{15}} = \frac{x_{15} - x_{16}}{x_{16} - x_4} \tag{8.6}$$

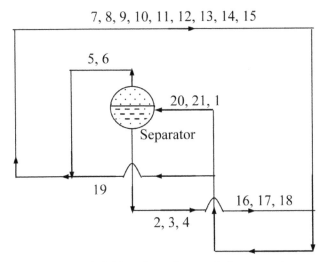

FIGURE 8.3   Looping of KCS fluids to frame the mass balance equations.

Equation 8.6 is simplified to

$$m_2 = \frac{x_{13} - x_1}{x_1 - x_2} \times m_{13}$$ (8.7)

Similarly from the mixing of two streams at the inlet of condenser,

$$m_6 + m_{19} = m_7$$ (8.8)

$$m_6 x_6 + m_{19} x_{19} = m_7 x_7$$ (8.9)

Combining eq 8.8 and eq 8.9,

$$\frac{m_5}{m_{19}} = \frac{x_7 - x_{19}}{x_5 - x_7} = \frac{x_{13} - x_1}{x_5 - x_{13}}$$ (8.10)

At the separator, to find m_1

$$\frac{m_1}{m_2} = \frac{x_5 - x_2}{x_5 - x_1}$$ (8.11)

On substituting m_2 in eq 8.7 in eq 8.11,

$$m_1 = \frac{x_5 - x_2}{x_5 - x_1} \times \frac{x_{13} - x_1}{x_1 - x_2} \times m_{13}$$ (8.12)

From eq 8.12,

$$m_1 = \frac{x_{13} - x_1}{DF(x_5 - x_1)} \times m_{13} \tag{8.13}$$

To find m_{19} between separation and mixing,

$$m_{19} = \frac{x_5 - x_{13}}{x_{13} - x_1} m_5 = \frac{x_5 - x_{13}}{x_{13} - x_1} m_1 DF \tag{8.14}$$

Substituting m_1 from eq 8.13,

$$m_{19} = \frac{x_5 - x_{13}}{x_{13} - x_1} \times \frac{x_{13} - x_1}{F(x_5 - x_1)} DF \times m_{13} \tag{8.15}$$

$$m_{19} = \frac{x_5 - x_{13}}{x_5 - x_1} \times m_{13} \tag{8.16}$$

From eqs 8.13 and 8.16, m_1 and m_{19} is calculated from x_1 at fixed mass, m_{13} and concentration x_{13} are functions of x_1. The vapor is diluted with addition of weak solution, m_4. Therefore, x_1 is less than x_{13}.

Similar to IP, LP is a function of temperature and concentration ($P_{17} = f$ (T_{17}, x_{17})), it is determined from the condenser temperature and vapor concentration.

Turbine inlet temperature from the hot gas temperature,

$$T_{13} = T_{26} - TTD_{SH} \tag{8.17}$$

Energy interactions in the plant components as follows,

Vapor turbine output,

$$W_t = m_{13}(h_{13} - h_{14})\eta_{m,t}\eta_{ge} \tag{8.18}$$

Work input to pump,

$$W_p = \frac{m_{18}(h_{18} - h_{17}) + m_9(h_9 - h_8)}{\eta_{m,p}} \tag{8.19}$$

Net output from Kalina cycle,

$$W_{net} = W_t - W_p \tag{8.20}$$

Heat supply in HRVG sections,

$$Q_{ECO} = m_{12}(h_{12} - h_{11}) \tag{8.21}$$

$$Q_{EVA} = m_{11}(h_{11} - h_{10}) \tag{8.22}$$

$$Q_{SH} = m_{13}(h_{13} - h_{12}) \tag{8.23}$$

$$Q_{Supply} = Q_{ECO} + Q_{EVA} + Q_{SH} \tag{8.24}$$

Kalina cycle efficiency,

$$\eta_{1,KC} = \frac{W_{net}}{Q_{Supply}} \times 100 \tag{8.25}$$

8.4   ANALYTICAL RESULTS AND INFERENCE

At change in operating conditions, the performance of high-temperature solar thermal power plant has been investigated parametrically. The plant performance is examined using the key parameters, that is, separator concentration, separator temperature, turbine inlet condition (pressure, temperature, and concentration), and hot gas supply temperature. The optimum efficiency and power generation have been aimed on conducting the parametric study (Figs. 8.4–8.9).

Figure 8.5 shows the effect of heat-source temperature with strong solution concentration on power generation and cycle efficiency with hot gas temperature changed from 400°C to 500°C and strong solution concentration varied from 0.5 to 0.6. The increase in heat source temperature increases the power generation and cycle efficiency. But the increase in concentration is dropping both output and efficiency. The expansion in the vapor turbine decreases as the LP increases with increase in the strong solution concentration.

Figure 8.6 shows the effect of separator temperature with strong solution concentration on power generation and cycle efficiency. The separator temperature is varied between 60°C and 100°C. The increase in separator temperature is enhancing the power generation and cycle efficiency. But as stated in the earlier section, the rise in strong solution concentration is not supporting the power and efficiency rise.

Figure 8.7 shows the involvement of turbine concentration with the strong solution concentration on power and thermal efficiency. The increase in turbine concentration is increasing the power and cycle efficiency. The dew-point temperature in HRVG depends on turbine concentration. So it plays a role on heat recovery and vapor generation from source.

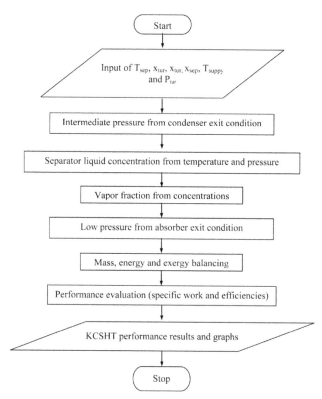

FIGURE 8.4 Summary of the methodology used to solve the KCS at high-temperature heat recovery.

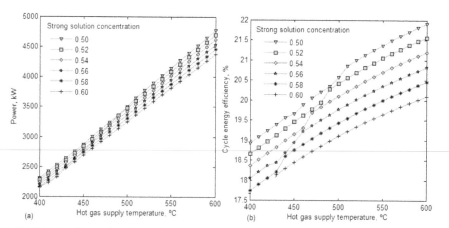

FIGURE 8.5 **(See color insert.)** Influence of hot gas inlet temperature with strong solution concentration on (a) power generation and (b) cycle thermal efficiency.

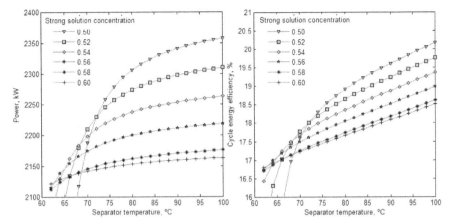

FIGURE 8.6 (See color insert.) Influence of separator temperature with strong solution concentration on (a) power generation and (b) cycle thermal efficiency.

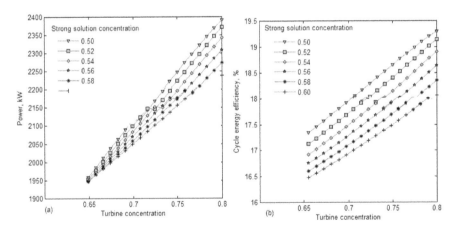

FIGURE 8.7 (See color insert.) Influence of turbine concentration with strong solution concentration on (a) plant's power generation and (b) cycle thermal efficiency.

Figure 8.8 shows the effect of turbine inlet pressure (50–100 bar) with strong solution concentration (0.5–0.6) on (a) power generation and (b) cycle thermal efficiency. At increase in turbine inlet pressure, the output and efficiency are increased. With increase in pressure, the turbine work increases with a small change in pump work.

Figure 8.9 compares the performance of KCS with the steam power plant at the same heat source conditions. In most of the cases, the steam power plant efficiency and power are dominating KCS performance. Only at lower source temperature, that is, up to around 450°C, the power generation of KCS is more than the steam power plant. Anyhow the efficiency of steam power plant is more than the KCS at all the selected temperature ranges.

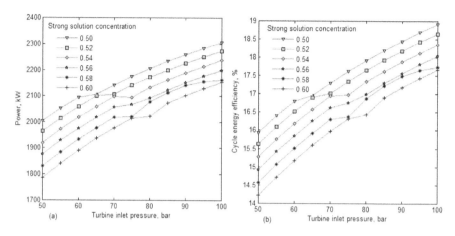

FIGURE 8.8 (See color insert.) Role of turbine inlet pressure with strong solution concentration on (a) power generation and (b) thermal efficiency of cycle.

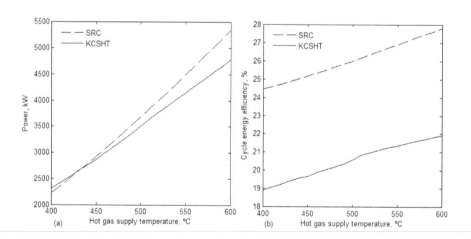

FIGURE 8.9 Comparison of performance of KCS suitable to high-temperature heat recovery with the steam power plant.

Table 8.2 presents the specifications of KCS at 400°C of hot gas supply and stated assumptions. The heat load in HRVG is the sum of economizer, evaporator, and superheater. The total heat supply from the external heat source is 12,189 kW. The net output from the plant at the hot gas flow rate of 100,000 N m³/h and 400°C is 2306.5 kW with the cycle thermal efficiency of 18.92%. At the exit of turbine, the internal heat recovery is 5872.7 kW which is the highest among all the internal heat recoveries.

The absorber heat load is more than the condenser as the fluid flow in absorber is much more than the condenser.

TABLE 8.2 Thermodynamic Results of KCS with High-temperature Heat Recovery at Hot Gas Inlet Temperature of 400°C.

Description	Result
Dryness fraction in separator (%)	18.00
Heat load in HRVG (ECO + EVA + SH) (kW)	12,189.00
Heat load in economizer 1 (kW)	1419.00
Heat load in economizer 2 (kW)	296.55
Heat load in economizer 3 (kW)	2603.40
Heat recovery	5872.70
Heat load in evaporator	6773.70
Heat load in superheater	2812.40
Heat load in absorber (kW)	5640.60
Heat load in condenser (kW)	3684.40
Work output of vapor turbine (kW)	2563.70
Work input to solution pumps (kW)	257.28
Net electricity output (kW)	2306.50
Kalina cycle energy efficiency (%)	18.92

8.5 SUMMARY

The source temperature, separator temperature, turbine pressure, turbine concentration, and strong solution concentration are identified as key parameters in KCS with HTHR. A lower strong solution concentration, higher turbine inlet concentration, higher separator temperature, and higher turbine inlet pressure are recommended to gain more power and high efficiency. Compared to steam power plant, KCS at HTHR, is competitive up to the source temperature of 450°C. Beyond this temperature, steam plant's power production is dominating with the heat recovery option. But the cycle efficiency is always more in SRC compared to the KCS at the selected temperature range. Therefore, the recommended source temperature of the KCS at HTHR is 400°C. This temperature is maintained fixed while the other operational conditions are varied.

KEYWORDS

- **Kalina cycle system**
- **vapor**
- **temperature**
- **Rankine cycle**
- **turbine**

CHAPTER 9

COOLING COGENERATION

ABSTRACT

The combined power and cooling generation is called as cooling cogeneration. A flexible Kalina cooling cogeneration cycle has been introduced and evaluated. Initially, the work is started with the assumption of four possible ways of operation of the proposed single cooling cogeneration. The final results concluded that only two possible ways can be selected from the proposed layout, namely, Kalina cycle system with once through or vapor absorption refrigeration with split operation.

9.1 INTRODUCTION

Small-capacity power generation suitable for rural people is also known as decentralized power generation as it has low transmission losses and possibility of renewable energy sources use. The self-generation of energy needs also decreases the load on the large-sector industry and decreases the global warming. Since the load on large-capacity plants relaxed, it indirectly drops the smoke, emission, thermal pollution, etc. from the large-scale sector. In addition to the renewable energy sources, waste heat recovery is also an attractive and economic option to operate the decentralized power systems. The waste heat from cement factories, thermal power plants, hotel's kitchen, steel plants, geothermal, sugar industry, etc. can be used as a source to operate the plants at the low- and medium-capacity level, may be up to 10 MW. The power plant name in the vapor absorption family is Kalina cycle system (KCS), and refrigeration cycle which works on vapor absorption principle is vapor absorption refrigeration (VAR). The features of KCS and VAR may be clubbed together to result new combined power and cooling cycle also known as Kalina cooling cogeneration cycle (KCCC). This chapter is focused on the evaluation and characterization of KCCC. Four-in-one mode has been invented in KCCC. Four in one means four options in operation from one plant. By properly controlling the working fluid and selecting the flow lines, the required option of configuration can be generated. The resulted configurations in four-in-one plant are power only (KCS), cooling only (VAR), combined power and cooling with once through flow, and combined power and cooling with split mode of KCS and VAR. In once through cogeneration, the working fluid cannot be controlled and all the fluids will flow in both power lines as well as in cooling lines, whereas in the split option, the working fluid selected for KCS and VAR lines are different. The amount of fluid in KCS and VAR can be selected and controlled to meet the required power and cooling. The cooling cogeneration or combined power and cooling to be promoted after showing the advantage over the KCS and VAR generations. The power and cooling alone can be generated from a separate power plant (KCS) and a separate refrigeration plant (VAR). The production rate of combined power and cooling should be higher than the production of KCS and VAR. At this condition, only the combined power cooling can be promoted. Therefore to justify the proposed cooling cogeneration configurations, the results are compared with the individual plants outputs under the same operation conditions. Out of two cooling cogeneration, one may be recommended from the best operating conditions.

The base for the comparison of the cooling cogeneration is the out ratio under the same input conditions. The output energy ratio in cooling cogeneration and individual plants are the same. Two categories of KCCC are configured; the best one should be identified. The thermodynamic evaluation has been conducted for KCS, VAR, and two cogeneration cycles. The identified key operational parameters are strong solution concentration, source temperature, dephlegmator effectiveness, vapor concentration, circulating water inlet temperature, and evaporator exit temperature. The focused results are power only, cooling only, combined power and cooling from once through cogeneration, combined power and cooling from split cogeneration, and energy utilization factor (EUF) for cycle and plant. The higher EUF in cycle is observed with once through cogeneration. The split cycle has not generated much output and energy conversions as its operated conditions are inferior to the once through cogeneration. The once through cogeneration boiler pressure is higher than the split cogeneration pressure. But split cogeneration may be recommended at special cases such as flexibility requirement and low-pressure (LP) plant operation. On output and energy conversion point of view, once through cogeneration is best.

9.2   CYCLE DESCRIPTION

The current proposed and studied cooling cogeneration differs with the existing cooling cogeneration. The four-in-one option in the present cooling cogeneration added the flexibility in operation of the plant as per the selection. The aims of this chapter are (1) to justify the merit of proposed cooing cogeneration compared to the performance of sum of separate power and cooling plants, (2) to choose the efficiency cooling cogeneration option out of two derived cogenerations, and (3) to identify the best process conditions to yield the maximum energy conversion from the given supply of energy. To meet these aims, first thermodynamic model has been generated, simulated, analyzed, and optimized to recommend the best configuration and also better parameters. Four plants are compared under the common ground to identify the potentials and limitations of the individual operational options.

KCCC has the characteristics of KCS and VAR. The power and cooling may also met from these two separate plants. But the investment and space requirements are more compared to four-in-one option. But the four in one needs to generate higher output than the sum of the individual plants output. Then only, it is easy to commercialize the ideal of proposed cooling cogeneration. The basic idea behind the cooling cogeneration is the identification of common features and process components from the individual plants and clubbing together. The resulted invention is suitable to operate at low-temperature heat recovery or sources in the range of 150–180°C with a terminal temperature difference (TTD) of 25°C in source.

In the conventional cooling cogeneration cycle, for example, Goswami cogeneration, the plant operated between pressures, that is, high pressure (HP) and LP. In the proposed cycle, three pressures exist, namely, HP, intermediate pressure (IP), and LP. The vapor expands from HP to LP in conventional cycle. But in proposed cooling cogeneration, vapor expands from HP to IP only. Therefore, lower power from proposed cooling cogeneration compared to the conventional cooling cogeneration. But on cooling side, liquid throttles for refrigeration which is not so in the conventional where the expanded vapor in the turbine generated the cooling. The dryness fraction of vapor before the cooling evaporator in the proposed plant is much lower than the existing. It absorbs more amount of cooling. Therefore, even though there is a penalty in the power generation, the extra cooling generation compensates the losses and improves the overall energy conversion and output. In conventional, the vapor should be of enough high concentration to result in cooling at the exit of turbine. There is no such restriction in the proposed cogeneration plant.

In the existed cooling cogeneration, since the refrigerant should be high concentrated, it demands high effective dephlegmator. In the new cogeneration, these issues are solved by adding a condenser at the outlet of vapor turbine. The vapor at the exit of the turbine is completely condensed into a saturated liquid state. Now, this liquid is throttled from IP to LP. Therefore compared to the existing cogeneration, the additional components are condenser and throttling, and the remaining configuration is the same. It indicates that a little modification and adoption of components in a new and efficient version can be generated. The superheat will not influence the cooling operation unlike the existing method. Earlier, the exit condition of the turbine is linked to the degree of superheating. More superheat favors the power and suppresses the cooling. Now since the power and cooling lines are separated by adding condenser, there is no such difficulty. One more extra component may be included in the new cycle is subcooler. It is not mandatory like condenser and throttling. The addition of subcooler saves the evaporator and absorber loads. Finally, the new one is cooling-favored cogeneration and the existing is the power-favored cogeneration. But the cooling-favored cogeneration gives much higher total output than the power-favored cogeneration and compensated the power penalty. In addition to the extract process components, in a new cogeneration, some control valves are also to be incorporated to choose the required flow lines and output.

A new plant is introduced having the qualities of KCS, VAR, and KCS–VAR characteristics. Ammonia–water mixture is taken as fluid in the all the cycles for the study. Figure 9.1 shows the schematic flow details of hybrid plant with flexibility to choose the required option. When a binary absorbent/refrigerant mixture changes its state from liquid to vapor, more volatile refrigerant vaporizes first and then absorbent. It ensures a better match with hot fluid temperature profile. To meet the power needs at small capacity, that is, domestic and industrial, Kalina developed a low-grade heat recovery power plant. The KCS is working on low heat temperature using vapor absorption refrigerant pair which is having a condenser above the atmospheric pressure.

In KCS configuration as shown in Figure 9.2, the strong solution of aqua–ammonia is pumped to HP of generator pressure to the boiler/generator via low-temperature regenerator (LTR) and high-temperature regenerator (HTR). Due to its high boiling point temperature difference, it is separated into ammonia vapor and weak solution. To achieve the pure ammonia vapor at the turbine, dephlegmator has been located at the exit of the separator. Dephlegmator/Rectifier removes the water content mixed with ammonia vapor by rejecting heat. The dephlegmator allows the operation of

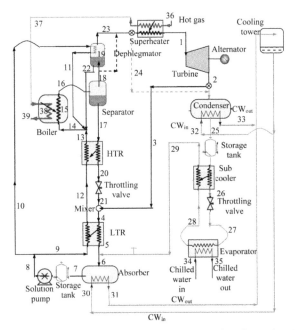

FIGURE 9.1   Proposed new cooling cogeneration cycle having the options in the operation of power plant, cooling plant, once through cogeneration, and split cogeneration.

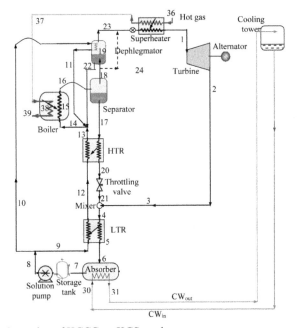

FIGURE 9.2   Operation of KCCC on KCS mode.

generator at LP; otherwise, it demands HP at higher concentration. But the use of dephlegmator in KCS drops the power and improves the coefficient of performance (COP) of VAR plant. Some of fluid from the pump is circulated in dephlegmator to cool the vapor. The ammonia vapor with little percentage of water is again heated in the superheater and the superheated ammonia vapor is used to run the turbine from HP to sink pressure. The weak solution moves to the absorber by sharing its heat to the strong solution in HTR and LTR. The weak solution after throttling and vapor from the turbine is mixed together to form strong solution concentration. As the exit of turbine is above atmospheric temperature, still there is a possible heat recovery at LTR. The mixture moves to the absorber where the heat of mixing is rejected to circulating water in absorber and the exhausted vapor absorbs into the liquid solution. The dilution of turbine vapor with weak solution allows the vapor to condense at a reasonable sink temperature; otherwise, too low temperature is required to condense the vapor without dilution. The low boiling temperature and condensation above the atmospheric pressure allows the operation of plant at different heat source, such as, geothermal, solar thermal, waste heat recovery, etc.

In VAR as shown in Figure 9.3, the operation of pump, LTR, HTR, throttling, generator, separator, and dephlegmator is same as in KCS plant. After the dephlegmator, the strong refrigerant is moved to the condenser for the condensation process. There is a drop in temperature when the HP-condensed liquid refrigerant is throttled to LP/sink pressure. It is used as cooling in the evaporator by absorbing the heat from the atmosphere. A subcooler is used to transfer the heat from the condensate to vapor and improve the evaporator and absorber performance. The weak solution from the generator (after throttling) absorbs the vapor from the evaporator and repeats the cycle. NH_3/H_2O and $H_2O/LiBr$ are the most widely used refrigerant/absorbent pair in VAR system. NH_3/H_2O VAR system has low COP compared to $H_2O/LiBr$ pair due to more heat load consumption in the generator for the separation process as well as the ammonia having less latent heat. Even though $H_2O/LiBr$ VAR system has high COP, it is unable to use for the cooling temperature less than 4°C application because the ice will be formed in the evaporator and also the complete plant is under vacuum.

In KCCC once through model as shown in Figure 9.4, the whole fluid circulates in power lines and cooling lines. The superheated vapor expands in the mixture turbine to generate power from HP to IP. The IP is determined from the concentration and sink temperature. It is condensed in a condenser from wet vapor to saturated liquid condition. The condensate cools after rejecting heat to low-temperature vapor at subcooler. The subcooled liquid

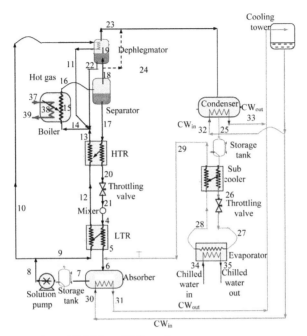

FIGURE 9.3   Operation of KCCC on VAR mode.

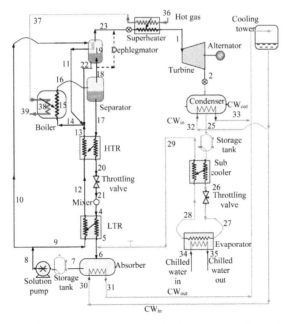

FIGURE 9.4   Operation of KCCC on combined power cooling with once through configuration.

throttles to low-temperature wet mixture. Depending upon the operational conditions, the exit temperature of throttling may be negative or positive. Since the state is above the saturated liquid condition, there is no practical issue of crystal formation at negative temperature which is the severe problem in case of LiBr–water mixture. The throttled and low-temperature liquid mixture can absorb the heat from the surroundings of the evaporator coils. At absorber inlet, a dry vapor receives the heat from subcooler and mixed with the weak solution. The mixer is condensed to a saturated liquid condition at the absorber. The liquid solution is pumped to boiler via LTR and HTR. Thus, the cycle repeats.

Another option in KCCC is split cycle shown in Figure 9.5. The vapor is generated in a common boiler for KCS lines and VAR lines. The vapor is shared and separated for KCS and VAR. The fluid can be controlled at the inlet of the KCS and VAR to meet the demand. The KCS and VAR cycles operate on separate lines. After the completion of these KCS and VAR operations, the two fluids mix together. The thermal compressor, LTR, HTR, and dephlegmator are the common components for KCS and VAR.

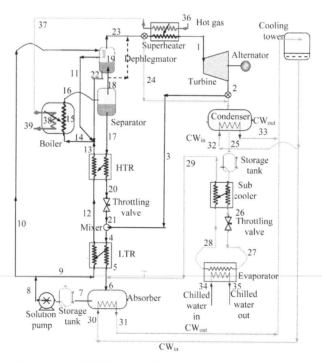

FIGURE 9.5 Operation of KCCC on combined power and cooling with shared working fluid.

Table 9.1 shows the flexible operation of single plant having multiple qualities on different modes of operation with proper control in splitters. The invention of all the plants in a single system to get the flexible operation consists of KCS, VAR, and two KCCC plants. Therefore, four modes are evolved in fully flexible KCCC and they are (1) KCS (power only), (2) VAR (cooling only), (3) KCCC once through, and (4) KCCC split. Splitter 1 (S_1) is located at the exit of dephlegmator and operated in association of splitter 2 (S_2).

TABLE 9.1   Operation of First Splitter at Exit of Dephlegmator and Second Splitter at Exit of Turbine for the Flexible Operation of KCCC.

Splitter	KCS	VAR	KCCC once through	KCCC split
S_1	Opened to SH	Opened to condenser	Opened to SH	Partially opened
S_2	Opened to KCS	Closed	Opened to condenser	Opened to KCS

9.3   THERMODYNAMIC EVALUATION

The simulation of combined power and cooling begins with the thermodynamic properties development for ammonia–water zeotropic mixture. After generating thermodynamic properties of plant fluids, evaluation of processes, cycle, and plant begins with the assumptions involved in the simulation. The circulating cooling water inlet temperature in condenser and absorber is 30°C. TTD in heat exchangers is 5°C. Pinch point in vapor generator's evaporator is 15°C. The hot gas flow in vapor generator is 100,000 N m³/h at 175°C. The isentropic efficiency of pump and vapor turbine is 75%. Similarly, the mechanical efficiency of machines is 96%. The electrical generator efficiency is 98%. Degree of superheat is 10°C. For comparative purpose, a suitable strong solution concentration is maintained in individual KCS, separate VAR, and two cooling cogeneration cycles. The strong solution concentration in VAR and split cycle is 0.3 due to its similar process conditions. The processes conditions in KCS and once through are similar to some extent. The strong solution concentration in KCS is 0.45 and 0.5 in once through cycle. The saturated vapor concentration in separator after the evaporation is 0.85. Effectiveness of dephlegmator is 0.5. The power-to-total output energy ratio, resulted in KCCC once through, is maintained in KCCC split model and individual KCS and VAR.

The degree of dephlegmator can be designed to choose the required vapor concentration. The degree of dephlegmator can be selected from its effectiveness. The effectiveness of dephlegmator is defined as the actual concentration rise to the maximum possible concentration rise.

Effectiveness of dephlegmator is

$$\varepsilon_{dephlegmator} = \frac{x_{23} - x_{18}}{1 - x_{18}} \tag{9.1}$$

In the dephlegmator or rectifier, the saturated vapor is distillated to increase its concentration. The increased concentration (x_{23}) can be determined from eq 9.1 with the dephlegmator effectiveness and vapor concentration (x_{18}). The distilled vapor improves the performance of cooling by decreasing the heat supply in the generator.

After heating the strong solution in boiler, the fluid is a liquid and vapor mixture. The lever rule solves the weak solution concentration and mass of separated fluids. The dryness fraction of the mixer is defined as the mass ratio of saturated vapor and total mass, that is, sum of liquid and vapor.

From the mass balance equations, dryness fraction or vapor fraction in separator is

$$DF_{separator} = \frac{x_{16} - x_{17}}{x_{18} - x_{17}} \tag{9.2}$$

Equation 9.2 shows that the vapor fraction is as a function of strong solution concentration, weak solution concentration, and vapor concentration. The weak solution concentration can be determined from the HP and at the saturated liquid condition. After partial condensation of fluid in dephlegmator, it results in liquid–vapor mixture. The dryness fraction in dephlegmator is more than the dryness fraction in boiler or separator.

The dryness fraction in dephlegmator is

$$DF_{dephlegmator} = \frac{x_{18} - x_{22}}{x_{23} - x_{22}} \tag{9.3}$$

In eq 9.3, x_{22} can be determined from HP and dephlegmator temperature at saturated liquid condition.

Due to condensation of vapor and removal (separation) of condensate from dephlegmator, the refrigerant mass will decrease at the exit of dephlegmator.

Vapor mass at the exit of dephlegmator is

$$m_{23} = \frac{DF_{separator} DF_{dephlegmator}}{1 + DF_{separator} DF_{dephlegmator} - DF_{separator}} m_6 \tag{9.4}$$

The above equation is resulted from the simplification of mass balance equations in separator and dephlegmator. The mass m_6 is the strong solution mass either unit mass or determined from the source fluid. At the boiler inlet, three fluids mix together and form liquid or liquid–vapor mixture. The three fluids are pumped strong solution, return liquid from dephlegmator, and heat exchanger fluid from dephlegmator. After mixing, the concentration slightly dilutes. The fluid mass in boiler increases due to mixing and determined from the simplification of mass balance equations.

Strong solution at the inlet of boiler and after mixing is

$$m_{16} = \frac{m_{23}}{\text{DF}_{\text{separator}}\text{DF}_{\text{dephlegmator}}} \tag{9.5}$$

From the simplification of mass balance equations, the weak solution flow is

$$m_{17} = \frac{\left(1 - \text{DF}_{\text{separator}}\right)}{\text{DF}_{\text{separator}}\text{DF}_{\text{dephlegmator}}} m_{23} \tag{9.6}$$

The fluid at the inlet of dephlegmator is

$$m_{18} = \frac{m_{23}}{\text{DF}_{\text{dephlegmator}}} \tag{9.7}$$

KCS results in only power from heat supply at boiler and superheater.

Cycle energy utilization factor (ENUF) of KCS (thermal efficiency) is

$$\text{ENUF}_{\text{KCS cycle}} = \frac{W_{\text{net}}}{Q_{\text{generator}} + Q_{\text{superheater}}} \tag{9.8}$$

VAR results in only cooling from the heat supply at boiler and superheater.

Cycle ENUF of VAR (COP) is

$$\text{ENUF}_{\text{VAR cycle}} = \frac{Q_{\text{evaporator}}}{Q_{\text{generator}} + Q_{\text{superheater}}} \tag{9.9}$$

KCCC results in both power and cooling from the heat supply in boiler and superheater.

Cycle ENUF of KCCC is

$$\text{ENUF}_{\text{KCCC cycle}} = \frac{W_{\text{net}} + Q_{\text{evaporator}}}{Q_{\text{generator}} + Q_{\text{superheater}}} \tag{9.10}$$

The plant efficiency can be determined from the heat content of hot gas. Plant ENUF of KCS (thermal efficiency) is

$$ENUF_{KCS\,plant} = \frac{W_{net}}{m_{hot\,gas}h_{hot\,gas}} \tag{9.11}$$

Plant ENUF of VAR (COP) is

$$ENUF_{VAR\,plant} = \frac{Q_{evaporator}}{m_{hot\,gas}h_{hot\,gas}} \tag{9.12}$$

Plant ENUF of KCCC is

$$ENUF_{KCCC\,plant} = \frac{W_{net}+Q_{evaporator}}{m_{hot\,gas}h_{hot\,gas}} \tag{9.13}$$

The power-to-total output ratio is made common in cogeneration cycles for the best comparison. The energy ratio resulted in KCCC once through is maintained in split cycle.

$$r_{PTTO,\,cogen} = \frac{W_{cogen}}{W_{cogen}+Q_{evaporator\,cogen}} \tag{9.14}$$

The KCS power and VAR cooling should be shared at the same heat input and same energy ratio (eq 9.14) for the comparison purpose. Therefore, a power-sharing factor (PSF) or cooling-sharing factor, that is, (1-PSF), is to be determined to meet the two constraints, that is, (1) same power to total output ratio and (2) same heat supply to the plant. The power to total output ratio in KCS and VAR should be equal to the $r_{PTTO,cogen}$. Therefore, KCS power and VAR cooling can be selected as per the ratio.

The following analogy is made to find the PSF:

$$\frac{PSF\,W_{KCS}}{PSF\,W_{KCS}+(1-PSF)Q_{VAR\,cooling}} = \frac{W_{cogen}}{W_{cogen}+Q_{evaporator\,cogen}} = r_{PTTO,\,cogen} \tag{9.15}$$

After simplification of eq 9.15,

$$PSF = \frac{r_{PTTO,\,cogen}Q_{VAR\,cooling}}{r_{PTTO,\,cogen}\left(Q_{VAR\,cooling}+W_{KCS}\right)+W_{KCS}} \tag{9.16}$$

Now, the cycle ENUF from the individual KCS and VAR cycles is

$$\text{ENUF}_{\text{KCS+VAR cycle}} = \frac{\text{PSF}\,W_{\text{KCS}} + (1-\text{PSF})Q_{\text{VAR}}}{\text{PSF}\,Q_{\text{KCS}} + (1-\text{PSF})Q_{\text{VAR}}} \qquad (9.17)$$

Similarly, the plant ENUF for the individual KCS and VAR plants is

$$\text{ENUF}_{\text{KCS plant}} = \frac{\text{PSF}\,W_{\text{KCS}} + (1-\text{PSF})Q_{\text{VAR}}}{m_{\text{hot gas}}\,h_{\text{hot gas}}} \qquad (9.18)$$

The thermodynamic work is outlined in Figures 9.6 and 9.7, respectively, for KCS and VAR. The methodology adopted in KCCC is similar to the procedure outlined in KCS and VAR. The thermodynamic evaluation of KCS involves mainly listing the assumptions, finding concentrations from mass balance, framing lever rules, determining properties at all the states from the simplification of mass and energy balance equations, and finally performance indices. After the thermodynamic evaluation, the key parameters and its feasible range are identified. The operational conditions are varied to analyze the cycle and plant performance. It will recommend the best operation environment.

Figure 9.7 outlines the methodology designed for VAR solution. The determination of HP in VAR differs from KCS. The HP in KCS is determined from boiler condition, that is, from one constraint. The vapor concentration in boiler is independent. In VAR, condenser and boiler are two constraints to determine the HP. Therefore, vapor concentration is dependent on VAR.

Tables 9.2–9.5 show the material balance results of KCS, VAR, KCCC, once through, and KCCC split plants. Table 9.2 shows the KCS plant data developed as per the assumptions listed. The strong solution concentration and vapor concentration are 0.45 and 0.85, respectively, in KCS. Since the HP is determined from the boiler conditions, that is, saturated vapor concentration and temperature, the resulted weak solution concentration is 0.3. The separator temperature is 150°C at the source temperature of 175°C. HP and LP in KCS are 28.79 and 3.43 bar, respectively. The dryness fraction of vapor at the exit of turbine is 93% obtained from dephlegmator effectiveness. Since the cooling lines are blocked, the property information in these lines is dotted in the table. In dephlegmator, the vapor concentration is increased from 0.85 to 0.93.

Similarly Table 9.3 shows the material flow data of VAR plant determined as per the assumptions. The HP in VAR boiler is less than the KCS boiler. In VAR, the HP and LP are 9.87 and 1.33 bar, respectively, at the strong solution concentration of 0.3. The saturated vapor concentration has been iterated from the boiler and condenser conditions, respectively, at saturated vapor and saturated liquid. In generator, the weak solution concentration is 0.1 as per the lever rule and mass balances. The resulted vapor concentration is 0.52 and it is distilled in the rectifier or dephlegmator to 0.76. In VAR, the power lines are blocked as shown. The refrigerant temperature is dropped from 31°C to −18.63°C by expanding the fluid from HP to LP.

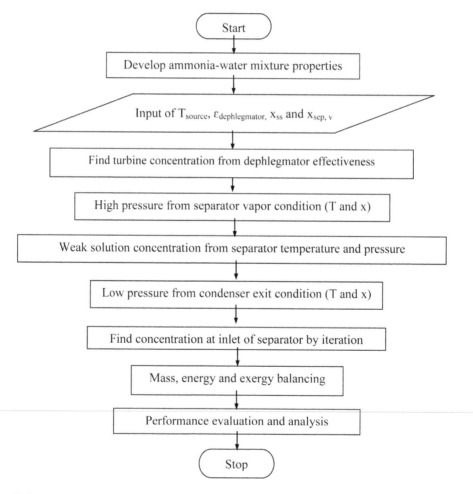

FIGURE 9.6   Methodology adopted for KCS plant.

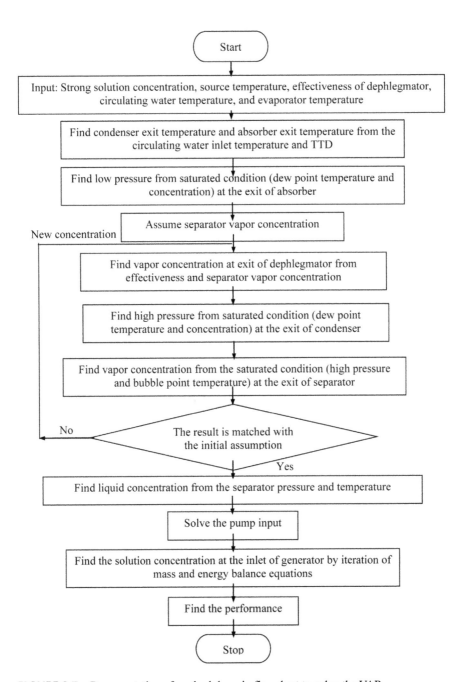

FIGURE 9.7   Representation of methodology in flowchart to solve the VAR.

TABLE 9.2   Material Flow Data of KCS Plant.

State	P (bar)	T (°C)	x	m (kg/s)	h (kJ/kg)	s (kJ/kg K)
1	28.79	160.00	0.93	0.67	1661.60	4.91
2	3.43	58.11	0.93	0.67	1398.83	5.17
3	3.43	58.11	0.93	0.67	1398.83	5.17
4	3.43	71.82	0.45	2.78	577.71	2.40
5	3.43	61.83	0.45	2.78	412.93	1.91
6	3.43	61.83	0.45	2.78	412.93	1.91
7	3.43	35.00	0.45	2.78	−80.82	0.38
8	28.79	37.02	0.45	2.78	−69.69	0.40
9	28.79	37.02	0.45	2.52	−69.69	0.40
10	28.79	37.02	0.45	0.27	−69.69	0.40
11	28.79	135.00	0.45	0.27	608.59	2.23
12	28.79	77.49	0.45	2.52	112.59	0.95
13	28.79	107.05	0.45	2.52	252.23	1.34
14	28.79	114.61	0.45	2.89	289.22	1.43
15	28.79	117.96	0.45	2.89	305.85	1.48
16	28.79	150.00	0.45	2.89	854.88	2.81
17	28.79	150.00	0.30	2.11	484.35	1.87
18	28.79	150.00	0.85	0.78	1704.37	4.99
19	28.79	129.12	0.85	0.78	1472.13	4.44
20	28.79	115.14	0.30	2.11	317.57	1.46
21	3.43	76.34	0.30	2.11	320.01	1.51
22	28.79	129.12	0.39	0.11	363.77	1.61
23	28.79	129.12	0.93	0.67	1571.95	4.70
24	–	–	–	–	–	–
25	–	–	–	–	–	–
26	–	–	–	–	–	–
27	–	–	–	–	–	–
28	–	–	–	–	–	–
29	–	–	–	–	–	–
30	1.01	30.00	0.00	12.25	20.90	0.07
31	1.01	56.83	0.00	12.25	133.03	0.42
32	–	–	–	–	–	–
33	–	–	–	–	–	–
34	–	–	–	–	–	–
35	–	–	–	–	–	–
36	1.01	175.00	0.00	36.89	157.21	0.43
37	1.01	173.50	0.00	36.89	155.60	0.42
38	1.01	132.96	0.00	36.89	112.53	0.32
39	1.01	131.75	0.00	36.89	111.25	0.32

TABLE 9.3   Material Flow Data of VAR Plant.

State	P (bar)	T (°C)	x	m (kg/s)	h (kJ/kg)	s (kJ/kg K)
1	–	–	–	–	–	–
2	–	–	–	–	–	–
3	–	–	–	–	–	–
4	1.33	15.38	0.10	1.12	0.30	0.28
5	1.33	15.38	0.10	1.12	0.30	0.28
6	1.33	50.50	0.30	1.62	239.66	1.33
7	1.33	35.00	0.30	1.62	−38.86	0.44
8	9.87	36.75	0.30	1.62	−30.54	0.47
9	9.87	36.75	0.30	0.90	−30.54	0.47
10	9.87	36.75	0.30	0.73	−30.54	0.47
11	9.87	135.00	0.30	0.73	909.24	2.97
12	9.87	53.91	0.30	0.90	43.53	0.70
13	9.87	83.20	0.30	0.90	171.72	1.08
14	9.87	117.28	0.28	1.97	487.29	1.90
15	9.87	117.28	0.28	1.97	487.29	1.90
16	9.87	150.00	0.28	1.97	1335.58	3.96
17	9.87	150.00	0.10	1.12	574.52	1.90
18	9.87	150.00	0.52	0.85	2115.62	6.12
19	9.87	127.40	0.52	0.85	1311.93	4.19
20	9.87	127.10	0.10	1.12	473.45	1.66
21	1.33	87.41	0.10	1.12	475.85	1.70
22	9.87	127.40	0.18	0.35	423.29	1.64
23	9.87	127.40	0.76	0.50	1800.68	5.63
24	9.87	127.40	0.76	0.50	1800.68	5.63
25	9.87	35.00	0.76	0.50	0.17	0.45
26	9.87	31.00	0.76	0.50	−18.66	0.38
27	1.33	−18.63	0.76	0.50	−16.35	0.48
28	1.33	15.00	0.76	0.50	758.21	3.38
29	1.33	16.78	0.76	0.50	778.02	3.45
30	1.01	30.00	0.00	13.53	20.90	0.07
31	1.01	38.00	0.00	13.53	54.34	0.18
32	1.01	30.00	0.00	2.97	20.90	0.07
33	1.01	102.40	0.00	2.97	323.55	0.96
34	1.01	−14.63	0.00	23.12	−165.66	−0.60
35	1.01	−18.63	0.00	23.12	−182.38	−0.66
36	1.01	175.00	0.00	36.89	157.21	0.43
37	1.01	175.00	0.00	36.89	157.21	0.43
38	1.01	132.28	0.00	36.89	111.80	0.32
39	1.01	132.30	0.00	36.89	111.82	0.32

Table 9.4 results from the material balance data of KCCC once through cycle. It has three pressures, namely, 28.79, 12.44, and 4.41 bar, respectively, for HP, IP, and LP. In dephlegmator, the concentration is increased from 0.85 to 0.93. The weak solution concentration is 0.3 determined at HP, separator temperature, and saturated liquid state. The considered strong solution concentration is 0.5.

TABLE 9.4 Material Flow Data of KCCC Once Through Plant.

State	P (bar)	T (°C)	x	m (kg/s)	h (kJ/kg)	s (kJ/kg K)
1	28.79	160.00	0.93	0.83	1661.60	4.91
2	12.44	97.97	0.93	0.83	1547.65	5.00
3	0.00	0.00	0.00	0.00	0.00	0.00
4	4.41	79.90	0.30	1.75	272.18	1.37
5	4.41	73.64	0.30	1.75	162.92	1.05
6	4.41	61.83	0.50	2.58	433.18	1.97
7	4.41	35.00	0.50	2.58	−82.98	0.37
8	28.79	37.04	0.50	2.58	−71.80	0.39
9	28.79	37.04	0.50	2.30	−71.80	0.39
10	28.79	37.04	0.50	0.28	−71.80	0.39
11	28.79	135.00	0.50	0.28	731.44	2.54
12	28.79	55.15	0.50	2.30	9.84	0.65
13	28.79	90.19	0.50	2.30	172.86	1.12
14	28.79	104.04	0.49	2.71	239.53	1.30
15	28.79	109.45	0.49	2.71	266.30	1.37
16	28.79	150.00	0.49	2.71	969.11	3.10
17	28.79	150.00	0.30	1.75	484.35	1.87
18	28.79	150.00	0.85	0.96	1704.37	4.99
19	28.79	129.12	0.85	0.96	1472.13	4.44
20	28.79	104.73	0.30	1.75	269.75	1.34
21	4.41	79.90	0.30	1.75	272.18	1.37
22	28.79	129.12	0.39	0.14	363.77	1.61
23	28.79	129.12	0.93	0.83	1571.95	4.70
24	–	–	–	–	–	–
25	12.44	35.00	0.93	0.83	110.89	0.56
26	12.44	30.00	0.93	0.83	86.77	0.48
27	4.41	3.04	0.93	0.83	87.69	0.51
28	4.41	15.00	0.93	0.83	981.26	3.72

TABLE 9.4   *(Continued)*

State	P (bar)	T (°C)	x	m (kg/s)	h (kJ/kg)	s (kJ/kg K)
29	4.41	16.68	0.93	0.83	1005.66	3.81
30	1.01	0.00	0.00	11.86	−104.50	−0.37
31	1.01	0.00	0.00	11.86	−104.50	−0.37
32	1.01	0.00	0.00	6.61	−104.50	−0.37
33	1.01	0.00	0.00	6.61	−104.50	−0.37
34	1.01	0.00	0.00	35.33	−104.50	−0.37
35	1.01	0.00	0.00	35.33	−104.50	−0.37
36	1.01	175.00	0.00	36.89	157.21	0.43
37	1.01	173.14	0.00	36.89	155.23	0.42
38	1.01	124.45	0.00	36.89	103.54	0.30
39	1.01	122.61	0.00	36.89	101.59	0.29

Table 9.5 is the material balance values of KCCC split cycle. Since it is the combination of KCS and VAR, the split cycle consists of two pressures. The HP is designed as per the VAR conditions. The split cycles HP and LP are same as in VAR. The strong solution concentration and weak solution concentration in generator separator are 0.3 and 0.1, respectively. The turbine concentration in split cycle is 0.76 with a saturated vapor concentration of 0.52. Similar to VAR, the saturated vapor concentration in boiler is iterated from the condenser and vapor generator conditions.

TABLE 9.5   Material Flow Data of KCCC Split Plant.

State	P (bar)	T (°C)	x	m (kg/s)	h (kJ/kg)	s (kJ/kg K)
1	9.87	160.00	0.76	0.14	1882.02	5.83
2	1.33	68.91	0.76	0.14	1621.57	6.07
3	1.33	68.91	0.76	0.14	1621.57	6.07
4	1.33	81.64	0.17	1.26	580.72	2.14
5	1.33	79.72	0.17	1.26	527.84	1.99
6	1.33	68.32	0.30	1.61	583.98	2.36
7	1.33	35.00	0.30	1.61	−38.86	0.44
8	9.87	36.75	0.30	1.61	−30.54	0.47
9	9.87	36.75	0.30	0.89	−30.54	0.47
10	9.87	36.75	0.30	0.72	−30.54	0.47
11	9.87	135.00	0.30	0.72	909.21	2.97
12	9.87	53.90	0.30	0.89	43.52	0.70

TABLE 9.5   *(Continued)*

State	P (bar)	T (°C)	x	m (kg/s)	h (kJ/kg)	s (kJ/kg K)
13	9.87	90.00	0.30	0.89	201.99	1.16
14	9.87	117.32	0.28	1.96	488.19	1.90
15	9.87	117.32	0.28	1.96	488.19	1.90
16	9.87	150.00	0.28	1.96	1335.51	3.96
17	9.87	150.00	0.10	1.12	574.51	1.90
18	9.87	150.00	0.52	0.85	2115.62	6.12
19	9.87	127.40	0.52	0.85	1311.88	4.19
20	9.87	121.21	0.10	1.12	447.86	1.59
21	1.33	86.47	0.10	1.12	450.28	1.63
22	9.87	127.40	0.18	0.35	423.28	1.64
23	9.87	127.40	0.76	0.50	1800.66	5.63
24	9.87	127.40	0.76	0.35	1800.66	5.63
25	9.87	35.00	0.76	0.35	0.17	0.45
26	9.87	30.00	0.76	0.35	−23.34	0.37
27	1.33	−18.69	0.76	0.35	−21.05	0.46
28	1.33	15.00	0.76	0.35	758.25	3.38
29	1.33	17.15	0.76	0.35	782.06	3.47
30	1.01	30.00	0.00	7.21	20.90	0.07
31	1.01	63.32	0.00	7.21	160.18	0.51
32	1.01	30.00	0.00	2.10	20.90	0.07
33	1.01	102.40	0.00	2.10	323.55	0.96
34	1.01	20.00	0.00	13.16	−20.90	−0.07
35	1.01	15.00	0.00	13.16	−41.80	−0.14
36	1.01	175.00	0.00	36.89	157.21	0.43
37	1.01	174.00	0.00	36.89	156.12	0.42
38	1.01	132.32	0.00	36.89	111.85	0.32
39	1.01	132.34	0.00	36.89	111.85	0.32

Figures 9.8–9.11 show the enthalpy–concentration and temperature–entropy diagrams for KCS, VAR, KCCC once through, and KCCC split cycles, respectively. These diagrams show the properties and process details to understand the nature of cycle. These two diagrams clearly show the separation process in generator and dephlegmator. The temperature–entropy diagram also depicts the heat addition or heat removal with variable temperature. The vapor absorption system also consists of mixing of streams as well as separation of one fluid into two.

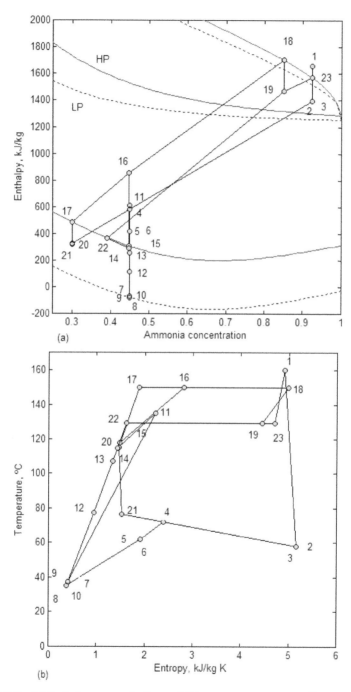

FIGURE 9.8 (a) Enthalpy–concentration and (b) temperature–entropy diagrams for KCS operation.

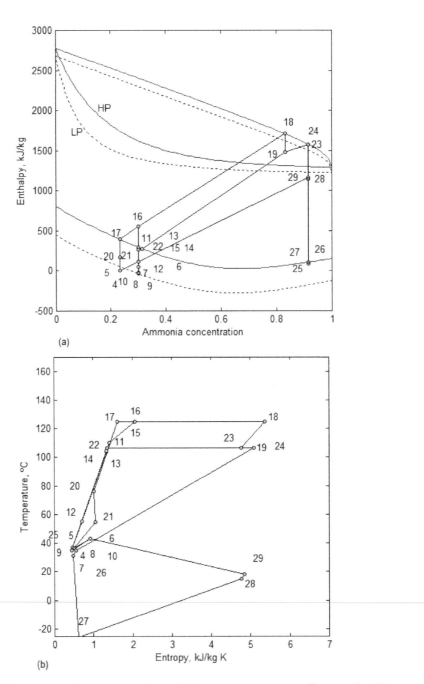

FIGURE 9.9   (a) Enthalpy–concentration and (b) temperature–entropy diagrams for VAR operation.

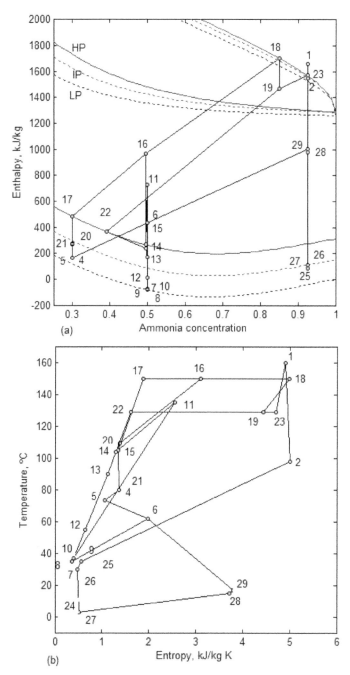

FIGURE 9.10 (a) Enthalpy–concentration and (b) temperature–entropy diagrams for KCCS once through mode.

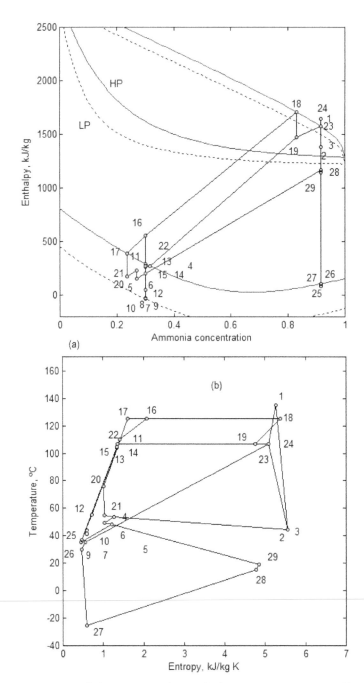

FIGURE 9.11 (a) Enthalpy–concentration and (b) temperature–entropy diagrams for KCCS split cycle.

9.4   ANALYTICAL RESULTS AND INTERPRETATION

In this section, the results of four-cycle configurations are compared and studied. Power only is the output from KCS and cooling only is from VAR. The KCCC generates both power and cooling from single plant setup. Power from KCS, KCCC once through, and KCCC split are compared under the same heat source and sink. Similarly, the cooling from VAR, KCCC once through, and KCCC split are also compared under common ground. The ENUF of cycle and plant are analyzed to identify the conditions for maximum energy conversion in cycle as well as plant. The power and cooling selection in split cycle can be obtained with fluid control. The energy ratio (power to total output) in once through and split cycle is the same. The share in working fluid is determined based on the energy ratio resulted in the once through cycle for comparison of the cycles under the similar situations. The identified independent operational variables are hot fluid supply temperature, strong solution concentration, separator vapor concentration, and effectiveness of dephlegmator. The focused results are power, cooling, cycle ENUF, and plant ENUF.

Figure 9.12 plots the comparative output results of plants with a change in strong solution concentration. The beginning value of strong solution in variations is determined from the resulted weak solution concentration. It is varied above the resulted weak solution concentration. Since the weak solution concentration in KCS and VAR are different, the lower limit of strong solution concentration differs with plant. The HP in KCS and KCCC once through is high compared to VAR and KCCC split; the lower limit of strong solution concentration is also high. Therefore in VAR and KCCC split, the lower limit of strong solution concentration is low. Strong solution concentration decides the sink pressure in absorber. It also influences the HP in VAR and KCCC split plant. In steam Rankine cycle, vacuum pressure is required to condense the steam at the exit of steam turbine. The addition of ammonia in water increases the condenser (absorber) pressure. A concentrated fluid in condenser demands HP to condenser. Therefore, the fluid is diluted at the inlet of absorber to condense at a suitable pressure. Approximately 25% of ammonia concentration is required to condensate the vapor at atmospheric pressure with a sink temperature of 35°C. It is with the circulating water inlet temperature of 30°C. The increase in strong solution concentration above 25% increases the sink pressure above the atmospheric. For example in KCS, at 45% concentration and 35°C sink's temperature, the absorber (sink) pressure is 3.5 bar which is above atmospheric pressure. Figure 9.12a shows that the power generation is increasing with increase in strong solution concentration in KCS and KCCC once through. In once through model, at the lower strong solution concentration, the pump work exceeds the turbine work. Therefore, the net power output in once through is negative below 0.35 concentration. But there is no such issue in KCS and split cycle. Due to fluid share in split cycle, the power generation is low compared to other plants. The power in split cycle is increasing with increase in strong solution concentration up to 0.4 and followed by a drop. From Figure 9.12b, it can be observed that the cooling is highest at an optimum strong solution concentration and differs with the plant. The optimum strong solution concentration is 0.32 in VAR for maximum cooling. The optimum strong solution concentration is 0.52 for KCCS once through based on maximum cooling (Fig. 9.12b) and maximum total output (Fig. 9.12c)

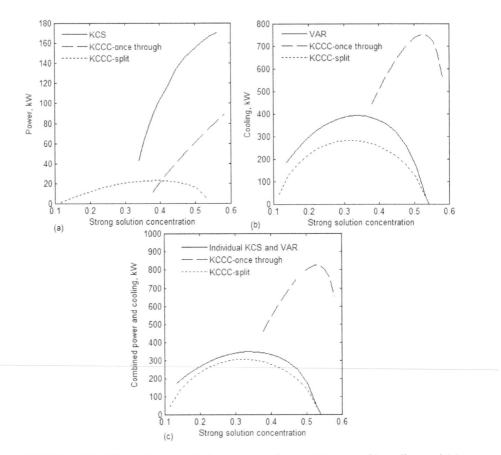

FIGURE 9.12 Effect of strong solution concentration on (a) power, (b) cooling, and (c) combined power and cooling from individual plants and KCCC groups.

condition. The combined power and cooling from the individual plants is the sum of power and cooling obtained, respectively, from KCS only and VAR only plants as shown in Figure 9.12c. The optimum strong solution concentration with a focus on combined power and cooling is a little more compared to optimum concentration from higher cooling. The increase in power with rise in concentration and cooling peak causes this difference in optimum strong solution concentration. The power from the power only is higher than the cogeneration plants as the configuration is designed only for power, and also it allows HP which is the same as the once through cycle. In the similar way, the VAR only is not resulting in the higher cooling compared to the cogeneration cycles. The VAR-iterated pressure is less than the once through HP. Therefore, once through is resulting higher cooling than VAR which is the surprising issue in this case.

Figure 9.13 depicts the cycle and plant ENUFs with a change in strong solution concentration. ENUF is the ratio of combined power and cooling to energy supply. The optimum strong solution concentration of KCCC once through is more than the individual plants and also KCCC split with a focus on combined power and cooling. The optimum strong solution concentration is 0.5 in KCCC once through and 0.3 in KCCC split to gain the maximum cycle ENUF and plant ENUF benefit. The optimum strong solution concentration resulted with maximum ENUF is less than the optimum concentration resulted with maximum combined power and cooling. Due to diversion of fluid to power, the cooling generation from split is less than the VAR at the same heat source conditions. Since the cooling efficiency (COP) is better than the power efficiency, which has Carnot's constraint, the split cycle efficiency falls below the sum of the individual cycle efficiency. Finally, the total output and ENUF of split cycle is less than the individual KCS and VAR systems. In fact, the total efficiency of KCS and VAR is between KCS and VAR efficiencies. In KCS, the generator pressure is more than the split cycle's generator's pressure (HP). In KCS, absorber works as a condenser and so there is no separate condenser. The KCS HP is designed on the basis of vapor concentration and separator temperature due to the absence of condenser. The heat rejects in KCS through absorber only. In split cycle which is the combination of KCS and VAR, heat rejection occurs at condenser and absorber. The HP is determined at condensate temperature and separator temperature on iteration mode. Since the generator pressure and condenser pressure are same, at assumed condensate temperature and generator temperature, the vapor concentration can be determined along with the HP. Therefore, the iteration results HP and vapor concentration. Split cycle's HP is equal to the VAR's HP. Even though there are difficulties in split cycle, it offers a great flexibility in control of power and cooling to meet the variable demand. The plant erection area and components cost decreases with split cycle compared to the individual cycles. As per the feedback from the strong solution analysis, in the subsequent analyses, the strong solution concentration in KCS, VAR, once through, and split are maintained are 0.45, 0.3, 0.5, and 0.3, respectively.

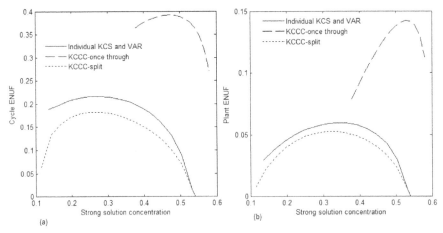

FIGURE 9.13 Effect of strong solution concentration on (a) cycle ENUF and (b) plant ENUF.

Figure 9.14 reveals the augmented performance of KCCC in the light of hot fluid supply temperature varying from 155°C to 178°C. The separator temperature is kept below the hot fluid source temperature by 25°C to accommodate the heat transfer in superheater and also to provide the TTD between hot fluid and superheated vapor. Figure 9.14a compares the power generated from KCS, KCCC once through, and KCCC. Since KCS is designed for power alone, it generates more power under the same source and sink conditions. In KCS, the sink temperature is determined at absorber temperature and strong solution concentration. In once through cycle, the turbine exit pressure (IP) is determined at condenser exit temperature and turbine concentration which is higher than the strong solution concentration. Therefore in once through, the expansion of vapor is limited over the KCS. In split cycle, due to diversion of fluid to cooling, the power generation is dropped compared to KCS. The HP in split cycle is determined based on condenser condition. In VAR and split cycle, it is not possible to design the HP from boiler saturated vapor condition. It leads a lower turbine pressure in split cycle compared to

KCS. All the four plants (KCS, VAR, KCCC once through, and KCCC split) have the same source with the flow rate of 100,000 liters per hour (LPH). The KCCC once through generates more cooling compared to the other plants. The cooling from split cycle is less than the other methods. The power from KCCC once through is more than the power from KCCC split with divide of fluid between power and cooling. Once through flow configuration is oriented with three pressures and they are HP, IP, and LP. These are designed at vapor generator, condenser, and absorber, respectively, for HP, IP, and LP. The selection of source temperature influences the HP. The IP and LP depend on sink temperature and the concentration. If the selected source temperature in once through is sufficiently low, the HP is less than IP. So, turbine has to work as a compressor to maintain HP in condenser than generator. It results in a negative output from turbine. In KCS and split, there is no such issue as they work between two pressures, that is, HP and LP. The output energy ratio in once through and split is maintained same for comparative purpose. The source temperature is started from 155°C in all the cycles on common base. However, the incremental rate in power with increase in source temperature is more in once through over others.

Figure 9.14b outlines the cooling trends in VAR, KCCC once through, and KCCC split. It shows that the cooling suffers with increase in source temperature in once through cycle. It is opposite to power changes with source temperature. In VAR, the cooling maximizes at 165°C of source. At high-source temperature, a drop in cooling is more in VAR and split cycle compared to once through. The VAR is more sensitive with source temperature and the cooling is drastically decreasing with increase in temperature. The cooling from once through is more than the VAR and split. Cooling cogeneration suffers with declined power but once through combined power and cooling is higher than the total generation from individual plants, that is, sum of KCS power and VAR cooling. To judge the energy conversion efficiency, ENUF has been evaluated at cycle and plant level. In KCCC once through, the generator vapor concentration is independent and it decides the HP from the generator temperature as per the assumption in the simulation. But in VAR and split cycle, vapor concentration is dependent on generator temperature and condenser temperature. Under the same generator or separator temperature, the iterated vapor concentration in VAR and split cycle is lower than the concentration assumed in KCS and once through cycle. Therefore, once through is generating more cooling over the other cooling methods.

Figure 9.14c compares the combined power and cooling from the individual plants and cogeneration plants with a change in source temperature.

Switching from the individual plants to cogeneration mode increases the total output with KCCC once through. The sum of power and cooling from individual plants maximizes with increase in plant inlet temperature. The combined power and cooling in KCCC once through is major compared to the others and not much fluctuation with source temperature. Maximum cooling over others and high power compared to split from once through lead the highest combined power and cooling.

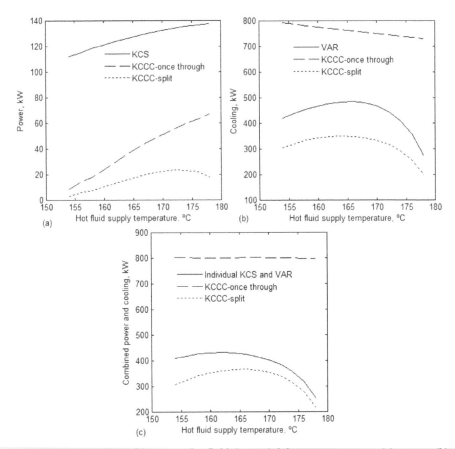

FIGURE 9.14 Effect of heat-transfer fluid (source) inlet temperature on (a) power, (b) cooling, and (c) combined power and cooling from individual plants (KCS + VAR) and KCCC plants.

Figure 9.15a compares the cycle ENUF and Figure 9.15b plots the plant ENUF for the combined power and cooling. The ENUFs are defined in the methodology section for all the plants under this comparison. The ENUF is the combined effect of power and cooling. The power increases with increase in source temperature as per the Carnot theorem and the refrigeration is decreasing with increase in source temperature after the optimum. The combined effect of the ENUF is decreasing with increase in source temperature in split and individual plants. It is not violating the first law of thermodynamics. Since the drop in cooling is dominating with rise in temperature, the total output and ENUF are decreasing with temperature. The plant ENUF is less than the cycle ENUF. The cycle and plant ENUF is increasing with increase in hot fluid supply temperature in KCCC once through. In other two plants, the ENUF is decreasing with increase in source temperature. A maximum feasible source temperature is recommended to once through as the power increases with increase in source temperature. The split cycle is suitable at low-temperature heat source compared to once through cycle. The overall once through plant efficiency is not increasing much with rise in temperature.

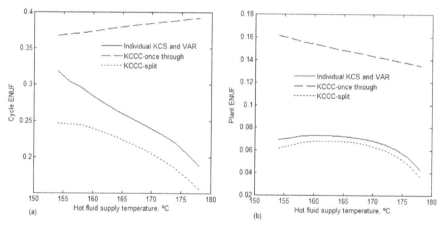

FIGURE 9.15 Effect of heat-transfer fluid (source) inlet temperature on (a) cycle ENUF and (b) plant ENUF of individual plants (KCS + VAR) and KCCC plants.

Figure 9.16 shows the influence of separator vapor concentration on power, cooling, and combined power and cooling for the plants. The vapor concentration depends on pressure and temperature of vapor in the generator which is controlled by the heat-transfer flow rate and temperature. The vapor concentration will increase with increase in pressure. The same will decrease with increase in temperature. In simulation, separator vapor fraction is made an independent variable so that generator pressure may be determined at the known vapor concentration and temperature. In

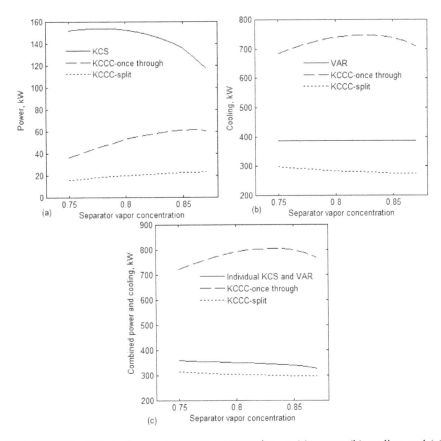

FIGURE 9.16 Effect of separator vapor concentration on (a) power, (b) cooling, and (c) combined power and cooling of KCS, VAR, and KCCC systems.

this section, vapor concentration is varied at a fixed separator temperature. So, an increase in vapor concentration indicates an increase in HP without altering the temperature. The power is increasing with declined rate with increase in separator vapor concentration. KCS power is increasing up to 0.78 concentration and followed by a drop in power with increase in concentration (Fig. 9.16a). Therefore, KCS may be operated at low vapor concentration to gain the higher power. In split cycle, the HP is less than the HP resulted in KCS and once through; therefore, the power decreases in split cycle with limited fluid flow. The increase in vapor concentration allows more expansion in the turbine but it also increases pumping power. In VAR and split cycle, the separator vapor concentration is dependent variable. The HP is iterated from condenser exit condition, that is, temperature and concentration. The separator vapor concentration satisfies the saturated vapor in the generator and saturated liquid in condenser. Therefore, the separator vapor concentration can be iterated between saturated vapor condition in generator and saturated liquid condition in condenser. Therefore, the cooling variations in VAR are zero as separator vapor concentration does not change at fixed source temperature and sink temperature. Due to change in fluid share in split, a little variation in cooling is observed. The HP in KCS and once through cycle is 28.7 bar and it is 10 bar in VAR and split cycle under the same separator temperature (150°C). The total output from the individual cycles, that is, from KCS and VAR is not just the sum of their direct outputs. The power from KCS and cooling from VAR are selected to maintain the same energy ratio resulted in cogeneration and common heat source. The combined power and cooling from the individual KCS and VAR is decreasing with increase in vapor concentration. Due to fall in KCS power, the total output is decreasing in split cycle and the sum of individual cycles.

Figure 9.17 shows the ENUF changes in cycle and plant with a change in separator vapor concentration applied to individual plants and KCCC plants. The cycle ENUF and plant ENUF are increasing with increase in separator vapor concentration in once through. The split cycle conditions are similar to VAR and so the variations are minor. As the once through generates more total output with increase in vapor concentration, the efficiencies are also increasing. For the individual KCS and VAR as well as split cycle, the cycle and plant efficiencies are dropping with vapor concentration by following the total generation trends.

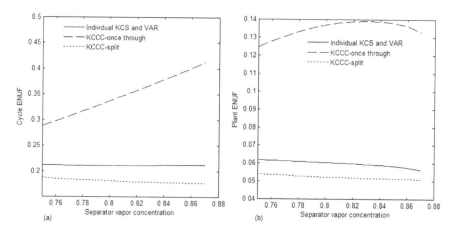

FIGURE 9.17   Effect of separator vapor concentration on (a) cycle ENUF and (b) plant ENUF individual plants and KCCC systems.

Figure 9.18 shows the influence of dephlegmator effectiveness on outputs of the plants under this study. Zero dephlegmator effectiveness indicates the plant operation without dephlegmator. A higher source temperature in VAR will not permit higher degree of dephlegmator. Therefore, the source temperature of VAR and split cycle is fixed at 150°C instead of 175°C to allow the effectiveness variations. Figure 9.18a shows that the power output from the KCS and two KCCC are decreasing with increasing dephlegmator effectiveness. Therefore, the dephlegmator will not support the power as the boiler pressure decreases with increase in dephlegmator effectiveness and also loss in turbine fluid. Figure 9.18b shows the significance of dephlegmator in once through cycle. The cooling increases with increase in dephlegmator effectiveness in KCCC once through. It is dropping in other plants. The cooling from VAR or split cycle decreases with drop in refrigerant in evaporator. The dephlegmator condenses some fluid and results in a loss in refrigerant. It increases the COP of cooling system.

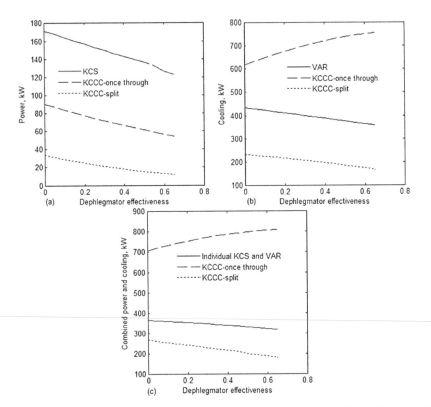

FIGURE 9.18 Effect of dephlegmator effectiveness on (a) power, (b) cooling, and (c) combined power and cooling applied to individual plants and cogeneration plants.

As shown in Figure 9.19, cycle and plant ENUF increase with increase in dephlegmator effectiveness in KCCC once through as the combined power and cooling is increasing. A rise and fall in ENUF is observed in split cycle and sum of KCS and VAR. Too high effectiveness in dephlegmator is not recommended in split cycle but there is no much constraint in once through.

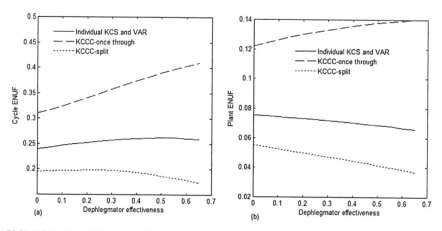

FIGURE 9.19   Effect of dephlegmator effectiveness on (a) cycle ENUF and (b) plant ENUF.

Figure 9.20 shows the influence of circulating water inlet temperature on outputs of KCS, VAR, and two cogenerations. It reveals the role of sink temperature on the output. As per the Carnot law, the output is decreasing with increase in sink temperature, that is, it shows the inverse relation. The power, cooling, and total production from the once through are decreasing with increase in sink temperature. The VAR and split cycle exhibit a small deviation at the beginning of the temperature rise. After that, it follows the same decreasing trend. At around 10–12°C of sink temperature, the VAR and split are exhibiting peaks in output and energy conversion. It can be noticed that split cycle is suffering to generate the cooling at higher sink temperature. Approximately after 35°C, the split cycle cannot generate the cooling at the stated operational conditions.

As per the earlier discussion, the ENUF of cycle and plant follows the outputs as shown in Figure 9.21.

Figure 9.22 reveals the relationship between evaporator exit temperature and outputs from KCS, VAR, once through, and split cycle. The operation of evaporator in cooling lines is independent of power flow lines. Therefore, the power generation from KCS, once through, and split cycle is not disturbed with a change in evaporator temperature. But as the evaporator is directly linked to cooling production, the operation at higher evaporator exit temperature provides more heat absorption by increasing its capacity. The flow is not affected but the temperature rise in the evaporator is increased. Therefore, the capacity of refrigerating effect is increased. The increasing cooling capacity in once through is much more than the VAR and split cycle due to its special design features. Because of this, the total generation from once through is also influenced much.

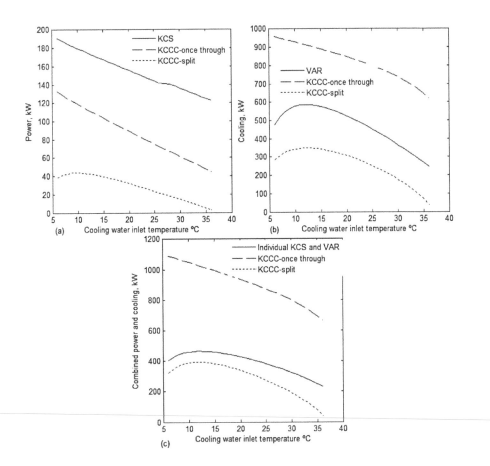

FIGURE 9.20 Effect of circulating water inlet temperature for sink on (a) power, (b) cooling, and (c) combined power and cooling applied to individual plants and cogeneration plants.

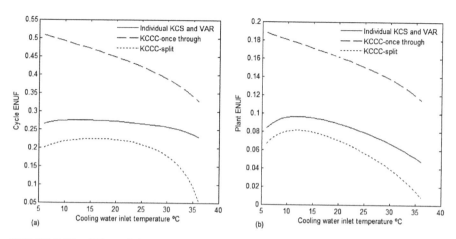

FIGURE 9.21 Effect of circulating water inlet temperature for sink on (a) cycle ENUF and (b) plant ENUF.

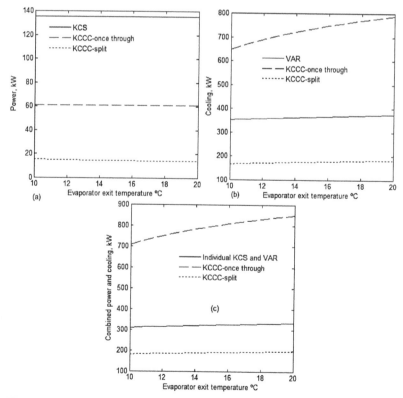

FIGURE 9.22 Effect of evaporator exit temperature for sink on (a) power, (b) cooling, and (c) combined power and cooling applied to individual plants and cogeneration plants.

The corresponding energy conversions are depicted in Figure 9.23 which follows the earlier quantities discussed. The evaporator temperature has certain limit. The required low temperature restricts the evaporator design temperature. It depends on the application such as air condition, cold storage, refrigeration, medicine, etc.

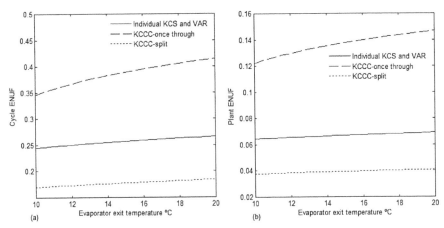

FIGURE 9.23   Effect of evaporator exit temperature on (a) cycle ENUF and (b) plant ENUF.

Table 9.6 overviews the specifications of the KCS, VAR, and two KCCC plants. In this comparison, a suitable strong solution concentration is maintained in the plants where 0.45, 0.3, 0.5, and 0.3 are for KCS, VAR, KCCC once through, and KCCC split, respectively. The fluid flow rates (working fluid, circulating water in absorber, circulating water in condenser, and chilled water in evaporator) are compared. The heat exchangers capacities are evaluated and tabulated. The performance parameters are compared to find the relative merits and demerits of the four plants. The source and sink conditions are common to all the plants. The heat-transfer fluid flow and temperature are 100,000 N m³/h and 175°C. The generated working fluid with the same heat source is high in KCS and once through and low in VAR and split cycle. It is due to the processes similarity between KCS and once through and VAR and split. The circulating water exit temperature absorber is designed as per the hot fluid inlet temperature by maintained TTD at 5°C. Relatively, the hot fluid inlet temperature in VAR is low compared to others. Therefore, the heat load in absorber is low compared to others. The drop in circulating water temperature difference increases the fluid flow as shown. Since more fluid is flowing in once through condenser compared to VAR and split, the condenser load is more in once through. The condenser is

TABLE 9.6   Specifications of the Plants at 100,000 N m³/h and 175°C of Hot Gas Supply.

Sr. no.	Description	KCS	VAR	KCCC once through	KCCC split
1	Working fluid in pump (kg/s)	2.78	1.62	2.58	1.61
2	Circulating water in absorber (kg/s)	12.25	13.53	11.86	7.21
3	Circulating water in condenser (kg/s)	–	2.97	6.61	2.10
4	Chilled water in evaporator (kg/s)	–	23.12	35.33	13.16
5	High pressure (bar)	28.80	9.87	28.80	9.87
6	Intermediate pressure (bar)	–	–	12.50	–
7	Low pressure (bar)	3.50	1.32	4.41	1.32
8	Absorber (kW)	1374.20	551.52	1329.90	878.18
9	Pump (kW)	29.38	31.63	27.32	28.78
10	LTR (kW)	458.60	264.69	187.64	261.30
11	HTR (kW)	352.38	723.37	375.57	715.82
12	Boiler (kW)	1636.40	1080.40	1978.50	1061.60
13	Dephlegmator (kW)	236.01	132.30	290.98	131.09
14	Superheater (kW)	60.09	–	74.09	12.68
15	Turbine (kW)	165.72	–	88.60	43.95
16	Condenser (kW)	–	525.27	1187.40	256.46
17	Subcooler (kW)	–	6.85	19.93	4.18
18	Evaporator (kW)	–	375.71	738.49	184.53
19	Power (net) (kW)	136.34	0.00	61.28	15.17
20	Cooling (kW)	0.00	375.71	738.49	184.53
21	Combined power and cooling (kW)	136.34	375.71	799.78	199.70
22	Cycle ENUF	0.08	0.34	0.39	0.18
23	Plant ENUF	0.03	0.08	0.14	0.04

located at the exit of turbine in KCS and once through. But it is at the exit of dephlegmator in VAR and split. In both the cases, the hot fluid temperature is high and so the circulating water exit temperature is maintained 25°C below the hot fluid inlet temperature. In KCS, there is no condenser and absorber is used to reject the heat. Similarly, evaporator is absent in KCS. Its capacity and chilled water flow are not applicable for KCS. The cooling from KCCC once through is high and it generates more chilled water with the same temperature difference. The bubble point temperature is 132.5°C and 127°C, respectively, in KCS and VAR with decrease in HP from KCS to VAR. The boiler pressure in KCS and VAR is 29 and 10 bar, respectively. The heat load in split cycle and VAR's boiler is more and generates less fluid in VAR compared to the others. The heat recovery in LTR in KCS is much more than the KCCC plants. The same boiler conditions in KCS and KCCC once through lead equal HTR load in two plants. Similarly, the boiler conditions in VAR and KCCC split are same and HTR load in two is also same. The hot fluid inlet temperature and flow rate is same in all the cycles and the generator load is nearly equal in all the plants. The highest power generation can be achieved through KCS and it drops in cooling cogeneration with power penalty by shifting from power to cooling cogeneration. The highest cooling is resulted in KCCC once through with maximum combined power and cooling. The high cycle ENUF (0.39) and plant ENUF (0.14) are achieved with KCCC once through plant. Finally, KCCC once through model is recommended for better results.

To justify the cogeneration performance, the individual KCS and VAR cycles and plants are considered. The total energy supply to KCS and VAR is equal the cogeneration plant. The energy ratio, that is, power-to-total output ratio in two cogeneration cycles and total output from individual plants, is kept fixed. For example, the power and combined power and cooling from KCCC once through are 61.28 and 799.78 kW, respectively. The power-to-total output ratio in this cogeneration is 0.076. The same ratio is maintained in split cycle. In split cycle, the power and combined power and cooling are, respectively, 15.17 and 199.70 kW, having the same energy ratio. The energy ratio obtained in cogeneration is to be maintained in KCS and VAR outputs taken from the separate plants with the same heat source. The power only from the KCS at the same heat source is 136.34 kW and cooling only from VAR at the same heat source is 375.71 kW. Now, we have to take these two outputs from the separate plants by maintaining the same energy ratio and heat supply concept. From the thermodynamic simulation, the resulted PSF is 0.19. Therefore, the required power from KCS is 0.19×136.34 kW = 25.90 kW. The required cooling from the VAR is $(1 - 0.19) \times 375.71$ kW = 304.32 kW. Now, the total power and cooling from the individual plants is 330.22 kW. The power-to-total output ratio from the individual plants is 0.076 which is same as the earlier. Now since the individual plants and cogeneration plants are normalized at the same heat source and energy ratio, they are fit for comparison without bias.

9.5 SUMMARY

KCS is extended to cooling cogeneration with possible combinations. Following are key conclusions drawn from the diagnosis of the individual plants and the combined cycles.

- Higher cooling and ENUF is found in KCCC once through.
- The optimum strong solution concentration is high in cogenerations than VAR.
- The optimum concentration is different with a focus on output and energy conversion efficiency.
- The critical strong solution concentration in KCCC is resulted as 0.55 at the separator temperature of 150°C or at the heat-transfer fluid inlet temperature of 175°C.
- The influence of dephlegmator in KCCC plant is more compared to VAR.
- The cycle ENUF is increased from 0.14 to 0.39 from individual cycles (KCS + VAR) to KCCC once through. Similarly, the plant ENUF is increased from 0.06 to 0.14 by the same KCCC.
- The boiler vapor concentration is dependent in VAR and KCCC split cycle.
- The operational conditions are similar in VAR and split cycle. Similarly, KCS states are matched with once through cycle.
- As per expectation, KCS gave higher power than cogeneration but surprisingly once through cogeneration resulted in higher cooling than VAR.
- Highest cooling (738.49 kW) and total output (799.78 kW) are resulted in KCCC once through and highest power (136.34 kW) in KCS at the hot gas flow of 100,000 N m^3/h at 175°C.
- Finally, KCCC once through is suggested for the best over the others.

KEYWORDS

- **small-capacity power generation**
- **renewable energy sources use**
- **waste heat recovery**
- **Kalina cycle system**
- **refrigeration cycle**

APPENDIX

AMMONIA–WATER MIXTURE PROPERTIES—TABLES AND GRAPHS

Generation of properties of pure substance is easy compared to the mixture properties. Zeotropic mixture properties are much complicated compared to azeotropic mixture. Refrigerant and steam properties are available as books. But ammonia–water mixture property tables are not available yet. An attempt has been made to develop the mixture properties with the variables in pressure, temperature, and concentration. The available ammonia–water mixture property charts are useful to read the saturation properties only. In this section, the wet mixture properties are plotter. The use of these wet properties avoids the use of lever rule which is the combination of mass and energy balance equations.

Tables A.1–A.13 show the properties of ammonia–water mixture in saturated and liquid–vapor mixture. The same properties are plotted in Figures A.1–A.13. At the given pressure, concentration, and temperature, the tabulated properties are bubble-point temperature, dew-point temperature, specific volume, specific enthalpy, specific entropy, liquid concentration, and vapor concentration. The temperature of the fluid is changed between bubble-point temperature and dew-point temperature. The temperature is increased in the steps of 10% increment up to the dew-point temperature.

The temperature steps are defined in the form of temperature fraction, F, and defined as follows:

$$F = \frac{T - T_{bp}}{T_{dp} - T_{bp}} \tag{A.1}$$

In the above equation, T_{bp} and T_{dp} are the bubble-point temperature and dew-point temperature, respectively. For example, at a step of 30%, that is, 0.1 of temperature fraction, the temperature can be determined as follows:

$$T = T_{bp} + F(T_{dp} - T_{bp}) \tag{A.2}$$

Since the vapor fraction is the function of mixture concentration, liquid concentration, and vapor concentration, the dryness fraction can be determined. For example from Table A.1, at the temperature fraction (F) of 0.3 and 0.5 concentration, the following properties can be listed.

$T = -16.30°C$

$v = 2.58 \ m^3/kg$

$h = 187.98 \ kJ/kg$

$s = 1.62 \ kJ/kg \ K$

$x_l = 0.33 \ kg/kg$

$x_v = 1.00 \ kg/kg$

The bubble-point temperature at the 0.5 concentration can be obtained from Table A.1 at 0 temperature fraction (saturated liquid) as $-40.96°C$. Table A.1 gives the dew-point temperature at the temperature fraction of 1 (saturated vapor) as $41.24°C$.

From eq A.2, the mixture temperature can be determined as follows:

$$T = -40.96 + 0.3 \ (41.24 - (-40.96)) = -16.30°C$$

The same temperature is shown at the 30% of temperature rise from the bubble-point temperature. The temperature increases from bubble-point temperature ($F = 0$) to dew-point temperature ($F = 1$).

The F value between 0 and 1 indicated the liquid–vapor mixture or wet mixture. The dryness fraction of wet mixture can be defined as the mass ratio of dry saturated vapor to mass of total mixture.

$$\text{Dryness fraction, } DF = \frac{m_{vapor}}{m_{liquid} + m_{vapor}}$$

$$m_{mixture} = m_{liquid} + m_{vapor} \tag{A.3}$$

The total ammonia portion before separation and after separation is same.

$$m_{mixture} x_{mixture} = m_{liquid} x_{liquid} + m_{vapor} x_{vapor} \tag{A.4}$$

Eliminating m_{liquid},

$$m_{mixture} x_{mixture} = (m_{mixture} - m_{vapor}) x_{liquid} + m_{vapor} x_{vapor}$$

$$m_{mixture} x_{mixture} = m_{mixture} x_{liquid} - m_{vapor} x_{liquid} + m_{vapor} x_{vapor}$$

$$m_{mixture} (x_{mixture} - x_{liquid}) = m_{vapor} (x_{vapor} - x_{liquid})$$

TABLE A.1 Ammonia–Water Property System at 0.15 Bar Under Saturated and Wet Conditions.

F, where T = T_bp + F(T_dp − T_bp)	Property	Ammonia concentration (mass fraction), x										
		0	0.1	0.2	0.3	0.4	0.5	0.6	0.7	0.8	0.9	1
0	T (°C)	54.31	23.38	4.45	−12.13	−27.61	−40.96	−51.13	−57.82	−61.60	−63.63	−65.09
	v (m³/kg)	0.00	0.00	0.00	0.00	0.00	0.00	0.00	0.00	0.00	0.00	0.00
	h (kJ/kg)	227.58	30.73	−116.34	−245.71	−356.25	−438.03	−481.38	−481.79	−442.71	−374.77	−291.36
	s (kJ/kg K)	0.76	0.39	0.06	−0.28	−0.64	−0.96	−1.20	−1.33	−1.35	−1.28	−1.20
	x_l (kg/kg)	0.00	0.10	0.20	0.30	0.40	0.50	0.60	0.70	0.80	0.90	1.00
	x_v (kg/kg)	0.00	0.00	0.00	0.00	0.00	0.00	0.00	0.00	0.00	0.00	0.00
0.1	T (°C)	54.27	26.16	8.88	−6.26	−20.44	−32.74	−42.27	−48.80	−52.95	−55.93	−64.38
	v (m³/kg)	1.00	0.96	0.91	0.87	0.82	0.78	0.75	0.73	0.71	0.70	0.67
	h (kJ/kg)	464.63	201.84	63.78	−50.78	−149.76	−227.27	−279.05	−306.71	−317.61	−319.27	−183.43
	s (kJ/kg K)	1.49	1.09	0.79	0.50	0.20	−0.07	−0.29	−0.44	−0.52	−0.58	−0.53
	x_l (kg/kg)	0.00	0.09	0.18	0.26	0.35	0.44	0.51	0.57	0.62	0.66	1.00
	x_v (kg/kg)	0.00	0.79	0.92	1.00	1.00	1.00	1.00	1.00	1.00	1.00	1.00
0.2	T (°C)	54.24	28.94	13.32	−0.40	−13.26	−24.52	−33.40	−39.79	−44.30	−48.22	−63.66
	v (m³/kg)	2.01	1.94	1.85	1.77	1.69	1.61	1.55	1.51	1.48	1.46	1.35
	h (kJ/kg)	701.69	379.42	243.00	140.21	52.63	−17.69	−68.49	−101.48	−122.13	−137.35	−64.35
	s (kJ/kg K)	2.21	1.80	1.51	1.26	1.01	0.79	0.60	0.47	0.37	0.29	0.15
	x_l (kg/kg)	0.00	0.08	0.15	0.23	0.31	0.38	0.44	0.49	0.53	0.57	1.00
	x_v (kg/kg)	0.00	0.76	0.90	0.97	1.00	1.00	1.00	1.00	1.00	1.00	1.00
0.3	T (°C)	54.21	31.72	17.75	5.47	−6.09	−16.30	−24.54	−30.77	−35.64	−40.52	−62.94
	v (m³/kg)	3.01	2.93	2.82	2.71	2.60	2.50	2.42	2.36	2.31	2.26	2.04
	h (kJ/kg)	938.77	566.20	424.23	328.33	250.52	187.98	141.00	107.64	83.08	60.28	68.22
	s (kJ/kg K)	2.94	2.52	2.24	2.01	1.80	1.62	1.46	1.34	1.25	1.15	0.83
	x_l (kg/kg)	0.00	0.07	0.13	0.19	0.26	0.33	0.38	0.42	0.46	0.50	1.00
	x_v (kg/kg)	0.00	0.72	0.86	0.94	1.00	1.00	1.00	1.00	1.00	1.00	1.00

TABLE A.1 *(Continued)*

F, where $T = T_{bp}$ $+ F(T_{dp} - T_{bp})$	Property	Ammonia concentration (mass fraction), x										
		0	0.1	0.2	0.3	0.4	0.5	0.6	0.7	0.8	0.9	1
0.4	T (°C)	54.17	34.51	22.19	11.33	1.08	−8.08	−15.67	−21.75	−26.99	−32.82	−62.23
	v (m³/kg)	4.01	3.92	3.81	3.69	3.56	3.44	3.34	3.26	3.19	3.12	2.72
	h (kJ/kg)	1175.87	765.23	611.86	516.29	444.71	388.91	346.28	314.06	287.55	259.54	212.36
	s (kJ/kg K)	3.66	3.26	2.97	2.76	2.58	2.42	2.29	2.19	2.10	1.99	1.54
	x_l (kg/kg)	0.00	0.06	0.11	0.16	0.22	0.28	0.32	0.36	0.40	0.44	1.00
	x_v (kg/kg)	0.00	0.67	0.83	0.91	0.96	1.00	1.00	1.00	1.00	1.00	1.00
0.5	T (°C)	54.14	37.29	26.62	17.20	8.25	0.14	−6.81	−12.73	−18.33	−25.11	−61.51
	v (m³/kg)	5.01	4.93	4.82	4.69	4.56	4.43	4.32	4.22	4.13	4.02	3.42
	h (kJ/kg)	1412.98	979.84	811.86	709.56	638.23	586.05	546.69	515.66	487.86	455.95	364.64
	s (kJ/kg K)	4.39	4.01	3.73	3.52	3.35	3.21	3.10	3.00	2.91	2.80	2.25
	x_l (kg/kg)	0.00	0.05	0.09	0.13	0.18	0.23	0.27	0.30	0.34	0.38	1.00
	x_v (kg/kg)	0.00	0.62	0.79	0.87	0.93	0.96	1.00	1.00	1.00	1.00	1.00
0.6	T (°C)	54.10	40.07	31.06	23.06	15.42	8.36	2.05	−3.71	−9.68	−17.41	−60.79
	v (m³/kg)	6.01	5.94	5.84	5.73	5.60	5.47	5.35	5.24	5.13	4.98	4.12
	h (kJ/kg)	1650.11	1213.63	1031.64	917.17	838.46	783.82	744.23	712.73	683.13	647.78	522.39
	s (kJ/kg K)	5.11	4.79	4.51	4.30	4.12	3.99	3.89	3.80	3.71	3.59	2.98
	x_l (kg/kg)	0.00	0.04	0.07	0.10	0.14	0.18	0.21	0.25	0.29	0.33	1.00
	x_v (kg/kg)	0.00	0.55	0.73	0.82	0.88	0.93	0.96	1.00	1.00	1.00	1.00
0.7	T (°C)	54.07	42.85	35.49	28.93	22.60	16.58	10.92	5.31	−1.02	−9.71	−60.08
	v (m³/kg)	7.02	6.96	6.88	6.78	6.67	6.56	6.44	6.32	6.18	5.98	4.82
	h (kJ/kg)	1887.25	1470.46	1279.82	1151.85	1059.36	993.73	946.52	909.36	874.79	834.73	683.85
	s (kJ/kg K)	5.84	5.58	5.32	5.11	4.94	4.79	4.68	4.58	4.49	4.36	3.70
	x_l (kg/kg)	0.00	0.03	0.05	0.08	0.10	0.13	0.16	0.20	0.23	0.29	1.00
	x_v (kg/kg)	0.00	0.46	0.65	0.76	0.83	0.87	0.91	0.94	0.97	1.00	1.00

TABLE A.1 (Continued)

F, where T = T_bp + F(T_dp − T_bp)	Property	Ammonia concentration (mass fraction), x										
		0	0.1	0.2	0.3	0.4	0.5	0.6	0.7	0.8	0.9	1
0.8	T (°C)	54.04	45.64	39.93	34.79	29.77	24.80	19.78	14.32	7.63	-2.01	-59.36
	v (m³/kg)	8.02	7.98	7.92	7.85	7.77	7.67	7.57	7.44	7.28	7.04	5.53
	h (kJ/kg)	2124.41	1754.45	1566.10	1429.42	1322.01	1238.16	1172.35	1117.97	1068.79	1018.14	847.90
	s (kJ/kg K)	6.56	6.38	6.17	5.98	5.80	5.65	5.51	5.39	5.26	5.12	4.43
	x_l (kg/kg)	0.00	0.02	0.04	0.05	0.07	0.09	0.12	0.15	0.18	0.24	1.00
	x_v (kg/kg)	0.00	0.36	0.55	0.67	0.75	0.81	0.85	0.89	0.93	0.97	1.00
0.9	T (°C)	54.00	48.42	44.36	40.66	36.94	33.02	28.65	23.34	16.28	5.70	-58.64
	v (m³/kg)	9.02	9.00	8.97	8.92	8.87	8.80	8.72	8.60	8.42	8.13	6.24
	h (kJ/kg)	2361.58	2070.04	1901.26	1768.14	1653.00	1550.09	1455.34	1365.57	1280.55	1202.60	1013.78
	s (kJ/kg K)	7.28	7.20	7.05	6.90	6.74	6.59	6.43	6.26	6.08	5.88	5.16
	x_l (kg/kg)	0.00	0.01	0.02	0.04	0.05	0.06	0.08	0.10	0.13	0.19	1.00
	x_v (kg/kg)	0.00	0.24	0.41	0.53	0.62	0.70	0.76	0.82	0.88	0.94	1.00
1.0	T (°C)	53.97	51.20	48.80	46.52	44.11	41.24	37.51	32.36	24.94	13.40	-57.93
	v (m³/kg)	10.02	10.00	9.98	9.97	9.96	9.93	9.87	9.76	9.57	9.25	6.96
	h (kJ/kg)	2598.77	2468.65	2340.55	2214.00	2088.46	1963.24	1837.47	1709.95	1578.84	1440.26	1179.70
	s (kJ/kg K)	8.01	8.02	7.96	7.87	7.75	7.61	7.45	7.26	7.03	6.74	5.88
	x_l (kg/kg)	0.00	0.00	0.00	0.00	0.00	0.00	0.00	0.00	0.00	0.00	0.00
	x_v (kg/kg)	0.00	0.10	0.20	0.30	0.40	0.50	0.60	0.70	0.80	0.90	1.00

$P = 0.15$ bar.

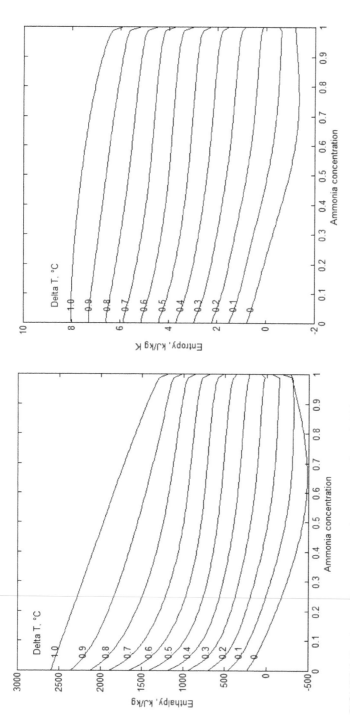

FIGURE A.1   Ammonia–water wet mixture properties at 0.15 bar.

TABLE A.2 Ammonia–Water Property System at 0.2 Bar Under Saturated and Wet Conditions.

F, where $T = T_{bp}$ $+ F(T_{dp} - T_{bp})$	Property	Ammonia concentration (mass fraction), x										
		0	0.1	0.2	0.3	0.4	0.5	0.6	0.7	0.8	0.9	1
0	T (°C)	60.36	29.44	10.11	−6.87	−22.66	−36.24	−46.58	−53.41	−57.30	−59.40	−60.93
	v (m³/kg)	0.00	0.00	0.00	0.00	0.00	0.00	0.00	0.00	0.00	0.00	0.00
	h (kJ/kg)	252.91	56.22	−91.94	−222.01	−332.62	−414.07	−457.14	−457.79	−419.71	−353.59	−272.86
	s (kJ/kg K)	0.84	0.47	0.15	−0.19	−0.54	−0.86	−1.09	−1.22	−1.24	−1.18	−1.12
	x_l (kg/kg)	0.00	0.10	0.20	0.30	0.40	0.50	0.60	0.70	0.80	0.90	1.00
	x_v (kg/kg)	0.00	0.00	0.00	0.00	0.00	0.00	0.00	0.00	0.00	0.00	0.00
0.1	T (°C)	60.34	32.23	14.58	−0.95	−15.42	−27.95	−37.65	−44.30	−48.53	−51.56	−60.28
	v (m³/kg)	0.77	0.74	0.70	0.66	0.63	0.60	0.57	0.56	0.55	0.54	0.52
	h (kJ/kg)	488.55	228.07	87.23	−28.66	−128.01	−205.44	−256.98	−284.43	−295.18	−296.72	−160.47
	s (kJ/kg K)	1.55	1.16	0.86	0.56	0.27	0.00	−0.21	−0.35	−0.44	−0.49	−0.45
	x_l (kg/kg)	0.00	0.09	0.18	0.27	0.35	0.44	0.51	0.57	0.62	0.67	1.00
	x_v (kg/kg)	0.00	0.78	0.92	0.98	1.00	1.00	1.00	1.00	1.00	1.00	1.00
0.2	T (°C)	60.33	35.03	19.05	4.97	−8.18	−19.66	−28.71	−35.20	−39.77	−43.71	−59.64
	v (m³/kg)	1.53	1.48	1.42	1.35	1.29	1.23	1.19	1.16	1.13	1.11	1.03
	h (kJ/kg)	724.21	406.86	266.13	161.10	72.72	2.21	−48.54	−81.38	−101.80	−116.71	−39.07
	s (kJ/kg K)	2.25	1.85	1.56	1.31	1.06	0.84	0.66	0.52	0.43	0.35	0.21
	x_l (kg/kg)	0.00	0.08	0.15	0.23	0.31	0.38	0.44	0.49	0.53	0.57	1.00
	x_v (kg/kg)	0.00	0.75	0.90	0.96	1.00	1.00	1.00	1.00	1.00	1.00	1.00
0.3	T (°C)	60.31	37.82	23.52	10.89	−0.95	−11.37	−19.77	−26.09	−31.00	−35.87	−58.99
	v (m³/kg)	2.30	2.23	2.16	2.07	1.99	1.91	1.85	1.80	1.76	1.73	1.55
	h (kJ/kg)	959.87	595.23	447.75	348.50	269.27	206.20	159.07	125.79	101.54	79.24	92.91
	s (kJ/kg K)	2.96	2.56	2.27	2.04	1.83	1.65	1.49	1.38	1.28	1.19	0.88
	x_l (kg/kg)	0.00	0.07	0.13	0.20	0.27	0.33	0.38	0.42	0.46	0.50	1.00
	x_v (kg/kg)	0.00	0.71	0.86	0.94	1.00	1.00	1.00	1.00	1.00	1.00	1.00

TABLE A.2 *(Continued)*

F, where T = T$_{bp}$ + F(T$_{dp}$ − T$_{bp}$)	Property	Ammonia concentration (mass fraction), x										
		0	0.1	0.2	0.3	0.4	0.5	0.6	0.7	0.8	0.9	1
0.4	T (°C)	60.29	40.61	27.99	16.82	6.29	−3.08	−10.83	−16.99	−22.24	−28.02	−58.34
	v (m³/kg)	3.06	3.00	2.91	2.81	2.72	2.63	2.55	2.49	2.44	2.38	2.08
	h (kJ/kg)	1195.54	796.03	636.55	536.52	462.63	405.81	362.79	330.57	304.40	277.02	234.66
	s (kJ/kg K)	3.67	3.28	2.99	2.77	2.59	2.43	2.30	2.20	2.11	2.01	1.56
	x$_l$ (kg/kg)	0.00	0.06	0.11	0.16	0.22	0.28	0.32	0.36	0.40	0.44	1.00
	x$_v$ (kg/kg)	0.00	0.66	0.83	0.91	0.96	1.00	1.00	1.00	1.00	1.00	1.00
0.5	T (°C)	60.28	43.40	32.46	22.74	13.53	5.21	−1.89	−7.88	−13.47	−20.18	−57.69
	v (m³/kg)	3.83	3.76	3.68	3.58	3.48	3.38	3.30	3.22	3.16	3.07	2.61
	h (kJ/kg)	1431.22	1012.36	838.38	730.74	656.13	602.23	562.09	530.88	503.41	472.18	383.94
	s (kJ/kg K)	4.37	4.02	3.73	3.51	3.34	3.20	3.09	2.99	2.91	2.80	2.25
	x$_l$ (kg/kg)	0.00	0.05	0.09	0.13	0.18	0.23	0.27	0.31	0.34	0.38	1.00
	x$_v$ (kg/kg)	0.00	0.60	0.78	0.87	0.93	0.96	0.98	1.00	1.00	1.00	1.00
0.6	T (°C)	60.26	46.19	36.93	28.66	20.77	13.50	7.05	1.22	−4.70	−12.33	−57.04
	v (m³/kg)	4.59	4.54	4.46	4.37	4.27	4.18	4.09	4.00	3.92	3.80	3.14
	h (kJ/kg)	1666.90	1247.49	1060.30	940.15	857.33	800.16	759.24	727.23	697.80	663.01	538.60
	s (kJ/kg K)	5.08	4.78	4.50	4.28	4.10	3.96	3.86	3.77	3.68	3.57	2.95
	x$_l$ (kg/kg)	0.00	0.04	0.07	0.10	0.14	0.18	0.22	0.25	0.29	0.33	1.00
	x$_v$ (kg/kg)	0.00	0.53	0.72	0.82	0.88	0.93	0.96	0.98	0.99	1.00	1.00
0.7	T (°C)	60.24	48.98	41.40	34.58	28.01	21.79	15.99	10.33	4.06	−4.49	−56.39
	v (m³/kg)	5.36	5.31	5.25	5.18	5.09	5.00	4.91	4.82	4.72	4.57	3.67
	h (kJ/kg)	1902.60	1504.88	1310.39	1177.03	1080.03	1011.25	962.14	924.04	889.27	849.36	697.08
	s (kJ/kg K)	5.79	5.55	5.30	5.08	4.90	4.75	4.64	4.54	4.44	4.32	3.65
	x$_l$ (kg/kg)	0.00	0.03	0.05	0.08	0.11	0.14	0.17	0.20	0.24	0.29	1.00
	x$_v$ (kg/kg)	0.00	0.44	0.64	0.75	0.83	0.88	0.92	0.95	0.97	1.00	1.00

TABLE A.2 (Continued)

F, where $T = T_{bp} + F(T_{dp} - T_{bp})$	Property	Ammonia concentration (mass fraction), x										
		0	0.1	0.2	0.3	0.4	0.5	0.6	0.7	0.8	0.9	1
0.8	T (°C)	60.23	51.77	45.87	40.50	35.24	30.08	24.93	19.44	12.83	3.36	−55.74
	v (m³/kg)	6.12	6.09	6.04	5.99	5.92	5.85	5.77	5.68	5.56	5.38	4.21
	h (kJ/kg)	2138.29	1788.17	1597.50	1456.17	1344.39	1257.24	1189.36	1133.94	1084.24	1032.90	858.29
	s (kJ/kg K)	6.50	6.34	6.13	5.93	5.75	5.59	5.45	5.33	5.20	5.06	4.36
	x_l (kg/kg)	0.00	0.02	0.04	0.06	0.08	0.10	0.12	0.15	0.19	0.24	1.00
	x_v (kg/kg)	0.00	0.34	0.54	0.66	0.74	0.81	0.85	0.89	0.93	0.97	1.00
0.9	T (°C)	60.21	54.57	50.34	46.42	42.48	38.37	33.87	28.54	21.59	11.20	−55.10
	v (m³/kg)	6.89	6.87	6.84	6.81	6.76	6.71	6.64	6.55	6.43	6.22	4.75
	h (kJ/kg)	2374.00	2101.19	1931.16	1794.13	1674.98	1569.13	1473.02	1383.27	1298.44	1218.79	1021.49
	s (kJ/kg K)	7.20	7.13	6.98	6.83	6.67	6.51	6.35	6.18	6.00	5.80	5.07
	x_l (kg/kg)	0.00	0.01	0.03	0.04	0.05	0.06	0.08	0.10	0.14	0.19	1.00
	x_v (kg/kg)	0.00	0.23	0.40	0.52	0.62	0.70	0.76	0.82	0.88	0.94	1.00
1.0	T (°C)	60.19	57.36	54.81	52.34	49.72	46.66	42.81	37.65	30.36	19.05	−54.45
	v (m³/kg)	7.65	7.63	7.62	7.61	7.60	7.57	7.52	7.44	7.31	7.07	5.29
	h (kJ/kg)	2609.72	2479.67	2351.49	2224.75	2098.97	1973.54	1847.68	1720.28	1589.57	1451.54	1185.78
	s (kJ/kg K)	7.91	7.92	7.86	7.77	7.65	7.51	7.34	7.15	6.93	6.64	5.77
	x_l (kg/kg)	0.00	0.00	0.00	0.00	0.00	0.00	0.00	0.00	0.00	0.00	0.00
	x_v (kg/kg)	0.00	0.10	0.20	0.30	0.40	0.50	0.60	0.70	0.80	0.90	1.00

$P = 0.2$ bar.

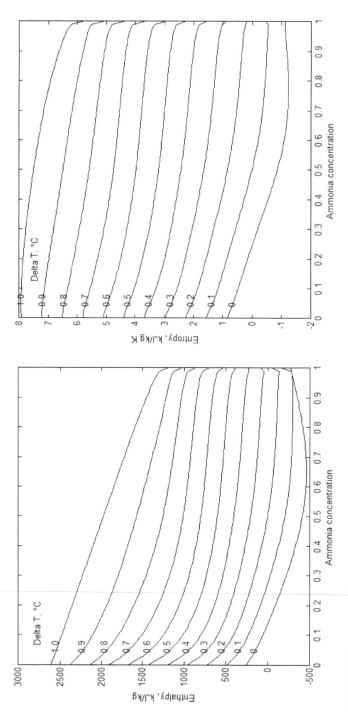

FIGURE A.2 Ammonia–water wet mixture properties at 0.2 bar.

TABLE A.3 Ammonia–Water Property System at 0.4 Bar Under Saturated and Wet Conditions.

F, where $T = T_{bp}$ $+ F(T_{dp} - T_{bp})$	Property	Ammonia concentration (mass fraction), x										
		0	0.1	0.2	0.3	0.4	0.5	0.6	0.7	0.8	0.9	1
0	T (°C)	75.96	45.11	24.80	6.86	−9.69	−23.85	−34.64	−41.82	−46.01	−48.37	−50.11
	v (m³/kg)	0.00	0.00	0.00	0.00	0.00	0.00	0.00	0.00	0.00	0.00	0.00
	h (kJ/kg)	318.27	121.85	−29.15	−161.16	−272.20	−353.12	−395.75	−397.09	−361.43	−299.54	−224.91
	s (kJ/kg K)	1.03	0.69	0.36	0.03	−0.31	−0.61	−0.83	−0.95	−0.97	−0.93	−0.90
	x_l (kg/kg)	0.00	0.10	0.20	0.30	0.40	0.50	0.60	0.70	0.80	0.90	1.00
	x_v (kg/kg)	0.00	0.00	0.00	0.00	0.00	0.00	0.00	0.00	0.00	0.00	0.00
0.1	T (°C)	75.97	47.91	29.34	12.90	−2.31	−15.41	−25.55	−32.55	−37.04	−40.28	−49.63
	v (m³/kg)	0.40	0.39	0.37	0.35	0.33	0.31	0.30	0.29	0.29	0.28	0.27
	h (kJ/kg)	550.19	295.95	148.31	28.67	−72.02	−149.58	−200.82	−227.94	−238.35	−239.34	−102.91
	s (kJ/kg K)	1.70	1.33	1.03	0.74	0.45	0.19	−0.01	−0.15	−0.23	−0.28	−0.26
	x_l (kg/kg)	0.00	0.09	0.18	0.27	0.35	0.44	0.51	0.58	0.63	0.67	1.00
	x_v (kg/kg)	0.00	0.75	0.92	0.98	1.00	1.00	1.00	1.00	1.00	1.00	1.00
0.2	T (°C)	75.99	50.71	33.88	18.94	5.07	−6.97	−16.45	−23.27	−28.07	−32.19	−49.14
	v (m³/kg)	0.80	0.77	0.74	0.71	0.68	0.65	0.62	0.61	0.59	0.58	0.54
	h (kJ/kg)	782.11	477.83	327.03	215.89	124.92	53.48	2.49	−30.32	−50.52	−64.98	23.96
	s (kJ/kg K)	2.36	1.99	1.69	1.43	1.19	0.97	0.79	0.67	0.58	0.50	0.37
	x_l (kg/kg)	0.00	0.08	0.15	0.23	0.31	0.38	0.44	0.49	0.53	0.57	1.00
	x_v (kg/kg)	0.00	0.71	0.89	0.96	0.99	1.00	1.00	1.00	1.00	1.00	1.00
0.3	T (°C)	76.01	53.51	38.42	24.98	12.45	1.47	−7.36	−13.99	−19.10	−24.10	−48.66
	v (m³/kg)	1.20	1.17	1.13	1.08	1.04	1.00	0.97	0.94	0.92	0.90	0.81
	h (kJ/kg)	1014.81	669.74	510.15	402.20	318.65	253.65	205.65	172.06	147.96	126.18	156.16
	s (kJ/kg K)	3.02	2.66	2.36	2.12	1.91	1.72	1.57	1.46	1.37	1.28	1.00

Flexible Kalina Cycle Systems

TABLE A.3 (Continued)

F, where $T = T_{bp}$ + $F(T_{dp} - T_{bp})$	Property	Ammonia concentration (mass fraction), x										
		0	0.1	0.2	0.3	0.4	0.5	0.6	0.7	0.8	0.9	1
0.4	x_l (kg/kg)	0.00	0.07	0.13	0.20	0.27	0.33	0.38	0.43	0.46	0.50	1.00
	x_v (kg/kg)	0.00	0.66	0.85	0.94	0.98	0.99	1.00	1.00	1.00	1.00	1.00
	T (°C)	76.03	56.31	42.96	31.02	19.83	9.91	1.74	−4.72	−10.12	−16.00	−48.17
	v (m³/kg)	1.60	1.56	1.52	1.47	1.42	1.37	1.33	1.30	1.27	1.25	1.08
	h (kJ/kg)	1247.29	874.09	701.99	590.97	510.65	450.52	405.84	372.91	346.80	319.97	293.80
	s (kJ/kg K)	3.69	3.34	3.05	2.82	2.62	2.46	2.33	2.23	2.14	2.04	1.63
	x_l (kg/kg)	0.00	0.06	0.11	0.17	0.23	0.28	0.33	0.37	0.40	0.44	1.00
	x_v (kg/kg)	0.00	0.61	0.81	0.91	0.96	0.98	0.99	1.00	1.00	1.00	1.00
0.5	T (°C)	76.04	59.12	47.51	37.06	27.21	18.35	10.83	4.56	−1.15	−7.91	−47.68
	v (m³/kg)	2.00	1.96	1.92	1.87	1.82	1.77	1.72	1.68	1.65	1.61	1.36
	h (kJ/kg)	1479.92	1093.35	907.87	787.94	704.79	645.83	602.92	570.40	542.72	511.89	436.51
	s (kJ/kg K)	4.35	4.04	3.75	3.52	3.33	3.18	3.07	2.97	2.89	2.79	2.27
	x_l (kg/kg)	0.00	0.05	0.09	0.14	0.19	0.24	0.28	0.31	0.35	0.39	1.00
	x_v (kg/kg)	0.00	0.55	0.76	0.86	0.93	0.96	0.98	0.99	0.99	1.00	1.00
0.6	T (°C)	76.06	61.92	52.05	43.11	34.59	26.79	19.93	13.84	7.82	0.18	−47.20
	v (m³/kg)	2.40	2.37	2.33	2.28	2.23	2.18	2.13	2.09	2.04	1.99	1.63
	h (kJ/kg)	1712.70	1330.05	1133.92	1001.54	908.83	844.89	799.78	765.53	735.27	700.36	583.60
	s (kJ/kg K)	5.01	4.76	4.48	4.25	4.06	3.91	3.80	3.70	3.62	3.51	2.91
	x_l (kg/kg)	0.00	0.04	0.07	0.11	0.15	0.19	0.23	0.26	0.29	0.34	1.00
	x_v (kg/kg)	0.00	0.48	0.69	0.81	0.88	0.93	0.96	0.97	0.98	0.99	1.00
0.7	T (°C)	76.08	64.72	56.59	49.15	41.98	35.24	29.02	23.11	16.80	8.28	−46.71
	v (m³/kg)	2.79	2.77	2.74	2.70	2.65	2.61	2.56	2.51	2.46	2.39	1.91
	h (kJ/kg)	1945.63	1586.78	1386.79	1242.61	1135.60	1059.09	1004.65	963.27	926.74	885.55	734.28

TABLE A.3 (Continued)

F, where $T = T_{bp} + F(T_{dp} - T_{bp})$	Property	Ammonia concentration (mass fraction), x										
		0	0.1	0.2	0.3	0.4	0.5	0.6	0.7	0.8	0.9	1
0.8	s (kJ/kg K)	5.67	5.48	5.24	5.01	4.82	4.66	4.54	4.43	4.34	4.22	3.55
	x_l (kg/kg)	0.00	0.03	0.06	0.08	0.11	0.15	0.18	0.21	0.24	0.29	1.00
	x_v (kg/kg)	0.00	0.40	0.60	0.74	0.82	0.88	0.92	0.95	0.97	0.98	1.00
	T (°C)	76.10	67.52	61.13	55.19	49.36	43.68	38.12	32.39	25.77	16.37	−46.22
	v (m³/kg)	3.20	3.17	3.15	3.12	3.09	3.05	3.00	2.96	2.90	2.81	2.18
	h (kJ/kg)	2173.45	1866.16	1673.37	1523.44	1402.59	1307.92	1234.73	1176.00	1124.06	1069.74	887.85
	s (kJ/kg K)	6.35	6.22	6.02	5.82	5.63	5.46	5.32	5.19	5.06	4.92	4.20
	x_l (kg/kg)	0.00	0.02	0.04	0.06	0.08	0.11	0.13	0.16	0.20	0.25	1.00
	x_v (kg/kg)	0.00	0.31	0.50	0.63	0.73	0.81	0.85	0.90	0.94	0.97	1.00
0.9	T (°C)	76.11	70.32	65.67	61.23	56.74	52.12	47.21	41.67	34.74	24.46	−45.74
	v (m³/kg)	3.59	3.58	3.56	3.54	3.52	3.49	3.46	3.41	3.35	3.24	2.46
	h (kJ/kg)	2405.31	2170.83	2000.68	1856.83	1729.75	1617.44	1517.72	1427.22	1342.37	1258.77	1043.71
	s (kJ/kg K)	7.01	6.96	6.82	6.66	6.50	6.33	6.17	6.00	5.83	5.63	4.85
	x_l (kg/kg)	0.00	0.01	0.03	0.04	0.06	0.07	0.09	0.12	0.15	0.20	1.00
	x_v (kg/kg)	0.00	0.21	0.37	0.50	0.60	0.69	0.76	0.83	0.88	0.94	1.00
1.0	T (°C)	76.13	73.12	70.22	67.27	64.12	60.56	56.31	50.95	43.72	32.56	−45.25
	v (m³/kg)	3.99	3.99	3.98	3.97	3.96	3.94	3.91	3.87	3.81	3.69	2.74
	h (kJ/kg)	2637.16	2507.36	2379.04	2251.89	2125.57	1999.60	1873.38	1745.98	1615.74	1478.26	1201.32
	s (kJ/kg K)	7.67	7.69	7.62	7.52	7.40	7.26	7.09	6.90	6.68	6.39	5.51
	x_l (kg/kg)	0.00	0.00	0.00	0.00	0.00	0.00	0.00	0.00	0.00	0.00	0.00
	x_v (kg/kg)	0.00	0.10	0.20	0.30	0.40	0.50	0.60	0.70	0.80	0.90	1.00

$P = 0.4$ bar

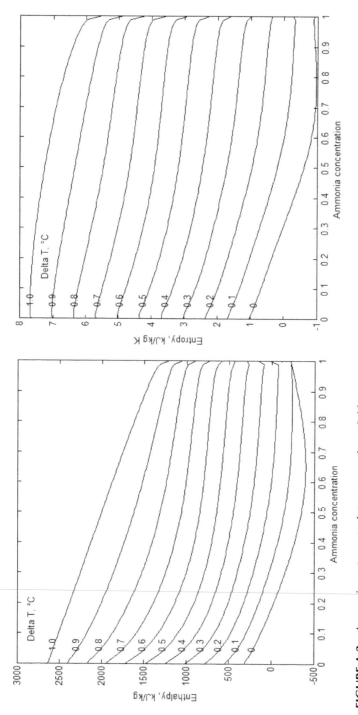

FIGURE A.3 Ammonia–water wet mixture properties at 0.4 bar.

TABLE A.4 Ammonia–Water Property System at 1 Bar Under Saturated and Wet Conditions.

F, where $T = T_{bp}$ $+ F(T_{dp} - T_{bp})$	Property	Ammonia concentration (mass fraction), x										
		0	0.1	0.2	0.3	0.4	0.5	0.6	0.7	0.8	0.9	1
0	T (°C)	99.38	68.64	47.03	27.79	10.20	-4.78	-16.24	-23.99	-28.69	-31.51	-33.67
	v (m³/kg)	0.00	0.00	0.00	0.00	0.00	0.00	0.00	0.00	0.00	0.00	0.00
	h (kJ/kg)	416.77	220.47	65.19	-70.02	-182.38	-263.35	-306.00	-308.58	-276.07	-219.35	-151.96
	s (kJ/kg K)	1.30	0.98	0.67	0.34	0.02	-0.26	-0.47	-0.58	-0.61	-0.59	-0.58
	x_l (kg/kg)	0.00	0.10	0.20	0.30	0.40	0.50	0.60	0.70	0.80	0.90	1.00
	x_v (kg/kg)	0.00	0.00	0.00	0.00	0.00	0.00	0.00	0.00	0.00	0.00	0.00
0.1	T (°C)	99.43	71.43	51.65	33.97	17.75	3.83	-6.98	-14.56	-19.55	-23.28	-33.38
	v (m³/kg)	0.17	0.16	0.16	0.15	0.14	0.13	0.13	0.12	0.12	0.12	0.12
	h (kJ/kg)	642.78	397.83	241.46	115.70	12.07	-66.60	-118.18	-145.37	-155.45	-155.10	-23.58
	s (kJ/kg K)	1.91	1.58	1.28	0.99	0.71	0.46	0.27	0.14	0.06	0.01	0.00
	x_l (kg/kg)	0.00	0.09	0.18	0.27	0.36	0.44	0.52	0.58	0.64	0.69	1.00
	x_v (kg/kg)	0.00	0.69	0.89	0.97	0.99	1.00	1.00	1.00	1.00	1.00	1.00
0.2	T (°C)	99.48	74.23	56.26	40.16	25.30	12.45	2.28	-5.12	-10.42	-15.04	-33.08
	v (m³/kg)	0.34	0.33	0.32	0.30	0.29	0.28	0.27	0.26	0.25	0.25	0.23
	h (kJ/kg)	869.62	583.35	420.91	300.47	204.46	130.53	78.22	44.62	23.97	9.56	106.83
	s (kJ/kg K)	2.51	2.19	1.89	1.62	1.38	1.16	0.99	0.87	0.78	0.70	0.57
	x_l (kg/kg)	0.00	0.08	0.16	0.24	0.31	0.39	0.45	0.50	0.54	0.59	1.00
	x_v (kg/kg)	0.00	0.65	0.86	0.95	0.98	0.99	1.00	1.00	1.00	1.00	1.00
0.3	T (°C)	99.53	77.02	60.88	46.34	32.85	21.06	11.54	4.32	-1.29	-6.81	-32.78
	v (m³/kg)	0.51	0.50	0.48	0.46	0.44	0.43	0.41	0.40	0.39	0.38	0.34
	h (kJ/kg)	1096.54	778.72	606.52	486.42	395.33	326.09	275.62	240.42	215.25	192.81	239.33
	s (kJ/kg K)	3.12	2.81	2.51	2.26	2.04	1.85	1.70	1.58	1.49	1.41	1.14
	x_l (kg/kg)	0.00	0.07	0.13	0.20	0.27	0.34	0.39	0.44	0.47	0.51	1.00
	x_v (kg/kg)	0.00	0.60	0.82	0.92	0.97	0.99	0.99	1.00	1.00	1.00	1.00

TABLE A.4 (Continued)

F, where $T = T_{bp}$ $+ F(T_{dp} - T_{bp})$	Property	Ammonia concentration (mass fraction), x										
		0	0.1	0.2	0.3	0.4	0.5	0.6	0.7	0.8	0.9	1
0.4	T (°C)	99.57	79.82	65.50	52.52	40.40	29.67	20.80	13.75	7.84	1.42	−32.49
	v (m³/kg)	0.68	0.66	0.65	0.63	0.60	0.58	0.57	0.55	0.54	0.53	0.46
	h (kJ/kg)	1323.78	985.69	802.04	677.12	586.71	520.26	471.66	436.18	408.38	380.26	373.96
	s (kJ/kg K)	3.72	3.44	3.14	2.90	2.69	2.52	2.39	2.28	2.19	2.10	1.71
	x_l (kg/kg)	0.00	0.06	0.11	0.17	0.23	0.29	0.34	0.38	0.41	0.46	1.00
	x_v (kg/kg)	0.00	0.54	0.77	0.89	0.95	0.97	0.99	0.99	1.00	1.00	1.00
0.5	T (°C)	99.62	82.62	70.12	58.70	47.95	38.29	30.06	23.19	16.97	9.66	−32.19
	v (m³/kg)	0.85	0.83	0.82	0.80	0.77	0.75	0.73	0.71	0.70	0.68	0.57
	h (kJ/kg)	1551.35	1206.00	1011.81	877.88	782.83	715.35	666.77	630.50	600.27	567.31	510.71
	s (kJ/kg K)	4.33	4.08	3.80	3.55	3.35	3.19	3.06	2.96	2.87	2.77	2.29
	x_l (kg/kg)	0.00	0.05	0.09	0.14	0.20	0.25	0.29	0.32	0.36	0.40	1.00
	x_v (kg/kg)	0.00	0.48	0.71	0.84	0.91	0.95	0.97	0.98	0.99	0.99	1.00
0.6	T (°C)	99.67	85.41	74.74	64.88	55.50	46.90	39.32	32.62	26.10	17.89	−31.90
	v (m³/kg)	1.02	1.00	0.99	0.97	0.95	0.92	0.90	0.88	0.87	0.84	0.69
	h (kJ/kg)	1779.24	1441.38	1240.57	1095.88	990.92	916.92	864.36	824.85	790.84	752.52	649.55
	s (kJ/kg K)	4.93	4.73	4.47	4.23	4.03	3.87	3.74	3.64	3.54	3.43	2.86
	x_l (kg/kg)	0.00	0.04	0.08	0.12	0.16	0.20	0.24	0.27	0.31	0.35	1.00
	x_v (kg/kg)	0.00	0.42	0.64	0.78	0.86	0.92	0.95	0.97	0.98	0.99	1.00
0.7	T (°C)	99.72	88.21	79.35	71.07	63.05	55.52	48.58	42.06	35.23	26.12	−31.60
	v (m³/kg)	1.18	1.17	1.16	1.14	1.12	1.10	1.08	1.06	1.04	1.01	0.80
	h (kJ/kg)	2007.46	1693.49	1493.19	1339.69	1221.76	1135.10	1072.26	1024.35	982.69	936.31	790.41
	s (kJ/kg K)	5.53	5.39	5.16	4.94	4.74	4.57	4.43	4.32	4.21	4.08	3.43
	x_l (kg/kg)	0.00	0.03	0.06	0.09	0.12	0.16	0.19	0.23	0.26	0.31	1.00
	x_v (kg/kg)	0.00	0.35	0.55	0.70	0.80	0.86	0.91	0.94	0.96	0.98	1.00

TABLE A.4 *(Continued)*

F, where T = T_bp + F(T_dp − T_bp)	Property	Ammonia concentration (mass fraction), x										
		0	0.1	0.2	0.3	0.4	0.5	0.6	0.7	0.8	0.9	1
0.8	T (°C)	99.76	91.00	83.97	77.25	70.60	64.13	57.84	51.49	44.36	34.35	−31.31
	v (m³/kg)	1.35	1.34	1.33	1.32	1.31	1.29	1.27	1.25	1.22	1.19	0.92
	h (kJ/kg)	2236.02	1963.98	1774.45	1618.46	1488.97	1385.56	1304.72	1239.71	1182.29	1121.08	933.21
	s (kJ/kg K)	6.14	6.06	5.88	5.68	5.49	5.31	5.16	5.02	4.89	4.73	4.01
	x_l (kg/kg)	0.00	0.02	0.04	0.07	0.09	0.12	0.15	0.18	0.21	0.27	1.00
	x_v (kg/kg)	0.00	0.28	0.45	0.59	0.70	0.79	0.85	0.89	0.93	0.97	1.00
0.9	T (°C)	99.81	93.80	88.59	83.43	78.16	72.74	67.10	60.93	53.49	42.59	−31.01
	v (m³/kg)	1.52	1.52	1.51	1.50	1.49	1.47	1.46	1.44	1.41	1.37	1.04
	h (kJ/kg)	2450.14	2254.41	2088.90	1941.12	1807.18	1687.92	1582.89	1489.20	1401.94	1312.11	1077.80
	s (kJ/kg K)	6.76	6.73	6.60	6.45	6.28	6.11	5.94	5.77	5.60	5.39	4.59
	x_l (kg/kg)	0.00	0.01	0.03	0.05	0.06	0.08	0.11	0.13	0.17	0.22	1.00
	x_v (kg/kg)	0.00	0.20	0.34	0.47	0.58	0.67	0.75	0.82	0.88	0.94	1.00
1.0	T (°C)	99.86	96.60	93.21	89.61	85.71	81.36	76.36	70.36	62.63	50.82	−30.71
	v (m³/kg)	1.69	1.69	1.69	1.68	1.67	1.66	1.65	1.63	1.61	1.56	1.15
	h (kJ/kg)	2675.96	2546.71	2418.43	2290.94	2163.99	2037.27	1910.31	1782.32	1651.67	1513.28	1224.04
	s (kJ/kg K)	7.36	7.37	7.31	7.21	7.09	6.94	6.77	6.57	6.35	6.06	5.16
	x_l (kg/kg)	0.00	0.00	0.00	0.00	0.00	0.00	0.00	0.00	0.00	0.00	0.00
	x_v (kg/kg)	0.00	0.10	0.20	0.30	0.40	0.50	0.60	0.70	0.80	0.90	1.00

$P = 1$ bar.

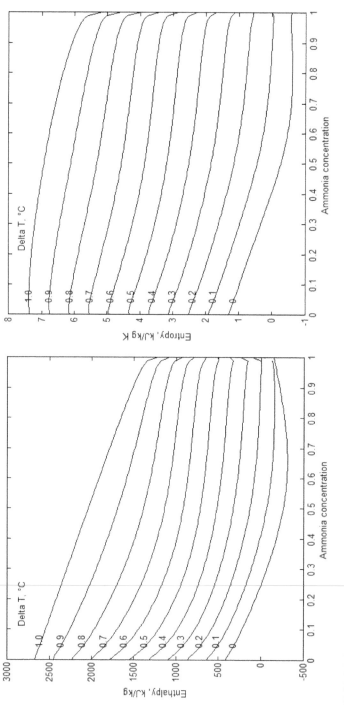

FIGURE A.4 Ammonia–water wet mixture properties at 1 bar.

TABLE A.5 Ammonia–Water Property System at 2 Bar Under Saturated and Wet Conditions.

F, where $T = T_{bp}$ $+ F(T_{dp} - T_{bp})$	Property	Ammonia concentration (mass fraction), x										
		0	0.1	0.2	0.3	0.4	0.5	0.6	0.7	0.8	0.9	1
0	T (°C)	119.85	89.17	66.56	46.30	27.86	12.21	0.18	−8.09	−13.26	−16.53	−19.12
	v (m³/kg)	0.00	0.00	0.00	0.00	0.00	0.00	0.00	0.00	0.00	0.00	0.00
	h (kJ/kg)	503.43	307.14	148.23	9.99	−104.08	−185.79	−229.03	−232.83	−202.67	−149.51	−86.90
	s (kJ/kg K)	1.53	1.23	0.92	0.60	0.29	0.02	−0.18	−0.29	−0.32	−0.31	−0.32
	x_l (kg/kg)	0.00	0.10	0.20	0.30	0.40	0.50	0.60	0.70	0.80	0.90	1.00
	x_v (kg/kg)	0.00	0.00	0.00	0.00	0.00	0.00	0.00	0.00	0.00	0.00	0.00
0.1	T (°C)	119.91	91.94	71.22	52.57	35.53	20.93	9.53	1.42	−4.09	−8.32	−18.94
	v (m³/kg)	0.09	0.09	0.08	0.08	0.07	0.07	0.07	0.07	0.06	0.06	0.06
	h (kJ/kg)	724.13	486.38	324.04	192.93	86.08	5.66	−46.88	−74.56	−84.50	−82.62	41.75
	s (kJ/kg K)	2.09	1.79	1.49	1.19	0.92	0.68	0.49	0.37	0.29	0.24	0.21
	x_l (kg/kg)	0.00	0.09	0.18	0.27	0.36	0.44	0.52	0.59	0.65	0.70	1.00
	x_v (kg/kg)	0.00	0.64	0.86	0.95	0.98	0.99	1.00	1.00	1.00	1.00	1.00
0.2	T (°C)	119.96	94.72	75.87	58.85	43.20	29.66	18.89	10.93	5.08	−0.12	−18.77
	v (m³/kg)	0.18	0.17	0.17	0.16	0.15	0.15	0.14	0.14	0.13	0.13	0.12
	h (kJ/kg)	945.13	673.51	504.16	376.31	275.32	198.35	144.13	109.22	87.70	73.07	171.33
	s (kJ/kg K)	2.65	2.36	2.06	1.79	1.54	1.33	1.16	1.03	0.94	0.87	0.74
	x_l (kg/kg)	0.00	0.08	0.16	0.24	0.32	0.39	0.45	0.51	0.55	0.60	1.00
	x_v (kg/kg)	0.00	0.59	0.83	0.93	0.97	0.99	1.00	1.00	1.00	1.00	1.00
0.3	T (°C)	120.01	97.49	80.53	65.12	50.86	38.39	28.25	20.44	14.25	8.08	−18.59
	v (m³/kg)	0.27	0.26	0.25	0.24	0.23	0.22	0.22	0.21	0.21	0.20	0.18
	h (kJ/kg)	1166.43	869.84	691.24	562.36	464.51	390.75	337.22	299.79	272.85	249.01	301.87
	s (kJ/Kg K)	3.21	2.93	2.64	2.38	2.15	1.96	1.81	1.69	1.60	1.51	1.26
	x_l (kg/kg)	0.00	0.07	0.14	0.21	0.28	0.34	0.40	0.44	0.48	0.53	1.00
	x_v (kg/kg)	0.00	0.55	0.78	0.90	0.96	0.98	0.99	0.99	1.00	1.00	1.00

TABLE A.5 (Continued)

F, where T = T_bp + F(T_dp − T_bp)	Property	Ammonia concentration (mass fraction), x										
		0	0.1	0.2	0.3	0.4	0.5	0.6	0.7	0.8	0.9	1
0.4	T (°C)	120.06	100.27	85.19	71.40	58.53	47.12	37.61	29.95	23.42	16.28	−18.42
	v (m³/kg)	0.35	0.35	0.34	0.33	0.32	0.31	0.30	0.29	0.28	0.28	0.24
	h (kJ/kg)	1388.03	1076.68	888.43	754.47	655.92	583.38	530.43	491.71	461.25	430.68	433.38
	s (kJ/kg K)	3.77	3.52	3.23	2.98	2.77	2.59	2.45	2.34	2.24	2.14	1.78
	x_l (kg/kg)	0.00	0.06	0.12	0.18	0.24	0.30	0.35	0.39	0.43	0.47	1.00
	x_v (kg/kg)	0.00	0.49	0.73	0.86	0.93	0.97	0.98	0.99	0.99	1.00	1.00
0.5	T (°C)	120.12	103.04	89.85	77.67	66.19	55.85	46.97	39.46	32.59	24.49	−18.24
	v (m³/kg)	0.44	0.44	0.43	0.42	0.41	0.39	0.38	0.37	0.37	0.36	0.30
	h (kJ/kg)	1609.93	1295.29	1099.25	957.32	853.64	778.80	724.52	683.83	649.94	613.37	565.86
	s (kJ/kg K)	4.32	4.11	3.84	3.60	3.39	3.22	3.08	2.97	2.87	2.76	2.30
	x_l (kg/kg)	0.00	0.05	0.10	0.15	0.20	0.25	0.30	0.34	0.37	0.42	1.00
	x_v (kg/kg)	0.00	0.44	0.67	0.81	0.89	0.94	0.97	0.98	0.99	0.99	1.00
0.6	T (°C)	120.17	105.81	94.51	83.94	73.86	64.58	56.33	48.97	41.76	32.69	−18.06
	v (m³/kg)	0.53	0.52	0.52	0.51	0.50	0.48	0.47	0.46	0.45	0.44	0.36
	h (kJ/kg)	1832.14	1526.92	1327.42	1176.89	1064.08	982.25	922.92	877.78	838.93	795.62	699.31
	s (kJ/kg K)	4.88	4.72	4.47	4.23	4.02	3.85	3.71	3.60	3.50	3.38	2.82
	x_l (kg/kg)	0.00	0.04	0.08	0.12	0.17	0.21	0.25	0.29	0.32	0.37	1.00
	x_v (kg/kg)	0.00	0.38	0.60	0.75	0.85	0.90	0.94	0.96	0.98	0.99	1.00
0.7	T (°C)	120.22	108.59	99.17	90.22	81.53	73.31	65.69	58.48	50.93	40.89	−17.89
	v (m³/kg)	0.62	0.61	0.61	0.60	0.59	0.58	0.57	0.55	0.54	0.53	0.42
	h (kJ/kg)	2054.65	1772.76	1576.66	1420.02	1296.15	1202.45	1132.60	1078.27	1030.66	977.77	833.72
	s (kJ/kg K)	5.44	5.33	5.11	4.89	4.69	4.51	4.36	4.24	4.12	3.98	3.34
	x_l (kg/kg)	0.00	0.03	0.06	0.10	0.13	0.17	0.20	0.24	0.28	0.33	1.00
	x_v (kg/kg)	0.00	0.32	0.52	0.66	0.77	0.85	0.90	0.93	0.96	0.98	1.00

TABLE A.5 *(Continued)*

F, where $T = T_{bp}$ $+ F(T_{dp} - T_{bp})$	Property	Ammonia concentration (mass fraction), x										
		0	**0.1**	**0.2**	**0.3**	**0.4**	**0.5**	**0.6**	**0.7**	**0.8**	**0.9**	**1**
0.8	T (°C)	120.28	111.36	103.83	96.49	89.19	82.03	75.05	67.99	60.10	49.10	−17.71
	v (m³/kg)	0.71	0.70	0.70	0.69	0.68	0.67	0.66	0.65	0.64	0.62	0.48
	h (kJ/kg)	2277.44	2033.93	1850.52	1693.83	1560.64	1451.95	1365.18	1294.16	1230.61	1161.88	969.10
	s (kJ/kg K)	6.00	5.94	5.77	5.58	5.39	5.21	5.05	4.90	4.76	4.59	3.87
	x_f (kg/kg)	0.00	0.02	0.05	0.07	0.10	0.13	0.16	0.19	0.23	0.29	1.00
	x_v (kg/kg)	0.00	0.26	0.42	0.56	0.68	0.77	0.84	0.88	0.92	0.96	1.00
0.9	T (°C)	120.33	114.14	108.49	102.77	96.86	90.76	84.40	77.50	69.27	57.30	−17.53
	v (m³/kg)	0.80	0.79	0.79	0.78	0.78	0.77	0.76	0.75	0.73	0.71	0.54
	h (kJ/kg)	2500.55	2311.50	2152.28	2005.04	1868.84	1745.99	1637.03	1539.59	1448.35	1352.12	1105.43
	s (kJ/kg K)	6.55	6.55	6.43	6.29	6.12	5.95	5.78	5.60	5.42	5.21	4.39
	x_f (kg/kg)	0.00	0.02	0.03	0.05	0.07	0.09	0.12	0.15	0.19	0.25	1.00
	x_v (kg/kg)	0.00	0.19	0.32	0.45	0.56	0.65	0.74	0.81	0.87	0.94	1.00
1.0	T (°C)	120.38	116.91	113.15	109.04	104.53	99.49	93.76	87.00	78.44	65.50	−17.36
	v (m³/kg)	0.88	0.88	0.88	0.88	0.87	0.87	0.86	0.85	0.83	0.81	0.60
	h (kJ/kg)	2706.89	2578.32	2450.35	2322.83	2195.58	2068.33	1940.69	1811.87	1680.15	1539.78	1242.67
	s (kJ/kg K)	7.13	7.14	7.08	6.98	6.85	6.70	6.53	6.33	6.10	5.81	4.91
	x_f (kg/kg)	0.00	0.00	0.00	0.00	0.00	0.00	0.00	0.00	0.00	0.00	0.00
	x_v (kg/kg)	0.00	0.10	0.20	0.30	0.40	0.50	0.60	0.70	0.80	0.90	1.00

$P = 2$ bar.

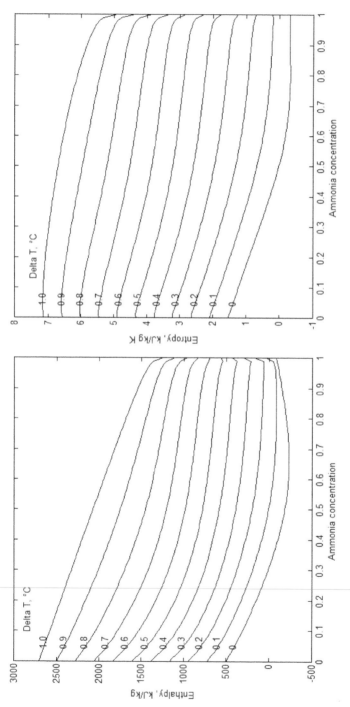

FIGURE A.5   Ammonia–water wet mixture properties at 2 bar.

TABLE A.6 Ammonia–Water Property System at 4 Bar Under Saturated and Wet Conditions.

F, where T = T$_{bp}$ + F(T$_{dp}$ − T$_{bp}$)	Property	Ammonia concentration (mass fraction), x										
		0	0.1	0.2	0.3	0.4	0.5	0.6	0.7	0.8	0.9	1
0	T (°C)	143.32	112.67	89.00	67.67	48.34	31.94	19.27	10.41	4.69	0.88	−2.23
	v (m³/kg)	0.00	0.00	0.00	0.00	0.00	0.00	0.00	0.00	0.00	0.00	0.00
	h (kJ/kg)	603.79	407.68	244.82	102.96	−13.56	−96.78	−141.19	−146.53	−118.68	−68.72	−10.24
	s (kJ/kg K)	1.78	1.50	1.19	0.88	0.58	0.32	0.13	0.02	−0.01	0.00	−0.03
	x$_l$ (kg/kg)	0.00	0.10	0.20	0.30	0.40	0.50	0.60	0.70	0.80	0.90	1.00
	x$_v$ (kg/kg)	0.00	0.00	0.00	0.00	0.00	0.00	0.00	0.00	0.00	0.00	0.00
0.1	T (°C)	143.35	115.39	93.68	74.02	56.10	40.76	28.70	19.95	13.82	8.93	−2.16
	v (m³/kg)	0.05	0.05	0.04	0.04	0.04	0.04	0.04	0.04	0.03	0.03	0.03
	h (kJ/kg)	817.27	587.40	419.90	283.10	172.14	89.07	34.83	6.17	−3.76	0.17	115.49
	s (kJ/kg K)	2.29	2.01	1.72	1.43	1.16	0.92	0.74	0.61	0.54	0.49	0.45
	x$_l$ (kg/kg)	0.00	0.09	0.18	0.27	0.36	0.44	0.52	0.59	0.66	0.72	1.00
	x$_v$ (kg/kg)	0.00	0.58	0.82	0.93	0.97	0.99	1.00	1.00	1.00	1.00	1.00
0.2	T (°C)	143.38	118.12	98.36	80.37	63.87	49.59	38.13	29.49	22.96	16.99	−2.08
	v (m³/kg)	0.09	0.09	0.09	0.08	0.08	0.08	0.07	0.07	0.07	0.07	0.06
	h (kJ/kg)	1031.13	774.36	600.00	464.97	358.21	277.18	220.13	183.20	160.32	145.44	241.53
	s (kJ/kg K)	2.80	2.54	2.25	1.97	1.73	1.51	1.34	1.22	1.13	1.05	0.92
	x$_l$ (kg/kg)	0.00	0.08	0.16	0.24	0.32	0.39	0.46	0.52	0.57	0.62	1.00
	x$_v$ (kg/kg)	0.00	0.54	0.78	0.90	0.96	0.98	0.99	1.00	1.00	1.00	1.00
0.3	T (°C)	143.41	120.84	103.03	86.71	71.63	58.41	47.56	39.03	32.10	25.04	−2.01
	v (m³/kg)	0.14	0.14	0.13	0.13	0.12	0.12	0.11	0.11	0.11	0.11	0.09
	h (kJ/kg)	1245.00	969.48	787.33	650.71	545.70	466.41	408.75	368.12	338.54	312.63	367.90
	s (kJ/kg K)	3.31	3.07	2.79	2.52	2.29	2.10	1.94	1.82	1.72	1.62	1.39

TABLE A.6 *(Continued)*

F, where $T = T_{bp}$ $+ F(T_{dp} - T_{bp})$	Property	0	0.1	0.2	0.3	0.4	0.5	0.6	0.7	0.8	0.9	1
						Ammonia concentration (mass fraction), x						
0.4	x_l (kg/kg)	0.00	0.07	0.14	0.21	0.28	0.35	0.40	0.45	0.50	0.55	1.00
	x_v (kg/kg)	0.00	0.50	0.74	0.87	0.94	0.97	0.98	0.99	1.00	1.00	1.00
	T (°C)	143.43	123.56	107.71	93.06	79.39	67.23	56.99	48.57	41.23	33.09	−1.93
	v (m³/kg)	0.19	0.18	0.18	0.17	0.17	0.16	0.16	0.15	0.15	0.14	0.12
	h (kJ/kg)	1458.99	1173.72	984.36	843.30	736.85	657.48	599.08	555.91	521.57	487.31	494.59
	s (kJ/kg K)	3.82	3.61	3.33	3.08	2.86	2.68	2.53	2.41	2.31	2.20	1.86
	x_l (kg/kg)	0.00	0.06	0.12	0.18	0.24	0.30	0.35	0.40	0.44	0.49	1.00
	x_v (kg/kg)	0.00	0.45	0.68	0.83	0.91	0.95	0.97	0.98	0.99	0.99	1.00
0.5	T (°C)	143.46	126.29	112.38	99.41	87.16	76.06	66.42	58.11	50.37	41.15	−1.86
	v (m³/kg)	0.23	0.23	0.22	0.22	0.21	0.21	0.20	0.20	0.19	0.19	0.16
	h (kJ/kg)	1673.12	1387.96	1193.81	1046.61	935.35	852.91	791.97	745.58	706.50	664.60	621.60
	s (kJ/kg K)	4.33	4.15	3.90	3.65	3.44	3.26	3.12	3.00	2.89	2.77	2.32
	x_l (kg/kg)	0.00	0.05	0.10	0.15	0.21	0.26	0.31	0.35	0.39	0.44	1.00
	x_v (kg/kg)	0.00	0.40	0.62	0.77	0.86	0.92	0.95	0.97	0.98	0.99	1.00
0.6	T (°C)	143.49	129.01	117.06	105.75	94.92	84.88	75.85	67.65	59.50	49.20	−1.78
	v (m³/kg)	0.28	0.27	0.27	0.26	0.26	0.25	0.25	0.24	0.24	0.23	0.19
	h (kJ/kg)	1887.39	1613.10	1418.49	1265.35	1146.52	1057.29	990.58	938.61	893.22	842.86	748.94
	s (kJ/kg K)	4.85	4.70	4.47	4.24	4.03	3.85	3.71	3.58	3.47	3.33	2.79
	x_l (kg/kg)	0.00	0.04	0.08	0.13	0.17	0.22	0.26	0.30	0.34	0.39	1.00
	x_v (kg/kg)	0.00	0.35	0.56	0.71	0.81	0.88	0.92	0.95	0.97	0.98	1.00
0.7	T (°C)	143.52	131.74	121.74	112.10	102.68	93.71	85.28	77.19	68.64	57.25	−1.71
	v (m³/kg)	0.32	0.32	0.32	0.31	0.31	0.30	0.30	0.29	0.28	0.27	0.22
	h (kJ/kg)	2101.78	1849.93	1661.18	1504.79	1377.35	1277.66	1200.69	1138.94	1083.73	1022.17	876.61

TABLE A.6 *(Continued)*

F, where T = T_bp + F(T_dp − T_bp)	Property	Ammonia concentration (mass fraction), x										
		0	0.1	0.2	0.3	0.4	0.5	0.6	0.7	0.8	0.9	1
	s (kJ/kg K)	5.36	5.26	5.06	4.85	4.65	4.47	4.31	4.18	4.05	3.89	3.26
	x_l (kg/kg)	0.00	0.03	0.07	0.10	0.14	0.18	0.22	0.25	0.30	0.35	1.00
	x_v (kg/kg)	0.00	0.30	0.48	0.63	0.74	0.82	0.88	0.92	0.95	0.97	1.00
0.8	T (°C)	143.55	134.46	126.41	118.45	110.45	102.53	94.71	86.73	77.78	65.31	−1.63
	v (m³/kg)	0.37	0.37	0.36	0.36	0.36	0.35	0.35	0.34	0.33	0.32	0.25
	h (kJ/kg)	2316.31	2099.25	1924.49	1770.31	1636.07	1523.64	1431.25	1353.47	1282.23	1203.98	1004.60
	s (kJ/kg K)	5.87	5.82	5.66	5.48	5.29	5.11	4.95	4.79	4.64	4.45	3.73
	x_l (kg/kg)	0.00	0.03	0.05	0.08	0.11	0.14	0.18	0.21	0.25	0.31	1.00
	x_v (kg/kg)	0.00	0.24	0.40	0.54	0.65	0.74	0.82	0.87	0.91	0.96	1.00
0.9	T (°C)	143.57	137.18	131.09	124.79	118.21	111.35	104.14	96.27	86.91	73.36	−1.56
	v (m³/kg)	0.42	0.41	0.41	0.41	0.40	0.40	0.40	0.39	0.38	0.37	0.28
	h (kJ/kg)	2524.74	2361.75	2210.77	2066.94	1931.20	1806.64	1694.39	1592.51	1495.63	1391.19	1132.92
	s (kJ/kg K)	6.39	6.37	6.27	6.12	5.96	5.79	5.62	5.44	5.25	5.02	4.20
	x_l (kg/kg)	0.00	0.02	0.04	0.06	0.08	0.11	0.13	0.17	0.21	0.27	1.00
	x_v (kg/kg)	0.00	0.18	0.31	0.43	0.54	0.64	0.72	0.80	0.87	0.93	1.00
1.0	T (°C)	143.60	139.91	135.77	131.14	125.98	120.18	113.57	105.81	96.05	81.41	−1.49
	v (m³/kg)	0.46	0.46	0.46	0.46	0.46	0.45	0.45	0.44	0.43	0.42	0.31
	h (kJ/kg)	2738.12	2610.62	2483.32	2356.09	2228.78	2101.15	1972.78	1842.86	1709.45	1565.95	1261.52
	s (kJ/kg K)	6.90	6.91	6.85	6.74	6.62	6.46	6.29	6.09	5.85	5.56	4.67
	x_l (kg/kg)	0.00	0.00	0.00	0.00	0.00	0.00	0.00	0.00	0.00	0.00	0.00
	x_v (kg/kg)	0.00	0.10	0.20	0.30	0.40	0.50	0.60	0.70	0.80	0.90	1.00

$P = 4$ bar.

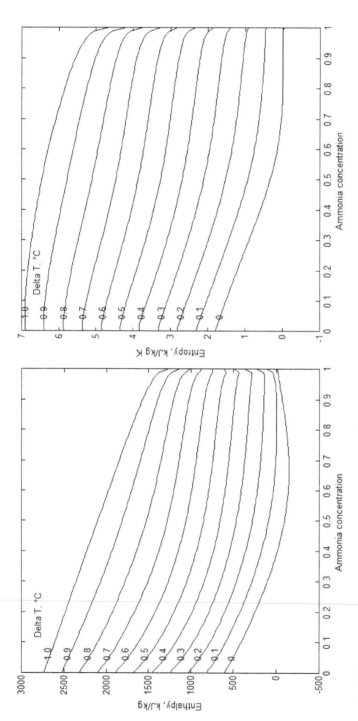

FIGURE A.6   Ammonia–water wet mixture properties at 4 bar.

TABLE A.7 Ammonia–Water Property System at 6 Bar Under Saturated and Wet Conditions.

F, where T = T_bp + F(T_dp − T_bp)	Property	Ammonia concentration (mass fraction), x										
		0	0.1	0.2	0.3	0.4	0.5	0.6	0.7	0.8	0.9	1
0	T (°C)	158.70	128.04	103.73	81.74	61.85	44.98	31.89	22.66	16.57	12.41	8.94
	v (m³/kg)	0.00	0.00	0.00	0.00	0.00	0.00	0.00	0.00	0.00	0.00	0.00
	h (kJ/kg)	670.28	474.51	309.26	165.00	46.65	−37.85	−83.27	−89.68	−63.18	−14.92	41.48
	s (kJ/kg K)	1.93	1.67	1.37	1.06	0.76	0.51	0.32	0.22	0.18	0.19	0.16
	x_l (kg/kg)	0.00	0.10	0.20	0.30	0.40	0.50	0.60	0.70	0.80	0.90	1.00
	x_v (kg/kg)	0.00	0.00	0.00	0.00	0.00	0.00	0.00	0.00	0.00	0.00	0.00
0.1	T (°C)	158.69	130.71	108.40	88.11	69.66	53.85	41.35	32.19	25.65	20.31	8.96
	v (m³/kg)	0.03	0.03	0.03	0.03	0.03	0.03	0.03	0.02	0.02	0.02	0.02
	h (kJ/kg)	878.80	653.49	483.42	343.24	229.51	144.43	88.81	59.31	49.29	54.66	164.54
	s (kJ/kg K)	2.42	2.16	1.87	1.58	1.31	1.07	0.89	0.77	0.69	0.64	0.60
	x_l (kg/kg)	0.00	0.09	0.18	0.27	0.36	0.44	0.52	0.60	0.66	0.73	1.00
	x_v (kg/kg)	0.00	0.55	0.80	0.91	0.97	0.99	0.99	1.00	1.00	1.00	1.00
0.2	T (°C)	158.68	133.38	113.07	94.49	77.46	62.71	50.80	41.72	34.72	28.22	8.98
	v (m³/kg)	0.06	0.06	0.06	0.06	0.06	0.05	0.05	0.05	0.05	0.05	0.04
	h (kJ/kg)	1087.34	839.16	662.81	523.86	413.49	329.63	270.47	231.98	208.07	193.10	287.59
	s (kJ/kg K)	2.90	2.65	2.37	2.10	1.85	1.63	1.46	1.34	1.24	1.16	1.03
	x_l (kg/kg)	0.00	0.08	0.16	0.24	0.32	0.39	0.46	0.52	0.57	0.63	1.00
	x_v (kg/kg)	0.00	0.51	0.75	0.88	0.95	0.98	0.99	0.99	1.00	1.00	1.00
0.3	T (°C)	158.67	136.06	117.73	100.86	85.27	71.58	60.26	51.26	43.80	36.12	9.00
	v (m³/kg)	0.10	0.09	0.09	0.09	0.08	0.08	0.08	0.08	0.07	0.07	0.06
	h (kJ/kg)	1295.88	1032.26	849.30	708.85	599.66	516.77	456.23	413.24	381.70	354.31	409.79
	s (kJ/kg K)	3.38	3.16	2.88	2.62	2.39	2.19	2.03	1.90	1.80	1.70	1.47
	x_l (kg/kg)	0.00	0.07	0.14	0.21	0.28	0.35	0.41	0.46	0.51	0.56	1.00
	x_v (kg/kg)	0.00	0.47	0.71	0.85	0.92	0.96	0.98	0.99	0.99	1.00	1.00

TABLE A.7 (Continued)

F, where $T = T_{bp} + F(T_{dp} - T_{bp})$	Property	Ammonia concentration (mass fraction), x										
		0	0.1	0.2	0.3	0.4	0.5	0.6	0.7	0.8	0.9	1
0.4	T (°C)	158.67	138.73	122.40	107.24	93.08	80.44	69.72	60.79	52.88	44.03	9.02
	v (m³/kg)	0.13	0.12	0.12	0.12	0.11	0.11	0.11	0.10	0.10	0.10	0.09
	h (kJ/kg)	1504.42	1233.56	1045.03	900.87	790.16	706.62	644.58	598.28	561.08	524.11	532.71
	s (kJ/kg K)	3.86	3.66	3.40	3.15	2.93	2.74	2.59	2.46	2.36	2.24	1.91
	x_l (kg/kg)	0.00	0.06	0.12	0.19	0.25	0.31	0.36	0.41	0.45	0.51	1.00
	x_v (kg/kg)	0.00	0.43	0.66	0.80	0.89	0.94	0.97	0.98	0.99	0.99	1.00
0.5	T (°C)	158.66	141.41	127.07	113.61	100.88	89.31	79.17	70.33	61.96	51.94	9.04
	v (m³/kg)	0.16	0.16	0.15	0.15	0.15	0.14	0.14	0.13	0.13	0.13	0.11
	h (kJ/kg)	1712.97	1443.77	1252.25	1103.29	988.31	901.52	836.38	786.09	743.28	697.48	655.70
	s (kJ/kg K)	4.35	4.18	3.93	3.69	3.48	3.30	3.15	3.02	2.91	2.78	2.34
	x_l (kg/kg)	0.00	0.05	0.10	0.16	0.21	0.27	0.31	0.36	0.40	0.46	1.00
	x_v (kg/kg)	0.00	0.38	0.60	0.75	0.85	0.91	0.94	0.96	0.98	0.99	1.00
0.6	T (°C)	158.65	144.08	131.74	119.98	108.69	98.17	88.63	79.86	71.04	59.84	9.06
	v (m³/kg)	0.19	0.19	0.18	0.18	0.18	0.17	0.17	0.17	0.16	0.16	0.13
	h (kJ/kg)	1921.52	1663.60	1473.30	1320.10	1198.74	1105.57	1034.43	977.98	928.04	872.62	778.77
	s (kJ/kg K)	4.83	4.70	4.48	4.25	4.04	3.86	3.71	3.58	3.46	3.31	2.78
	x_l (kg/kg)	0.00	0.04	0.09	0.13	0.18	0.22	0.27	0.31	0.35	0.41	1.00
	x_v (kg/kg)	0.00	0.34	0.53	0.68	0.79	0.86	0.91	0.94	0.96	0.98	1.00
0.7	T (°C)	158.64	146.76	136.41	126.36	116.50	107.04	98.08	89.39	80.11	67.75	9.08
	v (m³/kg)	0.22	0.22	0.22	0.21	0.21	0.21	0.20	0.20	0.19	0.19	0.15
	h (kJ/kg)	2130.08	1893.70	1710.49	1555.73	1427.36	1324.80	1243.77	1177.35	1116.98	1049.39	901.91
	s (kJ/kg K)	5.31	5.22	5.03	4.83	4.63	4.45	4.29	4.15	4.01	3.84	3.22
	x_l (kg/kg)	0.00	0.04	0.07	0.11	0.15	0.19	0.23	0.26	0.31	0.37	1.00
	x_v (kg/kg)	0.00	0.28	0.47	0.61	0.72	0.81	0.86	0.91	0.94	0.97	1.00

TABLE A.7 *(Continued)*

F, where T = T$_{bp}$ + F(T$_{dp}$ − T$_{bp}$)	Property	Ammonia concentration (mass fraction), x										
		0	0.1	0.2	0.3	0.4	0.5	0.6	0.7	0.8	0.9	1
0.8	T (°C)	158.63	149.43	141.08	132.73	124.30	115.90	107.54	98.93	89.19	75.66	9.10
	v (m³/kg)	0.25	0.25	0.25	0.25	0.24	0.24	0.24	0.23	0.23	0.22	0.17
	h (kJ/kg)	2338.64	2134.68	1965.99	1814.68	1681.07	1567.26	1471.87	1389.92	1313.55	1228.82	1025.75
	s (kJ/kg K)	5.80	5.74	5.60	5.42	5.24	5.06	4.89	4.73	4.57	4.37	3.65
	x$_l$ (kg/kg)	0.00	0.03	0.05	0.08	0.11	0.15	0.18	0.22	0.27	0.33	1.00
	x$_v$ (kg/kg)	0.00	0.23	0.39	0.52	0.63	0.73	0.80	0.86	0.91	0.95	1.00
0.9	T (°C)	158.62	152.11	145.75	139.10	132.11	124.77	117.00	108.46	98.27	83.56	9.12
	v (m³/kg)	0.28	0.28	0.28	0.28	0.28	0.27	0.27	0.27	0.26	0.25	0.19
	h (kJ/kg)	2547.20	2387.12	2241.75	2101.17	1966.98	1842.38	1728.67	1624.07	1523.25	1413.11	1148.75
	s (kJ/kg K)	6.28	6.27	6.16	6.03	5.87	5.70	5.52	5.34	5.15	4.91	4.09
	x$_l$ (kg/kg)	0.00	0.02	0.04	0.06	0.09	0.11	0.14	0.18	0.22	0.29	1.00
	x$_v$ (kg/kg)	0.00	0.17	0.30	0.42	0.53	0.63	0.72	0.80	0.86	0.93	1.00
1.0	T (°C)	158.61	154.78	150.41	145.48	139.92	133.63	126.45	117.99	107.35	91.47	9.14
	v (m³/kg)	0.31	0.31	0.31	0.31	0.31	0.31	0.31	0.30	0.29	0.28	0.21
	h (kJ/kg)	2755.76	2629.14	2502.46	2375.61	2248.44	2120.69	1991.93	1861.26	1726.56	1580.64	1271.74
	s (kJ/kg K)	6.76	6.78	6.71	6.61	6.48	6.33	6.15	5.95	5.71	5.41	4.52
	x$_l$ (kg/kg)	0.00	0.00	0.00	0.00	0.00	0.00	0.00	0.00	0.00	0.00	0.00
	x$_v$ (kg/kg)	0.00	0.10	0.20	0.30	0.40	0.50	0.60	0.70	0.80	0.90	1.00

P = 6 bar.

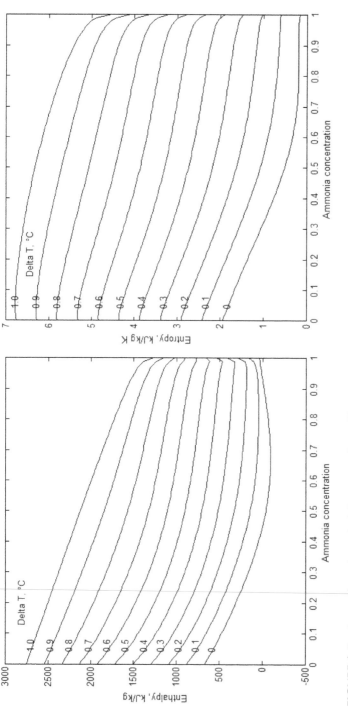

FIGURE A.7   Ammonia–water wet mixture properties at 6 bar.

TABLE A.8 Ammonia–Water Property System at 10 Bar Under Saturated and Wet Conditions.

F, where $T = T_{bp}$ $+ F(T_{dp} - T_{bp})$	Property	Ammonia concentration (mass fraction), x										
		0	0.1	0.2	0.3	0.4	0.5	0.6	0.7	0.8	0.9	1
0	T (°C)	180.08	149.36	124.21	101.36	80.73	63.22	49.57	39.81	33.22	28.56	24.61
	v (m³/kg)	0.00	0.00	0.00	0.00	0.00	0.00	0.00	0.00	0.00	0.00	0.00
	h (kJ/kg)	763.85	569.00	400.77	253.20	132.11	45.52	-1.56	-9.52	15.28	61.58	115.72
	s (kJ/kg K)	2.14	1.90	1.60	1.30	1.01	0.76	0.58	0.48	0.44	0.44	0.41
	x_l (kg/kg)	0.00	0.10	0.20	0.30	0.40	0.50	0.60	0.70	0.80	0.90	1.00
	x_v (kg/kg)	0.00	0.00	0.00	0.00	0.00	0.00	0.00	0.00	0.00	0.00	0.00
0.1	T (°C)	179.99	151.94	128.84	107.74	88.56	72.12	59.03	49.30	42.18	36.21	24.56
	v (m³/kg)	0.02	0.02	0.02	0.02	0.02	0.02	0.02	0.02	0.02	0.01	0.01
	h (kJ/kg)	964.91	745.55	572.81	428.41	310.87	222.78	164.96	134.08	123.84	131.23	232.27
	s (kJ/kg K)	2.59	2.35	2.07	1.78	1.51	1.28	1.10	0.98	0.90	0.86	0.81
	x_l (kg/kg)	0.00	0.09	0.18	0.27	0.36	0.45	0.53	0.60	0.67	0.75	1.00
	x_v (kg/kg)	0.00	0.51	0.76	0.89	0.95	0.98	0.99	1.00	1.00	1.00	1.00
0.2	T (°C)	179.90	154.52	133.47	114.12	96.40	81.01	68.50	58.80	51.13	43.87	24.51
	v (m³/kg)	0.04	0.04	0.04	0.04	0.03	0.03	0.03	0.03	0.03	0.03	0.03
	h (kJ/kg)	1166.02	927.96	750.07	606.65	491.62	403.79	341.48	300.61	275.14	260.17	348.84
	s (kJ/kg K)	3.03	2.81	2.53	2.26	2.01	1.80	1.63	1.50	1.40	1.32	1.20
	x_l (kg/kg)	0.00	0.08	0.16	0.24	0.32	0.40	0.47	0.53	0.59	0.65	1.00
	x_v (kg/kg)	0.00	0.47	0.72	0.85	0.93	0.97	0.98	0.99	0.99	1.00	1.00
0.3	T (°C)	179.81	157.10	138.10	120.50	104.24	89.91	77.96	68.29	60.09	51.52	24.46
	v (m³/kg)	0.06	0.06	0.06	0.05	0.05	0.05	0.05	0.05	0.05	0.04	0.04
	h (kJ/kg)	1367.17	1116.78	934.08	789.63	675.38	587.71	523.04	476.59	442.12	412.66	465.43
	s (kJ/kg K)	3.48	3.27	3.01	2.75	2.52	2.31	2.15	2.02	1.91	1.81	1.59
	x_l (kg/kg)	0.00	0.07	0.14	0.22	0.29	0.35	0.41	0.47	0.52	0.58	1.00
	x_v (kg/kg)	0.00	0.44	0.67	0.82	0.90	0.95	0.97	0.98	0.99	0.99	1.00

TABLE A.8 (Continued)

F, where T = T_{bp} + F(T_{dp} − T_{bp})	Property	Ammonia concentration (mass fraction), x										
		0	0.1	0.2	0.3	0.4	0.5	0.6	0.7	0.8	0.9	1
0.4	T (°C)	179.72	159.69	142.73	126.89	112.08	98.81	87.42	77.78	69.05	59.17	24.41
	v (m³/kg)	0.08	0.08	0.08	0.07	0.07	0.07	0.07	0.06	0.06	0.06	0.05
	h (kJ/kg)	1568.34	1312.58	1126.49	979.58	864.05	775.24	708.26	657.45	616.04	575.12	582.04
	s (kJ/kg K)	3.92	3.74	3.49	3.25	3.02	2.83	2.67	2.54	2.43	2.30	1.99
	x_l (kg/kg)	0.00	0.06	0.13	0.19	0.25	0.31	0.37	0.41	0.46	0.53	1.00
	x_v (kg/kg)	0.00	0.40	0.62	0.77	0.86	0.92	0.95	0.97	0.98	0.99	1.00
0.5	T (°C)	179.63	162.27	147.36	133.27	119.92	107.70	96.88	87.28	78.01	66.82	24.36
	v (m³/kg)	0.10	0.10	0.09	0.09	0.09	0.09	0.09	0.08	0.08	0.08	0.06
	h (kJ/kg)	1769.56	1515.87	1329.06	1179.24	1060.43	968.44	897.80	842.10	793.87	742.37	698.67
	s (kJ/kg K)	4.36	4.22	3.98	3.75	3.54	3.35	3.19	3.06	2.94	2.79	2.38
	x_l (kg/kg)	0.00	0.05	0.11	0.16	0.22	0.27	0.32	0.37	0.41	0.48	1.00
	x_v (kg/kg)	0.00	0.36	0.57	0.72	0.82	0.89	0.93	0.95	0.97	0.98	1.00
0.6	T (°C)	179.54	164.85	152.00	139.65	127.76	116.60	106.35	96.77	86.97	74.47	24.31
	v (m³/kg)	0.12	0.12	0.11	0.11	0.11	0.11	0.10	0.10	0.10	0.10	0.08
	h (kJ/kg)	1970.81	1727.19	1543.60	1391.75	1268.26	1170.68	1094.00	1031.51	975.09	912.37	815.32
	s (kJ/kg K)	4.81	4.69	4.49	4.27	4.06	3.88	3.72	3.58	3.45	3.28	2.77
	x_l (kg/kg)	0.00	0.05	0.09	0.14	0.19	0.23	0.28	0.32	0.37	0.43	1.00
	x_v (kg/kg)	0.00	0.31	0.51	0.66	0.77	0.84	0.89	0.93	0.95	0.98	1.00
0.7	T (°C)	179.45	167.43	156.63	146.04	135.59	125.49	115.81	106.26	95.93	82.13	24.26
	v (m³/kg)	0.14	0.14	0.13	0.13	0.13	0.13	0.12	0.12	0.12	0.11	0.09
	h (kJ/kg)	2172.10	1947.01	1771.89	1620.58	1492.22	1386.80	1300.89	1228.32	1160.80	1084.61	931.98
	s (kJ/kg K)	5.25	5.17	5.00	4.80	4.61	4.43	4.26	4.11	3.96	3.78	3.17
	x_l (kg/kg)	0.00	0.04	0.07	0.11	0.15	0.20	0.24	0.28	0.32	0.39	1.00
	x_v (kg/kg)	0.00	0.27	0.44	0.58	0.70	0.78	0.85	0.89	0.93	0.96	1.00

TABLE A.8 *(Continued)*

F, where $T = T_{bp} + F(T_{dp} - T_{bp})$	Property	Ammonia concentration (mass fraction), x										
		0	**0.1**	**0.2**	**0.3**	**0.4**	**0.5**	**0.6**	**0.7**	**0.8**	**0.9**	**1**
0.8	T (°C)	179.36	170.01	161.26	152.42	143.43	134.39	125.27	115.76	104.89	89.78	24.21
	v (m³/kg)	0.16	0.15	0.15	0.15	0.15	0.15	0.15	0.14	0.14	0.13	0.10
	h (kJ/kg)	2373.42	2175.79	2015.64	1869.25	1737.66	1623.00	1524.24	1436.92	1353.50	1259.64	1048.67
	s (kJ/kg K)	5.70	5.65	5.51	5.35	5.17	4.99	4.82	4.66	4.48	4.27	3.56
	x_l (kg/kg)	0.00	0.03	0.06	0.09	0.12	0.16	0.20	0.24	0.28	0.35	1.00
	x_v (kg/kg)	0.00	0.22	0.37	0.50	0.61	0.71	0.79	0.85	0.90	0.95	1.00
0.9	T (°C)	179.27	172.59	165.89	158.80	151.27	143.29	134.74	125.25	113.85	97.43	24.15
	v (m³/kg)	0.17	0.17	0.17	0.17	0.17	0.17	0.17	0.16	0.16	0.15	0.12
	h (kJ/kg)	2574.78	2413.96	2276.39	2141.13	2010.16	1886.56	1771.57	1663.53	1557.17	1438.84	1165.37
	s (kJ/kg K)	6.14	6.13	6.04	5.91	5.75	5.59	5.41	5.22	5.02	4.77	3.96
	x_l (kg/kg)	0.00	0.02	0.04	0.07	0.09	0.12	0.16	0.20	0.24	0.32	1.00
	x_v (kg/kg)	0.00	0.17	0.30	0.41	0.52	0.62	0.70	0.79	0.85	0.93	1.00
1.0	T (°C)	179.19	175.18	170.52	165.19	159.11	152.18	144.20	134.74	122.81	105.08	24.10
	v (m³/kg)	0.19	0.19	0.19	0.19	0.19	0.19	0.19	0.19	0.18	0.17	0.13
	h (kJ/kg)	2776.17	2651.00	2525.44	2399.38	2272.62	2144.90	2015.69	1883.99	1747.35	1597.65	1282.09
	s (kJ/kg K)	6.59	6.60	6.54	6.44	6.31	6.16	5.98	5.77	5.53	5.22	4.35
	x_l (kg/kg)	0.00	0.00	0.00	0.00	0.00	0.00	0.00	0.00	0.00	0.00	0.00
	x_v (kg/kg)	0.00	0.10	0.20	0.30	0.40	0.50	0.60	0.70	0.80	0.90	1.00

$P = 10$ bar.

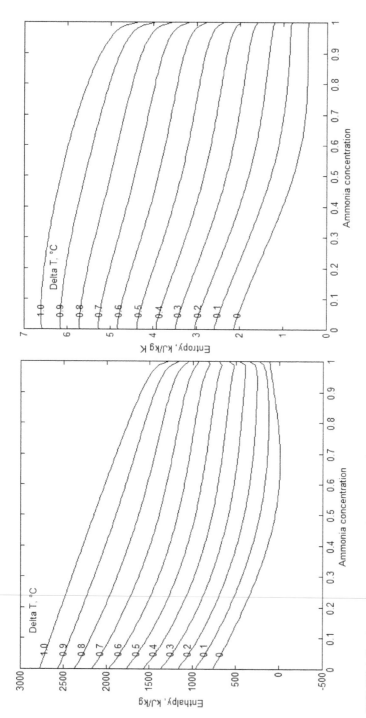

FIGURE A.8   Ammonia–water wet mixture properties at 10 bar.

TABLE A.9 Ammonia–Water Property System at 15 Bar Under Saturated and Wet Conditions.

F, where $T = T_{bp} + F(T_{dp} - T_{bp})$	Property	Ammonia concentration (mass fraction), x										
		0	0.1	0.2	0.3	0.4	0.5	0.6	0.7	0.8	0.9	1
0	T (°C)	198.81	168.03	142.17	118.60	97.34	79.30	65.17	54.96	47.93	42.84	38.47
	v (m³/kg)	0.00	0.00	0.00	0.00	0.00	0.00	0.00	0.00	0.00	0.00	0.00
	h (kJ/kg)	847.18	653.68	483.20	332.83	209.22	120.57	71.85	62.46	85.89	130.78	183.41
	s (kJ/kg K)	2.32	2.09	1.81	1.51	1.22	0.98	0.80	0.70	0.67	0.67	0.63
	x_l (kg/kg)	0.00	0.10	0.20	0.30	0.40	0.50	0.60	0.70	0.80	0.90	1.00
	x_v (kg/kg)	0.00	0.00	0.00	0.00	0.00	0.00	0.00	0.00	0.00	0.00	0.00
0.1	T (°C)	198.62	170.50	146.74	124.96	105.18	88.20	74.61	64.39	56.75	50.23	38.36
	v (m³/kg)	0.01	0.01	0.01	0.01	0.01	0.01	0.01	0.01	0.01	0.01	0.01
	h (kJ/kg)	1041.11	826.78	652.42	504.80	384.01	293.17	233.22	200.96	190.44	199.54	293.52
	s (kJ/kg K)	2.73	2.51	2.24	1.96	1.69	1.46	1.28	1.16	1.09	1.04	0.99
	x_l (kg/kg)	0.00	0.09	0.18	0.27	0.36	0.45	0.53	0.61	0.68	0.76	1.00
	x_v (kg/kg)	0.00	0.48	0.73	0.86	0.94	0.97	0.99	0.99	1.00	1.00	1.00
0.2	T (°C)	198.44	172.98	151.32	131.33	113.02	97.10	84.06	73.82	65.57	57.62	38.24
	v (m³/kg)	0.03	0.03	0.03	0.03	0.02	0.02	0.02	0.02	0.02	0.02	0.02
	h (kJ/kg)	1235.12	1005.01	826.69	680.13	561.40	470.14	404.96	361.84	334.91	320.08	403.68
	s (kJ/kg K)	3.15	2.94	2.67	2.41	2.16	1.94	1.77	1.64	1.54	1.46	1.35
	x_l (kg/kg)	0.00	0.08	0.16	0.25	0.32	0.40	0.47	0.53	0.60	0.67	1.00
	x_v (kg/kg)	0.00	0.45	0.69	0.83	0.91	0.96	0.98	0.99	0.99	1.00	1.00
0.3	T (°C)	198.25	175.45	155.89	137.69	120.86	105.99	93.50	83.25	74.39	65.02	38.13
	v (m³/kg)	0.04	0.04	0.04	0.04	0.04	0.03	0.03	0.03	0.03	0.03	0.03
	h (kJ/kg)	1429.20	1188.79	1007.24	860.31	742.30	650.70	582.45	532.84	495.67	464.43	513.87
	s (kJ/kg K)	3.56	3.37	3.12	2.86	2.63	2.43	2.26	2.13	2.02	1.91	1.71
	x_l (kg/kg)	0.00	0.07	0.15	0.22	0.29	0.36	0.42	0.48	0.53	0.60	1.00
	x_v (kg/kg)	0.00	0.41	0.64	0.79	0.88	0.93	0.96	0.98	0.99	0.99	1.00

TABLE A.9 *(Continued)*

F, where T = T$_{bp}$ + F(T$_{dp}$ − T$_{bp}$)	Property	Ammonia concentration (mass fraction), x										
		0	0.1	0.2	0.3	0.4	0.5	0.6	0.7	0.8	0.9	1
0.4	T (°C)	198.06	177.93	160.46	144.06	128.70	114.89	102.94	92.68	83.22	72.41	38.02
	v (m³/kg)	0.05	0.05	0.05	0.05	0.05	0.05	0.05	0.04	0.04	0.04	0.04
	h (kJ/kg)	1623.35	1378.55	1195.39	1047.21	928.36	835.45	764.31	709.52	664.27	619.81	624.11
	s (kJ/kg K)	3.97	3.81	3.57	3.33	3.11	2.91	2.75	2.61	2.49	2.36	2.06
	x$_l$ (kg/kg)	0.00	0.07	0.13	0.19	0.26	0.32	0.37	0.42	0.48	0.54	1.00
	x$_v$ (kg/kg)	0.00	0.38	0.60	0.75	0.85	0.91	0.94	0.96	0.98	0.99	1.00
0.5	T (°C)	197.88	180.41	165.04	150.42	136.54	123.79	112.39	102.11	92.04	79.80	37.90
	v (m³/kg)	0.07	0.07	0.06	0.06	0.06	0.06	0.06	0.06	0.05	0.05	0.04
	h (kJ/kg)	1817.58	1574.68	1392.56	1243.07	1121.92	1026.10	951.02	890.67	837.58	780.94	734.38
	s (kJ/kg K)	4.38	4.25	4.03	3.80	3.59	3.40	3.24	3.10	2.97	2.81	2.42
	x$_l$ (kg/kg)	0.00	0.06	0.11	0.17	0.22	0.28	0.33	0.38	0.43	0.50	1.00
	x$_v$ (kg/kg)	0.00	0.34	0.55	0.69	0.80	0.87	0.91	0.94	0.96	0.98	1.00
0.6	T (°C)	197.69	182.88	169.61	156.78	144.38	132.69	121.83	111.54	100.86	87.19	37.79
	v (m³/kg)	0.08	0.08	0.08	0.08	0.07	0.07	0.07	0.07	0.07	0.06	0.05
	h (kJ/kg)	2011.88	1777.58	1600.18	1450.43	1326.07	1225.47	1144.48	1076.93	1014.82	945.54	844.69
	s (kJ/kg K)	4.79	4.69	4.49	4.29	4.08	3.90	3.74	3.59	3.45	3.27	2.78
	x$_l$ (kg/kg)	0.00	0.05	0.09	0.14	0.19	0.24	0.29	0.33	0.38	0.45	1.00
	x$_v$ (kg/kg)	0.00	0.30	0.49	0.63	0.74	0.82	0.88	0.92	0.95	0.97	1.00
0.7	T (°C)	197.50	185.36	174.18	163.15	152.22	141.58	131.27	120.97	109.68	94.58	37.68
	v (m³/kg)	0.09	0.09	0.09	0.09	0.09	0.09	0.08	0.08	0.08	0.08	0.06
	h (kJ/kg)	2206.26	1987.64	1819.68	1672.07	1544.59	1437.47	1347.93	1270.35	1196.66	1112.84	955.05
	s (kJ/kg K)	5.21	5.13	4.97	4.78	4.59	4.41	4.25	4.09	3.93	3.73	3.14
	x$_l$ (kg/kg)	0.00	0.04	0.08	0.12	0.16	0.20	0.25	0.29	0.34	0.41	1.00
	x$_v$ (kg/kg)	0.00	0.26	0.43	0.57	0.68	0.77	0.83	0.88	0.92	0.96	1.00

TABLE A.9 (Continued)

F, where T = T$_{bp}$ + F(T$_{dp}$ − T$_{bp}$)	Property	Ammonia concentration (mass fraction), x										
		0	0.1	0.2	0.3	0.4	0.5	0.6	0.7	0.8	0.9	1
0.8	T (°C)	197.32	187.83	178.75	169.51	160.06	150.48	140.72	130.40	118.50	101.97	37.56
	v (m³/kg)	0.11	0.11	0.10	0.10	0.10	0.10	0.10	0.10	0.09	0.09	0.07
	h (kJ/kg)	2400.70	2205.19	2052.44	1910.86	1781.75	1667.07	1565.97	1474.36	1384.92	1282.99	1065.43
	s (kJ/kg K)	5.62	5.58	5.45	5.29	5.12	4.94	4.77	4.60	4.42	4.19	3.50
	x$_l$ (kg/kg)	0.00	0.03	0.06	0.09	0.13	0.17	0.21	0.25	0.30	0.38	1.00
	x$_v$ (kg/kg)	0.00	0.21	0.36	0.49	0.60	0.69	0.77	0.84	0.89	0.94	1.00
0.9	T (°C)	197.13	190.31	183.33	175.88	167.90	159.38	150.16	139.83	127.33	109.36	37.45
	v (m³/kg)	0.12	0.12	0.12	0.12	0.12	0.11	0.11	0.11	0.11	0.10	0.08
	h (kJ/kg)	2595.22	2430.60	2299.74	2169.57	2042.08	1920.03	1804.46	1693.72	1582.53	1456.83	1175.86
	s (kJ/kg K)	6.03	6.02	5.93	5.80	5.66	5.49	5.32	5.13	4.92	4.65	3.86
	x$_l$ (kg/kg)	0.00	0.03	0.05	0.07	0.10	0.13	0.17	0.21	0.26	0.34	1.00
	x$_v$ (kg/kg)	0.00	0.16	0.29	0.41	0.51	0.61	0.70	0.78	0.85	0.92	1.00
1.0	T (°C)	196.94	192.79	187.90	182.24	175.74	168.27	159.60	149.26	136.15	116.75	37.34
	v (m³/kg)	0.13	0.13	0.13	0.13	0.13	0.13	0.13	0.13	0.12	0.12	0.09
	h (kJ/kg)	2789.82	2666.16	2541.83	2416.68	2290.51	2162.98	2033.50	1900.91	1762.43	1609.01	1286.33
	s (kJ/kg K)	6.44	6.46	6.40	6.30	6.17	6.02	5.84	5.63	5.38	5.07	4.22
	x$_l$ (kg/kg)	0.00	0.00	0.00	0.00	0.00	0.00	0.00	0.00	0.00	0.00	0.00
	x$_v$ (kg/kg)	0.00	0.10	0.20	0.30	0.40	0.50	0.60	0.70	0.80	0.90	1.00

P = 15 bar.

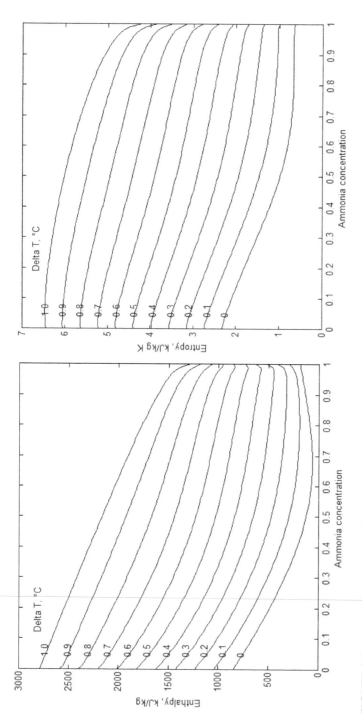

FIGURE A.9   Ammonia–water wet mixture properties at 15 bar.

TABLE A.10 Ammonia–Water Property System at 20 Bar Under Saturated and Wet Conditions.

F, where T = T$_{bp}$ + F(T$_{dp}$ − T$_{bp}$)	Property	Ammonia concentration (mass fraction), x										
		0	0.1	0.2	0.3	0.4	0.5	0.6	0.7	0.8	0.9	1
0	T (°C)	213.15	182.30	155.92	131.82	110.10	91.66	77.16	66.61	59.25	53.84	49.15
	v (m³/kg)	0.00	0.00	0.00	0.00	0.00	0.00	0.00	0.00	0.00	0.00	0.00
	h (kJ/kg)	911.87	719.80	547.89	395.47	269.88	179.55	129.46	118.95	141.40	185.36	237.05
	s (kJ/kg K)	2.46	2.24	1.96	1.66	1.38	1.14	0.97	0.87	0.83	0.83	0.80
	x$_l$ (kg/kg)	0.00	0.10	0.20	0.30	0.40	0.50	0.60	0.70	0.80	0.90	1.00
	x$_v$ (kg/kg)	0.00	0.00	0.00	0.00	0.00	0.00	0.00	0.00	0.00	0.00	0.00
0.1	T (°C)	212.87	184.68	160.43	138.15	117.92	100.54	86.58	75.98	67.95	61.01	48.99
	v (m³/kg)	0.01	0.01	0.01	0.01	0.01	0.01	0.01	0.01	0.01	0.01	0.01
	h (kJ/kg)	1099.94	889.46	714.28	564.47	441.29	348.31	286.64	253.24	242.43	252.77	341.67
	s (kJ/kg K)	2.84	2.64	2.37	2.09	1.82	1.60	1.42	1.30	1.22	1.18	1.13
	x$_l$ (kg/kg)	0.00	0.09	0.18	0.27	0.36	0.45	0.53	0.61	0.68	0.77	1.00
	x$_v$ (kg/kg)	0.00	0.45	0.70	0.85	0.92	0.96	0.98	0.99	0.99	1.00	1.00
0.2	T (°C)	212.59	187.06	164.95	144.48	125.75	109.42	95.99	85.34	76.65	68.18	48.82
	v (m³/kg)	0.02	0.02	0.02	0.02	0.02	0.02	0.02	0.02	0.02	0.02	0.01
	h (kJ/kg)	1288.13	1063.72	885.52	736.99	615.67	521.85	454.43	409.53	381.43	366.78	446.35
	s (kJ/kg K)	3.23	3.04	2.78	2.52	2.27	2.05	1.88	1.75	1.65	1.57	1.46
	x$_l$ (kg/kg)	0.00	0.08	0.16	0.25	0.33	0.40	0.47	0.54	0.60	0.68	1.00
	x$_v$ (kg/kg)	0.00	0.42	0.66	0.81	0.90	0.95	0.97	0.98	0.99	0.99	1.00
0.3	T (°C)	212.31	189.44	169.46	150.82	133.58	118.30	105.40	94.71	85.35	75.35	48.66
	v (m³/kg)	0.03	0.03	0.03	0.03	0.03	0.03	0.03	0.02	0.02	0.02	0.02
	h (kJ/kg)	1476.42	1242.90	1062.63	914.33	793.82	699.41	628.44	576.37	537.08	504.53	551.08
	s (kJ/kg K)	3.62	3.45	3.20	2.95	2.72	2.51	2.34	2.21	2.09	1.98	1.79
	x$_l$ (kg/kg)	0.00	0.08	0.15	0.22	0.29	0.36	0.42	0.48	0.54	0.61	1.00
	x$_v$ (kg/kg)	0.00	0.39	0.62	0.77	0.86	0.92	0.95	0.97	0.98	0.99	1.00

TABLE A.10 (Continued)

F, where $T = T_{bp}$ $+ F(T_{dp} - T_{bp})$	Property	Ammonia concentration (mass fraction), x										
		0	0.1	0.2	0.3	0.4	0.5	0.6	0.7	0.8	0.9	1
0.4	T (°C)	212.04	191.82	173.97	157.15	141.40	127.19	114.81	104.07	94.05	82.52	48.49
	v (m³/kg)	0.04	0.04	0.04	0.04	0.04	0.04	0.03	0.03	0.03	0.03	0.03
	h (kJ/kg)	1664.81	1427.34	1246.74	1098.11	977.18	881.45	807.26	749.43	701.20	654.02	655.87
	s (kJ/kg K)	4.01	3.86	3.63	3.39	3.17	2.97	2.81	2.67	2.54	2.41	2.13
	x_l (kg/kg)	0.00	0.07	0.13	0.20	0.26	0.32	0.38	0.43	0.49	0.56	1.00
	x_v (kg/kg)	0.00	0.36	0.58	0.73	0.83	0.89	0.93	0.96	0.97	0.99	1.00
0.5	T (°C)	211.76	194.20	178.48	163.49	149.23	136.07	124.23	113.44	102.75	89.69	48.33
	v (m³/kg)	0.05	0.05	0.05	0.05	0.05	0.05	0.04	0.04	0.04	0.04	0.03
	h (kJ/kg)	1853.32	1617.37	1439.02	1290.24	1167.80	1069.45	991.21	927.37	870.54	809.91	760.70
	s (kJ/kg K)	4.40	4.27	4.06	3.84	3.63	3.44	3.28	3.13	2.99	2.83	2.46
	x_l (kg/kg)	0.00	0.06	0.11	0.17	0.23	0.28	0.33	0.38	0.44	0.51	1.00
	x_v (kg/kg)	0.00	0.32	0.53	0.68	0.78	0.85	0.90	0.94	0.96	0.98	1.00
0.6	T (°C)	211.48	196.58	183.00	169.82	157.05	144.95	133.64	122.80	111.45	96.86	48.17
	v (m³/kg)	0.06	0.06	0.06	0.06	0.06	0.06	0.05	0.05	0.05	0.05	0.04
	h (kJ/kg)	2041.93	1813.32	1640.68	1492.88	1368.31	1265.82	1181.84	1110.58	1044.14	969.78	865.58
	s (kJ/kg K)	4.78	4.69	4.50	4.30	4.10	3.92	3.75	3.60	3.45	3.26	2.79
	x_l (kg/kg)	0.00	0.05	0.10	0.15	0.20	0.24	0.29	0.34	0.39	0.47	1.00
	x_v (kg/kg)	0.00	0.29	0.48	0.62	0.73	0.81	0.86	0.91	0.94	0.97	1.00
0.7	T (°C)	211.20	198.96	187.51	176.16	164.88	153.83	143.05	132.17	120.15	104.03	48.00
	v (m³/kg)	0.07	0.07	0.07	0.07	0.07	0.07	0.06	0.06	0.06	0.06	0.05
	h (kJ/kg)	2230.64	2015.48	1852.92	1708.37	1581.90	1473.88	1381.90	1300.70	1222.36	1132.67	970.51
	s (kJ/kg K)	5.17	5.11	4.95	4.77	4.58	4.40	4.23	4.07	3.90	3.69	3.12
	x_l (kg/kg)	0.00	0.04	0.08	0.12	0.17	0.21	0.25	0.30	0.35	0.43	1.00
	x_v (kg/kg)	0.00	0.25	0.42	0.55	0.67	0.75	0.82	0.87	0.92	0.96	1.00

TABLE A.10 *(Continued)*

F, where T = T_bp + F(T_dp − T_bp)	Property	Ammonia concentration (mass fraction), x										
		0	0.1	0.2	0.3	0.4	0.5	0.6	0.7	0.8	0.9	1
0.8	T (°C)	210.92	201.34	192.02	182.49	172.70	162.72	152.46	141.54	128.85	111.20	47.84
	v (m³/kg)	0.08	0.08	0.08	0.08	0.08	0.08	0.07	0.07	0.07	0.07	0.05
	h (kJ/kg)	2419.46	2224.16	2076.92	1939.16	1812.19	1697.80	1595.20	1500.53	1406.59	1298.44	1075.49
	s (kJ/kg K)	5.56	5.53	5.40	5.25	5.08	4.91	4.73	4.56	4.37	4.13	3.46
	x_l (kg/kg)	0.00	0.04	0.07	0.10	0.14	0.17	0.21	0.26	0.31	0.39	1.00
	x_v (kg/kg)	0.00	0.20	0.36	0.48	0.59	0.68	0.76	0.83	0.89	0.94	1.00
0.9	T (°C)	210.64	203.72	196.53	188.82	180.53	171.60	161.88	150.90	137.55	118.37	47.67
	v (m³/kg)	0.09	0.09	0.09	0.09	0.09	0.09	0.09	0.08	0.08	0.08	0.06
	h (kJ/kg)	2608.39	2439.65	2313.80	2187.63	2063.01	1942.39	1826.62	1713.96	1599.11	1467.63	1180.52
	s (kJ/kg K)	5.95	5.94	5.85	5.73	5.59	5.42	5.25	5.06	4.84	4.56	3.79
	x_l (kg/kg)	0.00	0.03	0.05	0.08	0.11	0.14	0.18	0.22	0.28	0.36	1.00
	x_v (kg/kg)	0.00	0.16	0.29	0.40	0.51	0.60	0.69	0.77	0.85	0.92	1.00
1.0	T (°C)	210.36	206.10	201.05	195.16	188.35	180.48	171.29	160.27	146.25	125.54	47.51
	v (m³/kg)	0.10	0.10	0.10	0.10	0.10	0.10	0.10	0.10	0.09	0.09	0.06
	h (kJ/kg)	2797.42	2675.08	2551.84	2427.54	2301.95	2174.68	2045.06	1911.79	1771.80	1615.22	1285.60
	s (kJ/kg K)	6.34	6.36	6.30	6.20	6.07	5.92	5.74	5.53	5.28	4.95	4.12
	x_l (kg/kg)	0.00	0.00	0.00	0.00	0.00	0.00	0.00	0.00	0.00	0.00	0.00
	x_v (kg/kg)	0.00	0.10	0.20	0.30	0.40	0.50	0.60	0.70	0.80	0.90	1.00

$P = 20$ bar.

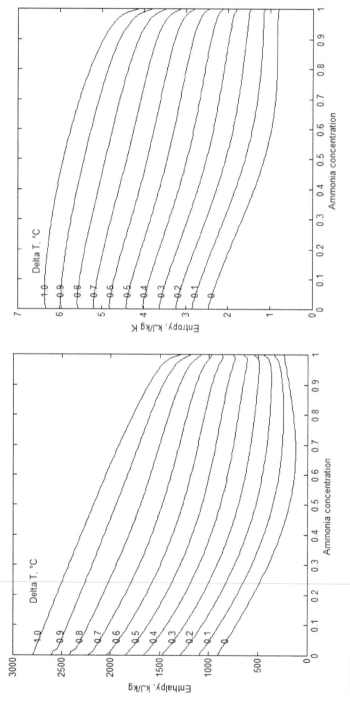

FIGURE A.10 Ammonia–water wet mixture properties at 20 bar.

TABLE A.11 Ammonia–Water Property System at 30 Bar Under Saturated and Wet Conditions.

F, where T = T_bp + F(T_dp − T_bp)	Property	Ammonia concentration (mass fraction), x										
		0	0.1	0.2	0.3	0.4	0.5	0.6	0.7	0.8	0.9	1
0	T (°C)	234.98	203.99	176.85	151.97	129.57	110.53	95.50	84.43	76.58	70.69	65.51
	v (m³/kg)	0.00	0.00	0.00	0.00	0.00	0.00	0.00	0.00	0.00	0.00	0.00
	h (kJ/kg)	1012.03	822.90	649.29	493.95	365.34	272.34	220.05	207.79	228.80	271.53	322.09
	s (kJ/kg K)	2.66	2.46	2.19	1.90	1.62	1.39	1.22	1.12	1.09	1.08	1.05
	x_l (kg/kg)	0.00	0.10	0.20	0.30	0.40	0.50	0.60	0.70	0.80	0.90	1.00
	x_v (kg/kg)	0.00	0.00	0.00	0.00	0.00	0.00	0.00	0.00	0.00	0.00	0.00
0.1	T (°C)	234.53	206.20	181.25	158.23	137.34	119.36	104.84	93.67	85.06	77.49	65.26
	v (m³/kg)	0.01	0.01	0.01	0.01	0.01	0.01	0.01	0.01	0.01	0.01	0.01
	h (kJ/kg)	1190.48	986.04	810.22	657.51	530.88	434.62	370.24	334.97	323.61	335.66	417.20
	s (kJ/kg K)	3.01	2.82	2.56	2.29	2.03	1.80	1.63	1.51	1.43	1.39	1.34
	x_l (kg/kg)	0.00	0.09	0.18	0.27	0.36	0.45	0.53	0.61	0.69	0.79	1.00
	x_v (kg/kg)	0.00	0.42	0.67	0.82	0.90	0.95	0.97	0.98	0.99	1.00	1.00
0.2	T (°C)	234.08	208.41	185.64	164.49	145.12	128.20	114.18	102.92	93.55	84.29	65.01
	v (m³/kg)	0.01	0.01	0.01	0.01	0.01	0.01	0.01	0.01	0.01	0.01	0.01
	h (kJ/kg)	1369.11	1153.00	975.60	824.68	699.80	602.22	531.37	483.65	453.70	439.42	512.39
	s (kJ/kg K)	3.36	3.19	2.94	2.68	2.44	2.22	2.05	1.92	1.82	1.74	1.64
	x_l (kg/kg)	0.00	0.09	0.17	0.25	0.33	0.41	0.48	0.55	0.61	0.70	1.00
	x_v (kg/kg)	0.00	0.39	0.63	0.79	0.87	0.93	0.96	0.98	0.99	0.99	1.00
0.3	T (°C)	233.63	210.62	190.03	170.75	152.89	137.03	123.52	112.16	102.03	91.10	64.76
	v (m³/kg)	0.02	0.02	0.02	0.02	0.02	0.02	0.02	0.02	0.02	0.02	0.01
	h (kJ/kg)	1547.90	1324.01	1146.20	996.49	872.72	774.32	699.32	643.47	600.87	566.40	607.63
	s (kJ/kg K)	3.71	3.56	3.33	3.08	2.85	2.64	2.47	2.33	2.22	2.10	1.93
	x_l (kg/kg)	0.00	0.08	0.15	0.22	0.30	0.36	0.43	0.49	0.55	0.64	1.00
	x_v (kg/kg)	0.00	0.36	0.59	0.75	0.85	0.90	0.94	0.96	0.98	0.99	1.00

TABLE A.11 *(Continued)*

F, where $T = T_{bp}$ + F($T_{dp} - T_{bp}$)	Property	Ammonia concentration (mass fraction), x										
		0	0.1	0.2	0.3	0.4	0.5	0.6	0.7	0.8	0.9	1
0.4	T (°C)	233.19	212.83	194.43	177.01	160.67	145.86	132.86	121.40	110.52	97.90	64.51
	v (m³/kg)	0.03	0.03	0.03	0.03	0.03	0.02	0.02	0.02	0.02	0.02	0.02
	h (kJ/kg)	1726.86	1499.29	1322.90	1174.24	1050.79	951.19	872.60	810.18	757.32	705.99	702.93
	s (kJ/kg K)	4.07	3.93	3.72	3.49	3.27	3.07	2.90	2.76	2.63	2.48	2.23
	x_l (kg/kg)	0.00	0.07	0.13	0.20	0.26	0.33	0.38	0.44	0.50	0.58	1.00
	x_v (kg/kg)	0.00	0.33	0.55	0.70	0.81	0.87	0.92	0.95	0.97	0.98	1.00
0.5	T (°C)	232.74	215.03	198.82	183.27	168.45	154.70	142.20	130.64	119.01	104.71	64.26
	v (m³/kg)	0.03	0.03	0.03	0.03	0.03	0.03	0.03	0.03	0.03	0.03	0.02
	h (kJ/kg)	1905.97	1679.09	1506.58	1359.40	1235.62	1133.95	1051.24	982.25	919.69	852.89	798.29
	s (kJ/kg K)	4.42	4.31	4.11	3.90	3.69	3.50	3.33	3.18	3.04	2.87	2.52
	x_l (kg/kg)	0.00	0.06	0.12	0.18	0.23	0.29	0.34	0.39	0.45	0.53	1.00
	x_v (kg/kg)	0.00	0.30	0.51	0.65	0.76	0.84	0.89	0.92	0.95	0.97	1.00
0.6	T (°C)	232.29	217.24	203.21	189.53	176.22	163.53	151.54	139.88	127.49	111.51	64.01
	v (m³/kg)	0.04	0.04	0.04	0.04	0.04	0.04	0.04	0.04	0.03	0.03	0.03
	h (kJ/kg)	2085.25	1863.64	1698.15	1553.65	1429.28	1324.47	1236.37	1159.70	1086.70	1004.50	893.70
	s (kJ/kg K)	4.77	4.69	4.51	4.32	4.12	3.94	3.77	3.61	3.45	3.26	2.82
	x_l (kg/kg)	0.00	0.06	0.10	0.15	0.20	0.25	0.30	0.35	0.41	0.49	1.00
	x_v (kg/kg)	0.00	0.27	0.46	0.60	0.71	0.79	0.85	0.89	0.93	0.96	1.00
0.7	T (°C)	231.84	219.45	207.61	195.79	184.00	172.36	160.88	149.12	135.98	118.32	63.76
	v (m³/kg)	0.05	0.05	0.05	0.05	0.05	0.04	0.04	0.04	0.04	0.04	0.03
	h (kJ/kg)	2264.70	2053.16	1898.56	1758.80	1634.22	1525.29	1430.03	1343.64	1258.33	1159.55	989.17
	s (kJ/kg K)	5.13	5.07	4.92	4.75	4.57	4.39	4.22	4.05	3.87	3.65	3.12
	x_l (kg/kg)	0.00	0.05	0.09	0.13	0.17	0.22	0.26	0.31	0.37	0.46	1.00
	x_v (kg/kg)	0.00	0.23	0.40	0.54	0.65	0.74	0.81	0.86	0.91	0.95	1.00

TABLE A.11 (Continued)

F, where T = T$_{bp}$ + F(T$_{dp}$ - T$_{bp}$)	Property	Ammonia concentration (mass fraction), x										
		0	0.1	0.2	0.3	0.4	0.5	0.6	0.7	0.8	0.9	1
0.8	T (°C)	231.39	221.66	212.00	202.05	191.77	181.20	170.22	158.36	144.46	125.12	63.51
	v (m³/kg)	0.05	0.05	0.05	0.05	0.05	0.05	0.05	0.05	0.05	0.05	0.03
	h (kJ/kg)	2444.30	2247.87	2108.70	1976.75	1853.22	1739.59	1635.10	1536.07	1435.39	1317.61	1084.70
	s (kJ/kg K)	5.48	5.45	5.33	5.18	5.02	4.85	4.68	4.50	4.30	4.04	3.41
	x$_l$ (kg/kg)	0.00	0.04	0.07	0.11	0.14	0.18	0.23	0.27	0.33	0.42	1.00
	x$_v$ (kg/kg)	0.00	0.19	0.34	0.47	0.58	0.67	0.75	0.82	0.88	0.94	1.00
0.9	T (°C)	230.95	223.87	216.40	208.31	199.55	190.03	179.56	167.61	152.95	131.92	63.26
	v (m³/kg)	0.06	0.06	0.06	0.06	0.06	0.06	0.06	0.06	0.05	0.05	0.04
	h (kJ/kg)	2624.06	2447.99	2329.49	2209.40	2089.29	1971.07	1855.21	1739.82	1619.40	1478.76	1180.28
	s (kJ/kg K)	5.84	5.83	5.74	5.62	5.48	5.33	5.15	4.96	4.73	4.44	3.71
	x$_l$ (kg/kg)	0.00	0.03	0.06	0.09	0.12	0.15	0.19	0.24	0.30	0.39	1.00
	x$_v$ (kg/kg)	0.00	0.15	0.28	0.39	0.50	0.59	0.68	0.77	0.85	0.92	1.00
1.0	T (°C)	230.50	226.08	220.79	214.57	207.32	198.86	188.90	176.85	161.43	138.73	63.01
	v (m³/kg)	0.07	0.07	0.07	0.07	0.07	0.07	0.07	0.06	0.06	0.06	0.04
	h (kJ/kg)	2803.99	2683.91	2562.60	2439.84	2315.34	2188.63	2058.88	1924.55	1782.04	1619.95	1275.91
	s (kJ/kg K)	6.19	6.21	6.15	6.06	5.93	5.78	5.60	5.38	5.13	4.80	4.01
	x$_l$ (kg/kg)	0.00	0.00	0.00	0.00	0.00	0.00	0.00	0.00	0.00	0.00	0.00
	x$_v$ (kg/kg)	0.00	0.10	0.20	0.30	0.40	0.50	0.60	0.70	0.80	0.90	1.00

P = 30 bar.

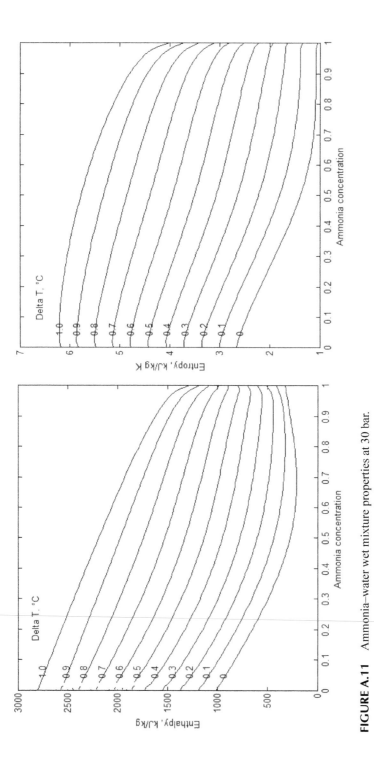

FIGURE A.11   Ammonia–water wet mixture properties at 30 bar.

TABLE A.12 Ammonia–Water Property System at 40 Bar Under Saturated and Wet Conditions.

F, where $T = T_{bp}$ $+ F(T_{dp} - T_{bp})$	Property	Ammonia concentration (mass fraction), x										
		0	0.1	0.2	0.3	0.4	0.5	0.6	0.7	0.8	0.9	1
0	T (°C)	251.69	220.59	192.88	167.42	144.51	125.03	109.59	98.14	89.92	83.66	78.13
	v (m³/kg)	0.00	0.00	0.00	0.00	0.00	0.00	0.00	0.00	0.00	0.00	0.00
	h (kJ/kg)	1090.21	904.01	729.56	572.18	441.28	346.16	292.12	278.48	298.44	340.34	390.20
	s (kJ/kg K)	2.80	2.62	2.36	2.07	1.80	1.57	1.40	1.31	1.28	1.28	1.24
	x_l (kg/kg)	0.00	0.10	0.20	0.30	0.40	0.50	0.60	0.70	0.80	0.90	1.00
	x_v (kg/kg)	0.00	0.00	0.00	0.00	0.00	0.00	0.00	0.00	0.00	0.00	0.00
0.1	T (°C)	251.09	222.64	197.16	173.60	152.23	133.81	118.85	107.27	98.22	90.16	77.80
	v (m³/kg)	0.01	0.01	0.01	0.01	0.01	0.01	0.01	0.01	0.01	0.01	0.00
	h (kJ/kg)	1260.71	1061.17	885.32	730.74	601.62	502.87	436.33	399.54	387.65	400.84	476.91
	s (kJ/kg K)	3.13	2.96	2.71	2.44	2.18	1.96	1.79	1.67	1.59	1.55	1.51
	x_l (kg/kg)	0.00	0.09	0.18	0.28	0.36	0.45	0.54	0.62	0.70	0.80	1.00
	x_v (kg/kg)	0.00	0.39	0.64	0.80	0.89	0.94	0.97	0.98	0.99	0.99	1.00
0.2	T (°C)	250.49	224.70	201.45	179.78	159.95	142.58	128.12	116.40	106.52	96.66	77.48
	v (m³/kg)	0.01	0.01	0.01	0.01	0.01	0.01	0.01	0.01	0.01	0.01	0.01
	h (kJ/kg)	1431.44	1221.59	1045.20	892.88	765.55	665.22	591.72	541.77	510.32	496.35	563.69
	s (kJ/kg K)	3.46	3.30	3.06	2.81	2.56	2.35	2.18	2.04	1.94	1.86	1.78
	x_l (kg/kg)	0.00	0.09	0.17	0.25	0.33	0.41	0.48	0.55	0.62	0.71	1.00
	x_v (kg/kg)	0.00	0.36	0.61	0.76	0.86	0.92	0.95	0.97	0.98	0.99	1.00
0.3	T (°C)	249.89	226.76	205.73	185.97	167.66	151.36	137.38	125.52	114.82	103.17	77.15
	v (m³/kg)	0.02	0.02	0.02	0.02	0.01	0.01	0.01	0.01	0.01	0.01	0.01
	h (kJ/kg)	1602.38	1385.44	1209.82	1059.47	933.56	832.34	754.31	695.53	650.31	614.41	650.51
	s (kJ/kg K)	3.78	3.65	3.42	3.18	2.95	2.75	2.57	2.43	2.31	2.20	2.04
	x_l (kg/kg)	0.00	0.08	0.15	0.23	0.30	0.37	0.43	0.50	0.56	0.65	1.00
	x_v (kg/kg)	0.00	0.33	0.57	0.72	0.83	0.89	0.93	0.96	0.97	0.98	1.00

TABLE A.12 *(Continued)*

F, where $T = T_{bp} + F(T_{dp} - T_{bp})$	Property	Ammonia concentration (mass fraction), x										
		0	0.1	0.2	0.3	0.4	0.5	0.6	0.7	0.8	0.9	1
0.4	T (°C)	249.29	228.81	210.01	192.15	175.38	160.13	146.65	134.65	123.12	109.67	76.83
	v (m³/kg)	0.02	0.02	0.02	0.02	0.02	0.02	0.02	0.02	0.02	0.02	0.01
	h (kJ/kg)	1773.53	1552.91	1379.88	1231.57	1106.59	1004.34	922.52	856.61	800.14	745.57	737.38
	s (kJ/kg K)	4.11	3.99	3.78	3.56	3.34	3.15	2.98	2.83	2.69	2.54	2.31
	x_l (kg/kg)	0.00	0.07	0.14	0.20	0.27	0.33	0.39	0.45	0.51	0.60	1.00
	x_v (kg/kg)	0.00	0.30	0.53	0.68	0.79	0.86	0.91	0.94	0.96	0.98	1.00
0.5	T (°C)	248.69	230.87	214.29	198.33	183.09	168.91	155.91	143.78	131.42	116.17	76.51
	v (m³/kg)	0.03	0.03	0.03	0.02	0.02	0.02	0.02	0.02	0.02	0.02	0.02
	h (kJ/kg)	1944.88	1724.17	1556.09	1410.40	1285.97	1182.11	1096.20	1023.34	956.33	884.68	824.30
	s (kJ/kg K)	4.44	4.34	4.15	3.94	3.74	3.55	3.38	3.23	3.07	2.90	2.58
	x_l (kg/kg)	0.00	0.07	0.12	0.18	0.24	0.29	0.35	0.40	0.46	0.55	1.00
	x_v (kg/kg)	0.00	0.27	0.48	0.63	0.74	0.82	0.87	0.91	0.94	0.97	1.00
0.6	T (°C)	248.09	232.93	218.57	204.52	190.81	177.68	165.18	152.90	139.72	122.68	76.18
	v (m³/kg)	0.03	0.03	0.03	0.03	0.03	0.03	0.03	0.03	0.03	0.02	0.02
	h (kJ/kg)	2116.45	1899.40	1739.20	1597.32	1473.37	1367.12	1276.14	1195.48	1117.44	1029.02	911.27
	s (kJ/kg K)	4.76	4.69	4.52	4.33	4.14	3.96	3.79	3.63	3.46	3.26	2.85
	x_l (kg/kg)	0.00	0.06	0.11	0.16	0.21	0.26	0.31	0.36	0.42	0.51	1.00
	x_v (kg/kg)	0.00	0.24	0.44	0.58	0.69	0.77	0.84	0.89	0.93	0.96	1.00
0.7	T (°C)	247.49	234.99	222.86	210.70	198.53	186.46	174.44	162.03	148.02	129.18	75.86
	v (m³/kg)	0.04	0.04	0.04	0.03	0.03	0.03	0.03	0.03	0.03	0.03	0.02
	h (kJ/kg)	2288.22	2078.76	1929.95	1793.83	1670.82	1561.45	1463.95	1373.76	1283.13	1177.15	998.27
	s (kJ/kg K)	5.09	5.04	4.90	4.73	4.56	4.38	4.21	4.04	3.85	3.62	3.12
	x_l (kg/kg)	0.00	0.05	0.09	0.13	0.18	0.22	0.27	0.32	0.38	0.48	1.00
	x_v (kg/kg)	0.00	0.21	0.38	0.52	0.63	0.72	0.79	0.85	0.90	0.95	1.00

TABLE A.12 *(Continued)*

F, where T = T_bp + F(T_dp − T_bp)	Property	Ammonia concentration (mass fraction), x										
		0	0.1	0.2	0.3	0.4	0.5	0.6	0.7	0.8	0.9	1
0.8	T (°C)	246.89	237.04	227.14	216.88	206.24	195.23	183.71	171.16	156.32	135.68	75.53
	v (m³/kg)	0.04	0.04	0.04	0.04	0.04	0.04	0.04	0.04	0.04	0.03	0.02
	h (kJ/kg)	2460.20	2262.45	2129.12	2001.48	1880.60	1767.68	1661.91	1559.71	1453.88	1328.39	1085.32
	s (kJ/kg K)	5.42	5.39	5.28	5.14	4.98	4.81	4.64	4.46	4.25	3.98	3.39
	x_l (kg/kg)	0.00	0.05	0.08	0.11	0.15	0.19	0.24	0.29	0.35	0.44	1.00
	x_v (kg/kg)	0.00	0.18	0.33	0.45	0.56	0.66	0.74	0.81	0.87	0.93	1.00
0.9	T (°C)	246.29	239.10	231.42	223.07	213.96	204.01	192.97	180.28	164.63	142.19	75.21
	v (m³/kg)	0.05	0.05	0.05	0.05	0.04	0.04	0.04	0.04	0.04	0.04	0.03
	h (kJ/kg)	2632.39	2450.61	2337.45	2221.86	2105.15	1988.79	1872.95	1755.52	1630.74	1482.61	1172.40
	s (kJ/kg K)	5.75	5.75	5.66	5.55	5.41	5.25	5.08	4.89	4.66	4.35	3.66
	x_l (kg/kg)	0.00	0.04	0.06	0.09	0.12	0.16	0.20	0.25	0.31	0.41	1.00
	x_v (kg/kg)	0.00	0.14	0.27	0.38	0.49	0.58	0.67	0.76	0.84	0.92	1.00
1.0	T (°C)	245.69	241.16	235.70	229.25	221.68	212.78	202.24	189.41	172.93	148.69	74.88
	v (m³/kg)	0.05	0.05	0.05	0.05	0.05	0.05	0.05	0.05	0.05	0.04	0.03
	h (kJ/kg)	2804.77	2686.64	2567.01	2445.63	2322.13	2195.96	2066.18	1930.98	1786.26	1619.13	1259.51
	s (kJ/kg K)	6.08	6.10	6.05	5.95	5.83	5.67	5.49	5.28	5.02	4.68	3.94
	x_l (kg/kg)	0.00	0.00	0.00	0.00	0.00	0.00	0.00	0.00	0.00	0.00	0.00
	x_v (kg/kg)	0.00	0.10	0.20	0.30	0.40	0.50	0.60	0.70	0.80	0.90	1.00

$P = 40$ bar.

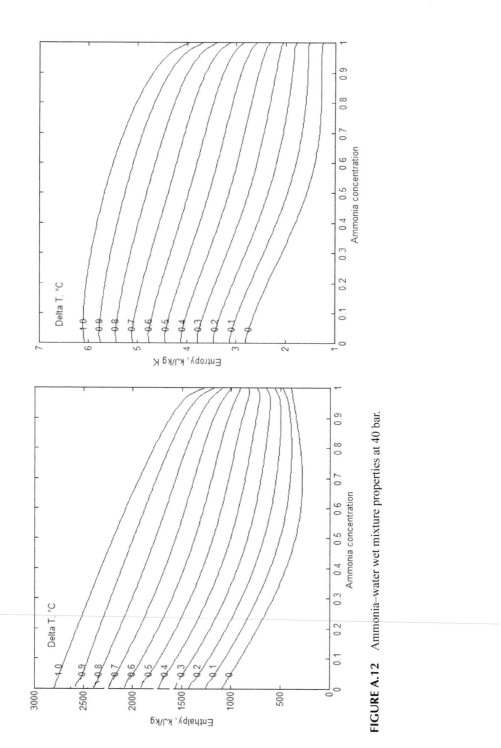

FIGURE A.12 Ammonia–water wet mixture properties at 40 bar.

TABLE A.13 Ammonia–Water Property System at 50 Bar Under Saturated and Wet Conditions.

F, where T = T_bp + F(T_dp − T_bp)	Property	Ammonia concentration (mass fraction), x										
		0	0.1	0.2	0.3	0.4	0.5	0.6	0.7	0.8	0.9	1
0	T (°C)	265.40	234.19	206.03	180.11	156.80	136.96	121.19	109.43	100.91	94.35	88.53
	v (m³/kg)	0.00	0.00	0.00	0.00	0.00	0.00	0.00	0.00	0.00	0.00	0.00
	h (kJ/kg)	1155.39	972.07	797.23	638.34	505.60	408.71	353.19	338.41	357.53	398.79	448.15
	s (kJ/kg K)	2.93	2.75	2.50	2.22	1.95	1.72	1.56	1.46	1.43	1.43	1.40
	x_l (kg/kg)	0.00	0.10	0.20	0.30	0.40	0.50	0.60	0.70	0.80	0.90	1.00
	x_v (kg/kg)	0.00	0.00	0.00	0.00	0.00	0.00	0.00	0.00	0.00	0.00	0.00
0.1	T (°C)	264.67	236.11	210.21	186.21	164.45	145.67	130.38	118.45	109.04	100.59	88.14
	v (m³/kg)	0.01	0.01	0.00	0.00	0.00	0.00	0.00	0.00	0.00	0.00	0.00
	h (kJ/kg)	1318.98	1123.69	948.12	792.21	661.16	560.36	492.03	453.92	441.53	455.49	527.12
	s (kJ/kg K)	3.23	3.07	2.83	2.56	2.31	2.09	1.92	1.80	1.72	1.69	1.65
	x_l (kg/kg)	0.00	0.09	0.18	0.28	0.37	0.45	0.54	0.62	0.70	0.80	1.00
	x_v (kg/kg)	0.00	0.37	0.63	0.78	0.87	0.93	0.96	0.98	0.98	0.99	1.00
0.2	T (°C)	263.93	238.03	214.39	192.32	172.11	154.39	139.57	127.47	117.18	106.84	87.74
	v (m³/kg)	0.01	0.01	0.01	0.01	0.01	0.01	0.01	0.01	0.01	0.01	0.01
	h (kJ/kg)	1482.83	1278.15	1102.84	949.60	820.43	717.89	642.22	590.37	557.63	543.88	606.11
	s (kJ/kg K)	3.53	3.39	3.16	2.91	2.67	2.46	2.28	2.15	2.05	1.97	1.89
	x_l (kg/kg)	0.00	0.09	0.17	0.25	0.33	0.41	0.49	0.56	0.63	0.73	1.00
	x_v (kg/kg)	0.00	0.35	0.59	0.75	0.85	0.91	0.94	0.96	0.98	0.99	1.00
0.3	T (°C)	263.19	239.95	218.57	198.43	179.76	163.10	148.76	136.49	125.32	113.08	87.35
	v (m³/kg)	0.01	0.01	0.01	0.01	0.01	0.01	0.01	0.01	0.01	0.01	0.01
	h (kJ/kg)	1646.93	1435.60	1261.91	1111.26	983.80	880.36	799.88	738.66	691.23	654.11	685.14
	s (kJ/kg K)	3.84	3.71	3.50	3.26	3.03	2.83	2.66	2.52	2.39	2.28	2.14
	x_l (kg/kg)	0.00	0.08	0.15	0.23	0.30	0.37	0.44	0.50	0.57	0.66	1.00
	x_v (kg/kg)	0.00	0.32	0.56	0.71	0.81	0.88	0.92	0.95	0.97	0.98	1.00

TABLE A.13   (Continued)

F, where $T = T_{bp}$ $+ F(T_{dp} - T_{bp})$	Property	Ammonia concentration (mass fraction), x										
		0	0.1	0.2	0.3	0.4	0.5	0.6	0.7	0.8	0.9	1
0.4	T (°C)	262.45	241.87	222.74	204.53	187.42	171.82	157.95	145.51	133.45	119.32	86.96
	v (m³/kg)	0.02	0.02	0.02	0.02	0.02	0.02	0.01	0.01	0.01	0.01	0.01
	h (kJ/kg)	1811.29	1596.16	1425.92	1278.07	1152.04	1047.77	963.38	894.58	835.06	777.75	764.19
	s (kJ/kg K)	4.14	4.04	3.84	3.62	3.40	3.21	3.04	2.89	2.75	2.60	2.39
	x_l (kg/kg)	0.00	0.07	0.14	0.21	0.27	0.33	0.39	0.45	0.52	0.61	1.00
	x_v (kg/kg)	0.00	0.30	0.52	0.67	0.78	0.85	0.89	0.93	0.95	0.97	1.00
0.5	T (°C)	261.71	243.79	226.92	210.64	195.07	180.53	167.14	154.53	141.59	125.57	86.57
	v (m³/kg)	0.02	0.02	0.02	0.02	0.02	0.02	0.02	0.02	0.02	0.02	0.01
	h (kJ/kg)	1975.89	1759.98	1595.47	1451.08	1326.28	1220.79	1132.37	1056.35	985.63	909.83	843.25
	s (kJ/kg K)	4.45	4.36	4.18	3.98	3.78	3.59	3.42	3.26	3.11	2.93	2.64
	x_l (kg/kg)	0.00	0.07	0.13	0.18	0.24	0.30	0.35	0.41	0.47	0.57	1.00
	x_v (kg/kg)	0.00	0.27	0.48	0.62	0.73	0.81	0.86	0.91	0.94	0.97	1.00
0.6	T (°C)	260.98	245.72	231.10	216.75	202.73	189.25	176.33	163.55	149.73	131.81	86.18
	v (m³/kg)	0.02	0.02	0.02	0.02	0.02	0.02	0.02	0.02	0.02	0.02	0.01
	h (kJ/kg)	2140.74	1927.19	1771.16	1631.45	1507.94	1400.64	1307.43	1223.54	1141.30	1047.55	922.33
	s (kJ/kg K)	4.76	4.69	4.53	4.35	4.16	3.98	3.81	3.64	3.47	3.26	2.89
	x_l (kg/kg)	0.00	0.06	0.11	0.16	0.21	0.26	0.31	0.37	0.43	0.53	1.00
	x_v (kg/kg)	0.00	0.24	0.43	0.57	0.68	0.77	0.83	0.88	0.92	0.96	1.00
0.7	T (°C)	260.24	247.64	235.28	222.85	210.38	197.96	185.52	172.57	157.87	138.05	85.79
	v (m³/kg)	0.03	0.03	0.03	0.03	0.03	0.03	0.03	0.03	0.02	0.02	0.02
	h (kJ/kg)	2305.83	2097.93	1953.65	1820.42	1698.72	1589.07	1489.84	1396.60	1301.55	1189.35	1001.42
	s (kJ/kg K)	5.06	5.02	4.88	4.72	4.55	4.38	4.20	4.03	3.84	3.60	3.14
	x_l (kg/kg)	0.00	0.06	0.10	0.14	0.18	0.23	0.28	0.33	0.39	0.49	1.00
	x_v (kg/kg)	0.00	0.21	0.38	0.52	0.63	0.71	0.79	0.85	0.90	0.94	1.00

TABLE A.13 *(Continued)*

F, where T = T_bp + F(T_dp − T_bp)	Property	Ammonia concentration (mass fraction), x										
		0	0.1	0.2	0.3	0.4	0.5	0.6	0.7	0.8	0.9	1
0.8	T (°C)	259.50	249.56	239.46	228.96	218.04	206.68	194.71	181.59	166.00	144.30	85.40
	v (m³/kg)	0.03	0.03	0.03	0.03	0.03	0.03	0.03	0.03	0.03	0.03	0.02
	h (kJ/kg)	2471.17	2272.35	2143.58	2019.33	1900.55	1788.22	1681.48	1576.71	1466.59	1334.41	1080.52
	s (kJ/kg K)	5.37	5.35	5.24	5.10	4.95	4.78	4.61	4.42	4.21	3.94	3.39
	x_l (kg/kg)	0.00	0.05	0.08	0.12	0.16	0.20	0.24	0.29	0.36	0.46	1.00
	x_v (kg/kg)	0.00	0.18	0.33	0.45	0.56	0.65	0.73	0.81	0.87	0.93	1.00
0.9	T (°C)	258.76	251.48	243.63	235.07	225.69	215.39	203.90	190.61	174.14	150.54	85.00
	v (m³/kg)	0.04	0.04	0.04	0.04	0.04	0.04	0.03	0.03	0.03	0.03	0.02
	h (kJ/kg)	2636.74	2450.57	2341.58	2229.55	2115.52	2000.63	1884.77	1765.64	1637.17	1482.39	1159.61
	s (kJ/kg K)	5.68	5.68	5.59	5.48	5.35	5.20	5.02	4.83	4.59	4.28	3.64
	x_l (kg/kg)	0.00	0.04	0.07	0.10	0.13	0.17	0.21	0.26	0.32	0.43	1.00
	x_v (kg/kg)	0.00	0.15	0.27	0.39	0.49	0.58	0.67	0.76	0.84	0.92	1.00
1.0	T (°C)	258.02	253.40	247.81	241.17	233.35	224.11	213.09	199.64	182.28	156.78	84.61
	v (m³/kg)	0.04	0.04	0.04	0.04	0.04	0.04	0.04	0.04	0.04	0.03	0.02
	h (kJ/kg)	2802.55	2686.17	2568.05	2447.91	2325.33	2199.68	2069.88	1933.88	1787.11	1615.09	1238.70
	s (kJ/kg K)	5.99	6.02	5.96	5.87	5.74	5.59	5.41	5.20	4.93	4.59	3.89
	x_l (kg/kg)	0.00	0.00	0.00	0.00	0.00	0.00	0.00	0.00	0.00	0.00	0.00
	x_v (kg/kg)	0.00	0.10	0.20	0.30	0.40	0.50	0.60	0.70	0.80	0.90	1.00

$P = 50$ bar.

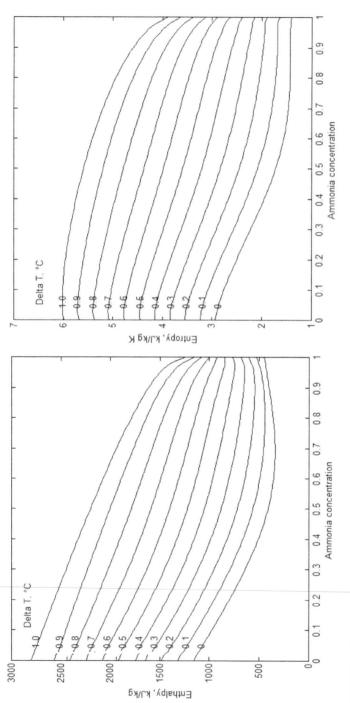

FIGURE A.13   Ammonia–water wet mixture properties at 50 bar.

Therefore, the dryness fraction of the mixture is defined from the simplification.

$$DF = \frac{m_{vapor}}{m_{mixture}} = \frac{x_{mixture} - x_{liquid}}{x_{vapor} - x_{liquid}} = \frac{x - x_l}{x_v - x_l} \tag{A.5}$$

For the above example, the dryness fraction of the mixture can be determined as follows:

$$DF = \frac{x - x_l}{x_v - x_l} = \frac{0.5 - 0.33}{1 - 0.33} = 0.25$$

It shows that at 0.15 bar, if the liquid–vapor mixture is at $-16.3°C$ (between bubble point and dew point), the dryness fraction is 0.25 (0.75 wet) with 0.5 liquid concentration and 100% pure vapor.

REFERENCES

Abovsky, V. Thermodynamics of Ammonia–Water Mixture. *Fluid Phase Equilibr.* **1996,** *116* (1), 170–176.

Ahmad, A.; Al-Dadah, R.; Mahmoud, S. Air Conditioning and Power Generation for Residential Applications Using Liquid Nitrogen. *Appl. Energy* **2016,** *184,* 630–640.

Ahmadzadeh, A.; Salimpour, M. R.; Sedaghat, A. Thermal and Exergoeconomic Analysis of a Novel Solar Driven Combined Power and Ejector Refrigeration (CPER) System. *Int. J. Refrig.* **2017,** *83,* 143–156.

Akasaka, R. A Rigorous Calculation of the Critical Point from the Fundamental Equation of State for the Water + Ammonia Mixture. *Int. J. Refrig.* **2009,** *32* (1), 95–101.

Alamdari, G. S. Simple Functions for Predicting the Thermodynamic Properties of Ammonia–Water Mixture (Research Note). *Int. J. Eng. Mater. Energy Res. Center* **2007,** *20* (1), 94–104.

Amano, Y.; Suzuki, T.; Hashizume, T.; Akiba, M.; Tanzawa, Y.; Usui, A. In *A Hybrid Power Generation and Refrigeration Cycle with Ammonia–Water Mixture,* IJPGC2000-15058, Proceedings of 2000 Joint Power Generation Conference, ASME, 2000.

Aneke, M.; Agnew, B.; Underwood, C. Approximate Analysis of the Economic Advantage of a Dual Source ORC System over Two Single ORC Systems in the Conversion of Dual Low and Mid-Grade Heat Energy to Electricity. *J. EUEC* **2011,** *5,* 1–17.

Antonopoulos, K. A.; Rogdakis, E. D. Nomographs for Optimum Solar Pond Driven LiBr/ZnBr$_2$/CH$_3$OH Absorption Refrigeration System. *Int. J. Energy Res.* **1992,** *16,* 413–429.

Aphornratana, S. Theoretical and Experimental Investigation of a Combined Ejector-Absorption Refrigerator. PhD Thesis, University of Sheffield, Sheffield, UK, 1995.

Aphornratana, S.; Eames, I. W. Experimental Investigation of a Combined Ejector-Absorption Refrigerator. *Int. J. Energy Res.* **1998,** *22,* 195–207.

Aphornratana, S.; Sriveerakul, T. Experimental Studies of a Single-Effect Absorption Refrigerator Using Aqueous Lithium Bromide: Effect of Operating Condition to System Performance. *Exp. Therm. Fluid Sci.* **2007,** *32* (2), 658–669.

Arabkoohsar, A.; Andresen, G. B. A Smart Combination of a Solar Assisted Absorption Chiller and a Power Productive Gas Expansion Unit for Cogeneration of Power and Cooling. *Renew. Energy* **2018,** *115,* 489–500.

Arslan, O. Exergoeconomic Evaluation of Electricity Generation by the Medium Temperature Geothermal Resources, Using a Kalina Cycle: Simav Case Study. *Int. J. Therm. Sci.* **2010,** *49* (9), 1866–1873.

Ashouri, M.; Razi Astaraei, F.; Ghasempour, R.; Ahmadi, M. H.; Feidt, M. Thermodynamic and Economic Evaluation of a Small-Scale Organic Rankine Cycle Integrated with a Concentrating Solar Collector. *Int. J. Low-Carbon Technol.* **2015,** *12* (1), 54–65.

Asou, H.; Yasunaga, T.; Ikegami, Y. In *Comparison between Kalina Cycle and Conventional OTEC System Using Ammonia–Water Mixtures as Working Fluid,* Proceedings of the Seventh International Offshore and Polar Engineering Conference, 2007.

Bai, O.; Nakamura, M.; Ikegami, Y.; Uehara, H. A Simulation Model for Hot Spring Thermal Energy Conversion Plant with Working Fluid of Binary Mixtures. *J. Eng. Gas Turbines Power.* **2004,** *126* (3), 445–454.

Baik, Y. J.; Kim, M.; Chang, K. C.; Kim, S. J. Power-Based Performance Comparison between Carbon Dioxide and R125 Transcritical Cycles for a Low-Grade Heat Source. *Appl. Energy* **2011,** *88* (3), 892–898.

Barhoumi, M.; Snoussi, A.; Ben Ezzine, N.; Mejbri, K.; Bellagi, A. Modelling of the Thermodynamic Properties of the Ammonia/Water Mixture. *Int. J. Refrig.* **2004,** *27* (3), 271–283.

Bennani, N.; Prevost, M.; Coronas, A. Absorption Heat Pump Cycle: Performance Analysis of Water Glycerol Mixture. *Heat Recov. Syst. CHP* **1989,** *9* (3), 257–63.

Best, R.; Porras, L.; Holland, F. A. Thermodynamic Design Data for Absorption Heat Pump System Operating on Ammonia-Nitrate: Part I Cooling. *Heat Recov. Syst. CHP* **1991,** *11* (1), 49–61.

Bliem, C. J. *The Kalina Cycle and Similar Cycles for Geothermal Power Production (No. EGG-EP-8132)*; EG and G Idaho, Inc.: Idaho Falls, ID, 1988.

Bombarda, P.; Invernizzi, C, M.; Pietra, C. Heat Recovery from Diesel Engines: A Thermodynamic Comparison between Kalina and ORC Cycles. *Appl. Therm. Eng.* **2010,** *30* (2), 212–219.

Borelli, S. J. S.; de Oliveira Junior, S. Exergy-Based Method for Analyzing the Composition of the Electricity Cost Generated in Gas-Fired Combined Cycle Plants. *Energy* **2008,** *33* (2), 153–162.

Borgert, J. A.; Velasquez, J. A. Exergoeconomic Optimisation of a Kalina Cycle for Power Generation. *Int. J. Exergy* **2004,** *1* (1), 18–28.

Brodyanskii, V. M. Improving the Efficiency of Nuclear and Geothermal Power Stations by Using Low Environment Temperatures. *Therm. Eng.* **2006,** *53* (3), 201–207.

Calise, F.; Libertini, L.; Vicidomini, M. Exergetic Analysis of a Novel Solar Cooling System for Combined Cycle Power Plants. *Entropy* **2016,** *18* (10), 1–31.

Cao, L.; Wang, J.; Dai, Y. Thermodynamic Analysis of a Biomass-Fired Kalina Cycle with Regenerative Heater. *Energy* **2014,** *77*, 760–770.

Cao, L.; Wang, J.; Wang, H.; Zhao, P.; Dai, Y. Thermodynamic Analysis of a Kalina-Based Combined Cooling and Power Cycle Driven by Low-Grade Heat Source. *Appl. Therm. Eng.* **2017,** *111*, 8–19.

Cayer, E.; Galanis, N.; Desilets, M.; Nesreddine, H.; Roy, P. Analysis of a Carbon Dioxide Transcritical Power Cycle Using a Low Temperature Source. *Appl. Energy* **2009,** *86* (7), 1055–1063.

Chacartegui, R.; Sanchez, D.; Munoz, J. M.; Sanchez, T. Alternative ORC Bottoming Cycles for Combined Cycle Power Plants. *Appl. Energy* **2009,** *86* (10), 2162–2170.

Chen, L. T. A New Ejector-Absorber Cycle to Improve the COP of an Absorption Refrigeration System. *Appl. Energy* **1988,** *30*, 37–51.

Chen, Y.; Guo, Z, Wu, J.; Zhang, Z and Hua, J. Energy and Exergy Analysis of Integrated System of Ammonia–Water Kalina–Rankine Cycle. *Energy* **2015,** *90*, 2028–2037.

Chen, Y.; Han, W.; Jin, H. Investigation of an Ammonia–Water Combined Power and Cooling System Driven by the Jacket Water and Exhaust Gas Heat of an Internal Combustion Engine. *Int. J. Refrig.* **2017,** *82*, 174–188.

Chew, J. M.; Reddy, C. C. S.; Rangaiah, G. P. Improving Energy Efficiency of Driving-Wall Columns Using Heat Pumps, Organic Rankine Cycle and Kalina Cycle. *Chem. Eng. Process.: Process Intensif.* **2014,** *76*, 7 45–59.

Chung, H.; Huor, M. H.; Prevost, M.; Bugarel, R. In *Domestic Heating Application of an Absorption Heat Pump, Directly Fired Heat Pumps*, Proceedings of International Conference, University of Bristol, City of Bristol, England, 1984; pp 19–21.

Conde-Petit, M. In *Thermophysical Properties of (NH₃ + H₂O) Mixtures for the Industrial Design of Absorption Refrigeration Equipment*, Manuel Conde Engineering, Zurich Switzerland, 2006.

Dejfors, C.; Svedberg, G. Second Law Analysis of Ammonia–Water Power Cycle for Direct-Fired Cogeneration Application. *Int. J. Thermodyn.* **2010,** *2* (3), 125–131.

Dejfors, C.; Thorin, E.; Svedberg, G. Ammonia–Water Power Cycles for Direct-Fired Cogeneration Applications. *Energy Convers. Manag.* **1998,** *39* (16), 1675–1681.

Deng, J.; Wang, R. Z; Han, G. Y. A Review of Thermally Activated Cooling Technologies for Combined Cooling, Heating and Power Systems. *Progr. Energy Combust. Sci.* **2011,** *37* (2), 172–203.

Desideri, U.; Bidini, G. Study of Possible Optimisation Criteria for Geothermal Power Plants. *Energy Convers. Manag.* **1997,** *38* (15), 1681–1691.

DeVault, B. *Integrated Energy Systems Cooling, Heating & Power Overview*; Oak Ridge National Laboratory: Oak Ridge, TN, 2005; p 8.

DiPippo, R. Second Law Assessment of Binary Plants Generating Power from Low-Temperature Geothermal Fluids. *Goethermics* **2004,** *33* (5), 565–586.

Dorj, P. *Thermoeconomic Analysis of a New Geothermal Utilization CHP Plant in Tsetserleg, Monglia (No. 2)*. United Nations University, 2005.

Ebrahimi, M.; Soleimanpour, M. Design and Evaluation of Combined Cooling, Heating and Power Using Micro Gas Turbine, Adsorption Chiller and a Thermal Damping Tank in Micro Scale. *Appl. Therm. Eng.* **2017,** *127*, 1063–1076.

Edison, T. A.; Sengers, J. V. Thermodynamic Properties of Ammonia in the Critical Region. *Int. J. Refrig.* **1999,** *22* (5), 365–378.

Ehsana, A.; Yilmazoglu, Z. M. Design and Exergy Analysis of a Thermal Power Plant Using Different Types of Turkish Lignite. *Int. J. Thermodyn.* **2011,** *14* (3), 125–133.

Eller, T.; Heberle, F.; Brüggemann, D. Second Law Analysis of Novel Working Fluid Pairs for Waste Heat Recovery by the Kalina Cycle. *Energy* **2017,** *119*, 188–198.

El-Sayed, Y. M.; Tribus. M. A Theoretical Comparison of Rankine and Kalina Cycles. *ASME Publ., AES* **1985,** *1*, 97–102.

Fallah, M.; Mahnoudi, S. M. S.; Yari, M.; Ghiasi, R. A. Advanced Exergy Analysis of the Kalina Cycle Applied for Low Temperature Enhanced Geothermal System. *Energy Convers. Manag.* **2016,** *109*, 190–201.

Farrokh Niae, A. H.; Moddarress, H.; Mohsen-Nia, M. A Three Parameter Cubic Equation of State for Prediction of Thermodynamic Properties of Fluids. *J. Chem. Thermodyn.* **2008,** *40* (1), 84–95.

Fontalvo, A.; Pinzon, H.; Duarte, J.; Bula, A.; Quiroga, A. G.; Padilla, R. V. Exergy Analysis of a Combined Power and Cooling Cycle. *Appl. Therm. Eng.* **2013,** *60*, 164–171.

Franco, A.; Casarosa, C. On Some Perspectives for Increasing the Efficiency of Combined Cycle Power Plants. *Appl. Therm. Eng.* **2002,** *22* (13), 1501–1518.

Franco, A.; Villani, M. Optimal Design of Binary Cycle Power Plants for Water-Dominated, Medium-Temperature Geothermal Fields. *Geothermics* **2009,** *38* (4), 379–391.

Fu, W.; Zhu, J.; Zhang, W.; Lu, Z. Performance Evaluation of Kalina Cycle Subsystem on Geothermal Power Generation in the Oilfield. *Appl. Therm. Eng.* **2013a,** *54* (2), 497–506.

Fu, W.; Zhu, J.; Zhang, W.; Lu, Z. Comparison of a Kalina Cycle Based Cascade Utilization System with an Existing Organic Rankine Cycle Based Geothermal Power System in an Oilfield. *Appl. Therm. Eng.* **2013b,** *58* (1–2), 224–233.

Galanis, N.; Cayer, E.; Roy, P.; Denis, E. S.; Desilets, M. Electricity Generation from Low Temperature Sources. *J. Appl. Fluid Mech.* **2009,** *2* (2), 55–67.

Ganapathy, T.; Alagumurthi, N.; Gakkhar, R. P.; Murugesan, K. Exergy Analysis of Operating Lignite Fired Thermal Power Plant. *J. Eng. Sci. Technol. Rev.* **2009,** *2* (1), 123–130.

Garland, P. W. *CHP for Buildings Integration: Test Centres at ORNL and University of Maryland*; Oak Ridge National Laboratory: Oak Ridge, MD, 2003; p 24.

Gholamian, E.; Zare, V. A Comparative Thermodynamic Investigation with Environmental Analysis of SOFC Waste Heat to Power Conversion Employing Kalina and Organic Rankine Cycles. *Energy Convers. Manag.* **2016,** *117*, 150–161.

Gosney, W. B. *Principle of Refrigeration*; Cambridge Univ. Press: Cambridge, United Kingdom, 1982; p 8.

Goswami, D. Y.; Vijayaraghavan, S.; Lu, S.; Tamm, G. New and Emerging Developments in Solar Energy. *Solar Energy* **2004,** *76* (1), 33–43.

Grossman, G. Absorption Heat Transformer for Process Heat Generation from Solar Ponds. *ASHRAE Trans.* **1991,** *97*, 420–427.

Grossman, G.; Gommed, K. A Computer Model for Simulation of Absorption System in Flexible and Modular Form. *ASHRAE Trans.* **1987,** *93*, 389–427.

Grover, G. S.; Eisa, M. A. R.; Holland, F. A. Thermodynamic Design Data for Absorption Heat Pump System Operating on Water–Lithium Chloride: Part I Cooling. *Heat Recov. Syst. CHP* **1988,** *8* (1), 33–41.

Guo, Z.; Zhang, Z.; Chen, Y.; Wu, J.; Dong, C. Dual-Pressure Vaporization Kalina Cycle for Cascade Reclaiming Heat Resource for Power Generation. *Energy Convers. Manag.* **2015,** *106*, 557–565.

Guzovic, Z.; Loncar, D.; Ferdelji, N. Possibilities of Electricity Generation in the Republic of Croatia by Means of Geothermal Energy. *Energy* **2010,** *35* (8), 3429–3440.

Hansen, P. L.; Kuczma, P. D.; Palsson, J. O.; Simon, J. S. Vapor Temperature Control in a Kalina Cycle Power Generation System. U.S. Patent No. 6,167,705B1. U.S. Patent and Trademark Office: Washington, DC, 2001a.

Hansen, P. L.; Kuczma, P. D.; Palsson, J. O.; Simon, J. S. Regenerative Subsystem Control in a Kalina Cycle Power Generation System. U.S. Patent No. 6,195,998 B1. U.S. Patent and Trademark Office: Washington, DC, 2001b.

Hansen, P. L.; Kuczma, P. D.; Palsson, J, S. Distillation and Condensation Subsystem (DCSS) Control in Kalina Cycle Power Generation System. U.S. Patent No. 6,158,220. U.S. Patent and Trademark Office: Washington, DC, 2003.

Hasan, A. A.; Goswami, D. Y. Exergy Analysis of a Combined Power and Refrigeration Thermodynamic Cycle Driven by a Solar Heat Source. *J. Sol. Energy Eng.* **2003,** *125* (1), 55–60.

Hasan, A. A.; Goswami, D. Y.; Vijayaraghavan, S. First and Second Law Analysis of a New Power and Refrigeration Thermodynamic Cycle Using a Solar Heat Source. *Sol. Energy* **2002,** *73* (5), 385–393.

Hatem, M. M. A. Experimental and Analytical Investigation of Ammonia-Vapor Absorption into Ammonia–Water Solution. Doctoral Dissertation, Saga University, 2007.

He, J.; Liu, C.; Xu, X.; Li, Y.; Wu, S.; Xu, J. Performance Research on Modified KCS (Kalina Cycle System) 11 Without Throttle Valve. *Energy* **2014,** *64*, 389–397.

Henry, A. M. *An Introduction to the Kalina Cycle*; American Society of Mechanical Engineers: New York, NY, United States, 1996 (ASME Int., No. CONF-961006).

Heppenstall, T. Advanced Gas Turbine Cycles for Power Generation: A Critical Review. *Appl. Therm. Eng.* **1998,** *18* (9), 837–846.

Herold, K. E.; Radermacher, L. Absorption Heat Pump. *Mech. Eng.* **1989,** *68,* 68–73.

Hettiarachchi, H. M.; Golubovic, M.; Worek, W. M.; Ikegami, Y. The Performance of the Kalina Cycle System 11 (KCS-11) with Low-Temperature Heat Sources. *ASME J. Energy Resour. Technol.* **2007,** *129* (3), 243–247.

Holcomb, C. D.; Outcalt, S. L. Near-Saturation (P,ρ,T) and Vapour-Pressure Measurements of NH_3 and Liquid-Phase Isothermal (P,ρ,T) and Bubble-Point-Pressure Measurements of NH_3 + H_2O Mixtures. *Fluid Phase Equilibr.* **1999,** *164* (1), 97–106.

Holmberg, P.; Berntsson, T. Alternative Working Fluids in Heat Transformers. *ASHRAE Trans.* **1990,** *96*, 1582–1589.

Horuza, I.; Callander, T. M. S. Experimental Investigation of a Vapor Absorption Refrigeration System. *Int. J. Refrig.* **2004,** *27* (1), 10–16.

Hu, E.; Yang, Y.; Nishimura, A.; Yilmaz, F.; Kouzani, A. Solar Thermal Aided Power Generation. *Appl. Energy* **2010,** *87* (9), 2881–2885.

Hua, J.; Chen, Y.; Wu, J.; Zhi, Z, and Ding, C. Waste Heat Supply—Side Power Generation with Variable Concentration for Turbine in Kalina Cycle. *Appl. Therm. Eng.* **2015,** *91*, 583–590.

Hussein, I. B.; Yusoff, M. B.; Boosroh, M, H. In *Exergy Analysis of a 120-MW Thermal Power Plant*, Proceedings of the BSME—ASME International Conference on Thermal Engineering, Dhaka, Bangladesh, 2001; pp 177–182.

Idema, P. D. In *Simulation of Stationary Operation and Control of a LiBr/ZnBr₂/CH₃OH Absorption Heat Pump System, Directly Fired Heat Pump*, Proceedings of International Conference, University of Bristol, United Kingdom, Paper 2.1, 1984.

Inoue, T.; Monde, M.; Teruya, Y. Pool Boiling Heat Transfer in Binary Mixtures of Ammonia/Water. *Int. J. Heat Mass Transf.* **2002,** *45* (22), 4409–4415.

International Energy outlook. 2013. http://www.eia.gov/forecasts/ieo/.

James Hartley, G. Power Generation Cycle Analysis. PhD Thesis, 2001.

Jawahar, C. P.; Saravanan, R.; Bruno, J. C.; Coronas, A. Simulation Studies on Gas Based Kalina Cycle for Both Power and Cooling Applications. *Appl. Therm. Eng.* **2013,** *50* (2), 1522–1529.

Jonsson, M. Advanced Power Cycles with Mixture as the Working Fluid. Doctoral Dissertation, KTH, 2003, ISBN 91-7283-443-9.

Kalina, A. I. Combined Cycle System with Novel Bottoming Cycle. *ASME J. Eng. Gas Turbine Power* **1984,** *106* (4), 737–742.

Kalina, A. I. Method and Apparatus for Implementing a Thermodynamic Cycle with Intercooling. U.S. Patent No. 4,604,867. U.S. Patent and Trademark Office: Washington, DC, 1986.

Kalina, A. I. Direct Fired Power Cycle. U.S. Patent No. 4,732,005. U.S. Patent and Trademark Office: Washington, DC, 1988a.

Kalina, A. I. Method and Apparatus for Implementing a Thermodynamic Cycle with Recuperative Preheating. U.S. Patent No. 4,763,480. U.S. Patent and Trademark Office: Washington, DC, 1988b.

Kalina, A. I. Method and Apparatus for Converting Heat from Geothermal Fluid to Electric Power. U.S. Patent No. 4,982,568. U.S. Patent and Trademark Office: Washington, DC, 1991a.

Kalina, A. I. Method and Apparatus for Converting Low Temperature Heat to Electric Power. U.S. Patent No. 5,029,444. U.S. Patent and Trademark Office: Washington, DC, 1991b.

Kalina, A. I. Method and Apparatus for Converting Thermal Energy into Electric Power. U.S. Patent No. 5,095,708. U.S. Patent and Trademark Office: Washington, DC, 1992.

Kalina, A. I. Method and Apparatus for Converting Heat from Geothermal Liquid and Geothermal Steam to Electric Power. U.S. Patent No. 5,440,882. U.S. Patent and Trademark Office: Washington, DC, 1995a.

Kalina, A. I. Multi-Stage Combustion System for Externally Fired Power Plants. U.S. Patent No. 5,450,821. U.S. Patent and Trademark Office: Washington, DC, 1995b.

Kalina, A. I. System and Apparatus for Conversion of Thermal Energy into Mechanical and Electrical Power. U.S. Patent No. 5,572,871. U.S. Patent and Trademark Office: Washington, DC, 1996.

Kalina, A. I.; Hillsborough, C. A. Dual Pressure Geothermal System. U.S. Patent No. 6,735,948. U.S. Patent and Trademark Office: Washington, DC, 2004a.

Kalina, A. I.; Hillsborough, C. A. Power Cycle and System for Utilizing Moderate and Low Temperature Heat Sources. U.S. Patent No. 6,769,256. U.S. Patent and Trademark Office: Washington, DC, 2004b.

Kalina, A. I.; Hillsborough, C. A. Low Temperature Geothermal System. U.S. Patent No. 6,820,421 B2. U.S. Patent and Trademark Office: Washington, DC, 2004c.

Kalina, A. I.; Hillsborough, C. A. Geothermal System. U.S. Patent No. 6,829,895 B2. U.S. Patent and Trademark Office: Washington, DC, 2004d.

Kalina, A. I.; Hillsborough, C. A. Power Cycle and System for Utilizing Moderate and Low Temperature Heat Sources. U.S. Patent No. 6,910,334 B2. U.S. Patent and Trademark Office: Washington, DC, 2005a.

Kalina, A. I.; Hillsborough, C. A. Power Cycle and System for Utilizing Moderate and Low Temperature Heat Sources. U.S. Patent No. 6,941,757 B2. U.S. Patent and Trademark Office: Washington, DC, 2005b.

Kalina, A. I.; Hillsborough, C. A. Dual Pressure Geothermal System. U.S. Patent No. 6,923,000. U.S. Patent and Trademark Office: Washington, DC, 2005c.

Kalina, A. I.; Hillsborough, C. A. Power System and Apparatus for Utilizing Waste Heat. U.S. Patent No. 6,968,690 B2. U.S. Patent and Trademark Office: Washington, DC, 2005d.

Kalina, A. I.; Hillsborough, C. A. Single Flow Cascade Power System. U.S. Patent No. 7,055,326 B1. U.S. Patent and Trademark Office: Washington, DC, 2006a.

Kalina, A. I.; Hillsborough, C. A. Process and Apparatus for Boiling and Vaporizing Multi-Component Fluids. U.S. Patent No. 7,065,967 B2. U.S. Patent and Trademark Office: Washington, DC, 2006b.

Kalina, A. I.; Hillsborough, C. A. Power Cycle and System for Utilizing Moderate and Low Temperature Heat Sources. U.S. Patent No. 7,065,969. U.S. Patent and Trademark Office: Washington, DC, 2006c.

Kalina, A. I.; Hillsborough, C. A. Power Cycle and System for Utilizing Moderate and Low Temperature Heat Sources. U.S. Patent No. 7,065,969 B2. U.S. Patent and Trademark Office: Washington, DC, 2006d.

Kalina, A. I.; Hillsborough, C. A. Combustion System with Recirculation of Flue Gas. U.S. Patent No. 7,350,471 B2. U.S. Patent and Trademark Office: Washington, DC, 2008.

Kalina, A. I.; Mirolli, M. D. Supplying Heat to an Externally Fired Power System. U.S. Patent No. 5,588,298. U.S. Patent and Trademark Office: Washington, DC, 1996.

Kalina, A. I.; Pelletier, R. I. Method and Apparatus for Implementing a Thermodynamic Cycle. U.S. Patent No. 5,649,426. U.S. Patent and Trademark Office: Washington, DC, 1997.

Kalina, A. I.; Rhodes, L. B. Converting Heat into Useful Energy Using Separate Closed Loops. U.S. Patent No. 5,822,990. U.S. Patent and Trademark Office: Washington, DC, 1998.

Kalina, A. I.; Pelletier, R. I.; Rhodes, L. B. Method and Apparatus of Converting Heat to Useful Energy. U.S. Patent No. 5,953,918. U.S. Patent and Trademark Office: Washington, DC, 1999.

Kandlikar, S. G. A New Absorber Heat Recovery Cycle to Improve COP of Aqua–Ammonia Absorption Refrigeration System. *ASHRAE Trans.* **1982,** *88,* 141–158.

Karaali, R. Exergy Analysis of a Combined Power and Cooling Cycle. *Acta Phys. Pol. A* **2016,** *8* (T9), 209–213.

Kaushik, S. C.; Kumar, R. A Comparative Study of an Absorber Heat Recovery Cycle for Solar Refrigeration Using NH_3-Refrigerant with Liquid/Solid Absorbents. *Energy Res.* **1987,** *11,* 123–132.

Kiselev, S. B.; Rainwater, J. C. Extended Law of Corresponding States and Thermodynamic Properties of Binary Mixtures in and beyond the Critical Region. *Fluid Phase Equilibr.* **1997,** *141* (1), 129–154.

Kohler, S.; Saadat, A. In *Thermodynamic Modeling of Binary Cycles—Looking for Best Case Scenarios,* International Geothermal Conference, 2004; pp 14–19.

Korobitsyn, M. A. New and Advanced Energy Conversion Technologies. Analysis of Cogeneration, Combined and Integrated Cycles. PhD Thesis, University of Twente, 1998. ISBN 90-365-11070.

Koroneos, C. J.; Rovas, D. C. In *Electricity from Geothermal Energy with the Kalina Cycle an Exergy Approach,* ICCEP'07, International Conference on Clean Electrical Power, IEEE, 2007; pp 423–428.

Kumar, R. Thermodynamic Analysis of Power and Ejector Cooling Cycle. *J. Mater. Sci. Mech. Eng.* **2015,** *2* (6), 32–34.

Larsen, U.; Nguyen, T. V.; Knudsen, T.; Haglind, F. System Analysis and Optimisation of a Kalina Split-Cycle for Waste Heat Recovery on Large Marine Diesel Engines. *Energy* **2014,** *64,* 484–494.

Leibowitz, H.; Mirolli, M. First Kalina Combined-Cycle Plant Tested Successfully. *Power Eng.* **1997,** 45–48.

Lemmon, E. W.; Tillner-Roth, R. A Helmholtz Energy Equation of State for Calculating the Thermodynamic Properties of Fluid Mixtures. *Fluid Phase Equilibr.* **1999,** *165* (1), 1–21.

Li, S.; Dai, Y. Thermo-Economic Comparison of Kalina and CO_2 Transcritical Power Cycle for Low Temperature Geothermal Sources in China. *Appl. Therm. Eng.* **2014,** *70* (1), 139–152.

Li, H.; Hu, D.; Wang, M.; Dai, Y. Off-Design Performance Analysis of Kalina Cycle for Low Temperature Geothermal Source. *Appl. Therm. Eng.* **2016,** *70* (107), 728–737.

Li, M.; Mu, H.; Li, H. Analysis and Assessments of Combined Cooling, Heating and Power Systems in Various Operation Modes for a Building in China, Dalian. *Energies* **2013,** *6* (5), 2446–2467.

Liu, M.; Zhang, N. Proposal and Analysis of a Novel, Ammonia–Water Cycle for Power and Refrigeration Cogeneration. *Energy* **2007,** *32* (6), 961–970.

Liu, M.; Shi, Y.; Fang, F. Combined Cooling, Heating and Power Systems: A Survey. *Renew. Sustain. Energy Rev.* **2014,** *35,* 1–22.

Lolos, P. A.; Rogdakis, E. D. A Kalina Power Cycle Driven by Renewable Energy Sources. *Energy* **2009a,** *34* (4), 457–464.

Lolos, P, A.; Rogdakis, E, D. Thermodynamic Analysis of a Kalina Power Unit Driven by Low Temperature Heat Sources. *Therm. Sci.* **2009b,** *13* (4), 21–31.

Lopez-Villada, J.; Ayou, D. S.; Bruno, J. C.; Coronas, A. Modelling, Simulation and Analysis of Solar Absorption Power-Cooling Systems. *Int. J. Refrig.* **2014,** *39,* 125–136.

Lu, S.; Goswami, D. Y. In *Theoretical Analysis of Ammonia-Based Combined Power/ Refrigeration Cycle at Low Refrigeration Temperatures,* Proceedings of SOLAR, 2002.

Lu, S.; Goswami, D. Y. Optimization of a Novel Combined Power/Refrigeration Thermodynamic Cycle. *J. Sol. Energy Eng.* **2003,** *125* (2), 125–213.

Mahmoudi, S. M. S.; Pourreza, A.; Akbari, A. D.; Yari, M. Exergoeconomic Evaluation and Optimization of a Novel Combined Augmented Kalina Cycle/Gas Turbine-Modular Helium Reactor. *Appl. Therm. Eng.* **2016,** *109,* 109–120.

Marcriss, R. A.; Gutraj, J. M.; Zawacki, T. S. *Absorption Fluid Data Survey: Final Report on Worldwide Data.* ORLN/sub/8447989/3, Institute of Gas Technology, 1988; p 3.

Marcuccilli, F.; Zouaghi, S. In *Radial Inflow Turbines for Kalina and Organic Rankine Cycles,* Proceedings European Geothermal Congress, Unterhaching, Germany, 2007.

Maria, S. M.; Miguel, M. C.; Mussati, S. F. NLP Model-Based Optimal Design of LiBr–H$_2$O Absorption Refrigeration Systems. *Int. J. Refrig.* **2014,** *38,* 58–70.

Marston, C. H. Parametric Analysis of the Kalina Cycle. *J. Eng. Gas Turbines Power* **1990,** *112* (1), 107–116.

Marston, C. H.; Hyre, M. Gas Turbines Bottoming Cycles: Triple—Pressure Steam Versus Kalina. *J. Eng. Gas Turbines Power* **1995,** *117* (1), 10–15.

Martin, C.; Goswami, D. Y. Effectiveness of Cooling Production with a Combined Power and Cooling Thermodynamic Cycle. *Appl. Therm. Eng.* **2006,** *26,* 576–582.

Mejbri, Kh.; Bellagi, A. Modelling of the Thermodynamic Properties of the Water–Ammonia Mixture by Three Different Approaches. *Int. J. Refrig.* **2006,** *29* (2), 211–218.

Mills, D.; Morrison, G. L.; Le Lievre, P. In *Design of a 240 MWe Solar Thermal Power Plant,* Proc. EuroSun Conference, Freiburg, 2004.

Mirolli, M.; Hjartarson, H.; Mlcak, H. A.; Ralph, M. In *Testing and Operating Experience of the 2MW Kalina Cycle Geothermal Power Plant in Husavik, Iceland,* Proceedings of the World Renewable Energy Congress VII, Cologne, Germany, 2002.

Mirolli, M. D. In *Commercialization of Kalina Cycle for Power Generation and its Potential Impact on CO$_2$ Emissions,* International Joint Power Generation Conference, 2001, pp 1–7.

Mirolli, M. D. In *The Kalina Cycle for Cement Kiln Waste Heat Recovery Power Plants,* Cement Industry Technical Conference, Conference Record, IEEE, 2005; pp 330–336.

Mirolli, M. D. Cementing Kalina Cycle Effectiveness. *IEEE Industry Applications Magazine,* 2006, 12 (4), pp 60–64.

Mishra, R. D.; Sahoo, P. K.; Gupta, A. Thermoeconomic Evaluation and Optimization of an Aqua–Ammonia Vapour–Absorption Refrigeration System. *Int. J. Refrig.* **2006,** *29* (1), 47–59.

Misra, R. D.; Sahoo, P. K.; Sahoo, S.; Gupta, A. Thermoeconomic Optimization of a Single Effect Water/LiBr Vapor Absorption Refrigeration System. *Int. J. Refrig.* **2003,** *26* (2), 158–169.

Mittelman, G.; Epstein, M. A Novel Power Block for CSP Systems. *Sol. Energy* **2010,** *84* (10), 1761–1771.

Mlcak, H.; Mirolli, M.; Hjartarson, H.; Husavikur, O.; Ralph, M. *Notes from the North: A Report on the Debut Year of the 2 MW Kalina Cycle Geothermal Power Plant in Husavik, Iceland*; Transactions-Geothermal Resources Council, 2002; pp 715–718.

Mlcak, H. A. In *Design and Start-up of the 2MW Kalina Cycle Orkuveita Husavikur Geothermal Power Plant in Iceland*, European Geothermal Energy Council 2nd Business Seminar, 2001.

Modi, A.; Haglind, F. Performance Analysis of a Kalina for a Central Receiver Solar Thermal Power Plant with Direct Steam Generation. *Appl. Therm. Eng.* **2014**, *65* (1), 201–208.

Modi, A.; Haglind, F. Thermodynamic Optimisation and Analysis of Four Kalina Cycle Layouts for High Temperature Applications. *Appl. Therm. Eng.* **2015**, *76*, 196–205.

Modi, A.; Andreasen, J. G.; Kaem, M. R.; Haglind, F. Part-Load Performance of a High Temperature Kalina Cycle. *Energy Convers. Manag.* **2015**, *105*, 453–461.

Mohan, G.; Dahal, S.; Kumar, U.; Martin, A.; Kayal, H. Development of Natural Gas Fired Combined Cycle Plant for Tri-Generation of Power, Cooling and Clean Water Using Waste Heat Recovery: Techno-Economic Analysis. *Energies* **2014**, *7* (10), 6358–6381.

Mohanty, M. *Sustainable Urban Energy: A Sourcebook for Asia*; UN-HABITAT: Kenya, 2012; pp 20–75.

Murugan, R, S.; Subbarao, P. M. V. Thermodynamic Analysis of Rankine–Kalina Combined Cycle. *Int. J. Thermodyn.* **2008**, *11* (3), 133–141.

Muthu, V.; Saravanan, R.; Renganarayanan, S. Experimental Studies on R134a–DMAC Hot Water Based Vapor Absorption Refrigeration Systems. *Int. J. Therm. Sci.* **2008**, *47*, 175–181.

Nag, P. K.; Gupta, A. V. S. S. K. S. Exergy Analysis of the Kalina Cycle. *Appl. Therm. Eng.* **1998**, *18* (6), 427–439.

Najjar, S. H. Determination of Thermodynamic Properties of Some Engineering Fluids Using Two-Constant Equations of State. *Thermochim. Acta* **1997**, *303* (2), 137–143.

Naveen, M. R.; Manavalan, S. Experimental Investigation of a Combined Power Refrigeration Transcritical CO_2 Cycle. *Indian J. Sci. Technol.* **2015**, *8* (31), 1–4.

Nemati, A.; Nami, H.; Ranjbar, F.; Yari, M. A Comparative Thermodynamic Analysis of ORC and Kalina Cycles for Waste Heat Recovery: A Case Study for CGAM Cogeneration System. *Case Stud. Therm. Eng.* **2017**, *9*, 1–13.

Nguyen, T. V.; Knudsen, T.; Larsen, U, and Haglind, F. Thermodynamic Evaluation of the Kalina Split-Cycle Concepts for Waste Heat Recovery Applications. *Energy* **2014**, *71*, 277–288.

Nowarski, A, and Friend, D. G. Application of the Extended Corresponding States Method to the Calculation of the Ammonia–Water Mixture Thermodynamic Surface. *Int. J. Thermophys.* **1998**, *19* (4), 1133–1142.

Ogriseck, S. Integration of Kalina Cycle in a Combined Heat and Power Plant, A Case Study, *Appl. Therm. Eng.* **2009**, *29* (14), 2843–2848.

Orbey, H.; Sandler, S. I. On the Combination of Equation of State and Excess Free Energy Models. *Fluid Phase Equilibr.* **1995**, *111* (1), 53–70.

Padilla, R. V.; Demirkaya, G.; Goswami, D. Y.; Stefanakos, E.; Rahman, M, M. Analysis of Power and Cooling Cogeneration Using Ammonia–Water Mixture. *Energy* **2010**, *35* (12), 4649–4657.

Panea, C.; Rosca, G. M.; Blaga, C. A. Power Generation from Low-Enthalpy Geothermal Resources. *J. Sustain. Energy* **2010**, *1* (2), 1–5.

Patek, J.; Klomfar, J. Simple Functions for Fast Calculations of Selected Thermodynamic Properties of the Ammonia–Water System. *Int. J. Refrig.* **1995,** *18* (4), 228–234.

Peletz Jr., L. J. Fluidized Bed for Kalina Cycle Power Generation System. U.S. Patent No. 6,253,552 B1. U.S. Patent and Trademark Office: Washington, DC, 2001.

Peletz Jr., L. J.; Tanca, M. C. Refurbishing Conventional Power Plants for Kalina Cycle Operation. U.S. Patent No. 6,035,642. U.S. Patent and Trademark Office: Washington, DC, 2000.

Pilatowsky, I.; Rivera, W.; Romero, J. R. Performance Evaluation of a Monomethylamine–Water Solar Absorption Refrigeration System for Milk Cooling Purposes. *Appl. Therm. Eng.* **2004,** *24*, 1103–1115.

Polikhronidi, N. G.; Abdulagatov, I. M.; Batyrova, R. G.; Stepanov, G. V. PVT Measurements of Water–Ammonia Refrigerant Mixture in the Critical and Supercritical Regions. *Int. J. Refrig.* **2009,** *32* (8), 1897–1913.

Pouraghaie, M.; Atashkari, K.; Besarati, S. M.; Nariman-Zadeh, N. Thermodynamic Performance Optimization of a Combined Power/Cooling Cycle. *Energy Convers. Manag.* **2010,** *51* (1), 204–211.

Prisyazhniuk, V. A. Alternate Trends in Development of Thermal Power Plants. *Appl. Therm. Eng.* **2008,** *28* (2), 190–194.

Ranasinghe, J.; Anand, A. K.; Smith, R. W. Integrated Gasification Combined Cycle Power Plant with Kalina Bottoming Cycle. U.S. Patent No. 6,216,436 B1. U.S. Patent and Trademark Office: Washington, DC, 2001.

Ranasinghe, J.; Smith, R. W.; Bjorge, R. W. Method for Kalina Combined Cycle Power Plant with District Heating Capability. U.S. Patent No. 6,347,520 B1. U.S. Patent and Trademark Office: Washington, DC, 2002.

Rashad, A.; El Maihy, A. In *Energy and Exergy Analysis of a Steam Power Plant in Egypt*, 13th International Conference on Aerospace Sciences and Aviation Technology, ASAT-13, 2009.

Rashidi, J.; Yoo, C. Exergetic and Exergoeconomic Studies of Two Highly Efficient Power-Cooling Cogeneration Systems Based on the Kalina and Absorption Refrigeration Cycles. *Appl. Therm. Eng.* **2017,** *124*, 1023–1037.

Reddy, V. S.; Kaushik, S. C.; Tyagi, S. K.; Panwar, N. L. An Approach to Analyse Energy and Exergy Analysis of Thermal Power Plants: A Review. *Smart Grid Renew. Energy* **2010,** *1* (3), 143–152.

Reid, R. C.; Prausnitz, J. M.; Poling, B. E. *The Properties of Gases and Liquids*, 4th ed.; McGraw-Hill: New York, NY, 1987; p 667. ISBN 0-07-051799-1.

Renon, H.; Guillevic, J. L.; Richon, D.; Boston, J.; Britt, H. A Cubic Equation of state Representation of Ammonia–Water Vapour-Liquid Equilibrium Data. *Int. J. Refrig.* **1986,** *9* (2), 70–73.

Rodriguez, C. E. C.; Palacio, J. C. E.; Rodríguez, C.; Cobas, M.; dos Santos, D. M.; Dotto, F. R. L.; Gialluca, V. *Exergetic and Economic Analysis of Kalina Cycle for Low Temperature Geothermal Sources in Brazil*; ECOS: Perugia, Italia, 2012.

Rogdakis, E. D. Thermodynamic Analysis, Parametric Study and Optimum Operation of the Kalina Cycle. *Int. J. Energy Res.* **1996,** *20* (4), 359–370.

Rogdakis, E. D.; Antonopoulos, K. A. A High Efficiency NH_3/H_2O Absorption Power Cycle. *Heat Recov. Syst. CHP* **1991,** *11* (4), 263–275.

Rosen, M. A. In *Exergy Concept and its Application*, Electrical Power Conference, EPC, IEEE Canada, IEEE, 2007; pp 473–478.

Rosen, M. A.; Dincer, I. Exergy as the Confluence of Energy, Environment and Sustainable Development. *Exergy Int. J.* **2001**, *1* (1), 3–13.

Roy, P.; Desilets, M.; Galanis, N.; Nesreddine, H.; Cayer, E. Thermodynamic Analysis of a Power Cycle Using a Low-Temperature Source and a Binary NH_3–H_2O Mixture as Working Fluid. *Int. J. Therm. Sci.* **2010**, *49* (1), 48–58.

Ruiter, J. P. Simplified Thermodynamic Description of Mixtures and Solutions. *Int. J. Refrig.* **1990**, *13* (4), 223–236.

Sabeti, V.; Boyaghchi, F. A. Multi-Objective Optimization of a Solar Driven Combined Power and Refrigeration System Using Two Evolutionary Algorithms Based on Exergoeconomic Concept. *J. Renew. Energy Environ.* **2016**, *3* (1), 43.

Sadeghi, S.; Saffari, H.; Bahadormanesh, N. Optimization of a Modified Double-Turbine Kalina Cycle by Using Artificial Bee Colony algorithm. *Appl. Therm. Eng.* **2015**, *91*, 19–32.

Sadrameli, S. M.; Goswami, D. Y. Optimum Operating Conditions for a Combined Power and Cooling Thermodynamic Cycle. *Appl. Energy* **2007**, *84* (3), 254–265.

Saffari, H.; Sadeghi, S.; Khoshzat, M.; Mehregan, P. Thermodynamic Analysis and Optimization of a Geothermal Kalina Cycle System Using Artificial Bee Colony Algorithm. *Renew. Energy* **2016**, *89*, 154–167.

Saidur, R.; Kazi, S. N.; Hossain, M. S.; Rahman, M. M.; Mohammed, H. A. A Review on the Performance of Nanoparticles Suspended with Refrigerants and Lubricating Oils in Refrigeration Systems. *Renew. Sustain. Energy Rev.* **2011**, *15*, 310–323.

Saravanan, R.; M. P. Maiya. Thermodynamic Comparison of Water-Based Working Fluid Combinations for a Vapor Absorption Refrigeration System. *Appl. Therm. Eng.* **1998**, *18* (7), 553–568.

Shankar, R.; Srinivas, T. Development and Analysis of a New Integrated Power and Cooling Plant Using LiBr–H_2O Mixture. *Sadhana* **2014**, *39* (6), 1547–1562.

Sharma, R.; Singhal, D.; Ghosh, R.; Dwivedi, A. Potential Applications of Artificial Neural Networks to Thermodynamics: Vapor–Liquid Equilibrium Predictions. *Comput. Chem. Eng.* **1999**, *23* (3), 385–390.

Shi, X.; Che, D. A Combined Power Cycle Utilizing Low-Temperature Waste Heat and LNG Cold Energy. *Energy Convers. Manag.* **2009**, *50* (3), 567–575.

Shokati, N.; Ranjbar, F.; Yaru, M, Exergoeconomic Analysis and Optimization of Basic, Dual-Pressure and Dual-Fluid ORCs and Kalina Geothermal Power Plants: A Comparative Study. *Renew. Energy* **2015**, *83*, 527–542.

Singh, O. K.; Kaushik, S. Energy and Exergy Analysis and Optimization of Kalina Cycle Coupled with a Coal Fired Steam Power Plant. *Appl. Therm. Eng.* **2012**, *5*, 787–800.

Singh, O. K.; Kaushik, S. Reducing CO_2 Emission and Improving Exergy Based Performance of Natural Gas Fired Combined Cycle Power Plants by Coupling Kalina Cycle. *Energy* **2013**, *55* (C), 1002–1013.

Som, S. K.; Dutta, A. Thermodynamic Irreversibilities and Exergy Balance in Combustion Processes. *Progr. Energy Combust. Sci.* **2008**, *34* (3), 351–376.

Spinks, A. H. In *Application of the Kalina Cycle to the Generation of Power from Geothermal Sources in New Zealand*, IPENZ Annual Conference, Proceedings of Choosing Our Future; 2: General, Electrical, Mechanical and Chemical, Institution of Professional Engineers New Zealand: Wellington, 1991; pp 303–314.

Srinivas, T.; Gupta, A. V. S. S. K. S.; Reddy, B. V. Performance Simulation of Combined Cycle with Kalina Bottoming Cycle. *Cogenerat. Distrib. Generat. J.* **2008**, *23* (1), 6–21.

Srinivas, T.; Reddy, B. V.; Gupta, A. V. S. S. K. S. Biomass-Fuelled Integrated Power and Refrigeration System. *Proc. Inst. Mech. Eng.* A **2011**, *225* (3), 249–258

Srinivas, T.; Reddy, B. V. Thermal Optimization of a Solar Thermal Cooling Cogeneration Plant at Low Temperature Heat Recovery. *ASME J. Energy Resour. Technol.* **2014**, *136* (2), 1–10.

Srinivas, T.; Vignesh, D. Performance Enhancement of GT–ST Power Plant with Inlet Air Cooling Using Lithium Bromide/Water Vapour Absorption Refrigeration System. *Int. J. Energy Technol. Policy* **2012**, *8* (1), 94–107.

Srinophakun, T.; Laowithayangkul, S.; Ishida, M. Simulation of Power Cycle with Energy Utilization Diagram. *Energy Convers. Manag.* **2001**, *42* (12), 1437–1456.

Stecco, S. S.; Desideri, U. Considerations on the Design Principles for a Binary Mixture Heat Recovery Boiler. *J. Eng. Gas Turbines Power.* **1992**, *114* (4), 701–706.

Styliaras, V. E. The Use of an Absorber-Generator Heat Exchanger in Power Production, *Appl. Therm. Eng.* **1996**, *16* (3), 273–275.

Sun, L.; Han, W.; Jing, X.; Zheng, D.; Jin, H. A Power and Cooling Cogeneration System Using Mid/Low-Temperature Heat Source. *Appl. Energy* **2013**, *112*, 886–897.

Talbia, M. M.; Agnew, B. Exergy Analysis: An Absorption Refrigerator Using Lithium Bromide and Water as the Working Fluids. *Appl. Therm. Eng.* **2000**, *20* (7), 619–630.

Tamm, G. O. Experimental Investigation of an Ammonia-Based Combined Power and Cooling Cycle, Doctoral Dissertation, University of Florida, 2003.

Tamm, G.; Goswami, D. Y. Novel Combined Power and Cooling Thermodynamic Cycle for Low Temperature Heat Sources, Part II: Experimental Investigation. *J. Sol. Energy Eng.* **2003**, *125* (2), 223–229.

Tamm, G.; Goswami, D. Y.; Lu, S.; Hasan, A. A. Theoretical and Experimental Investigation of an Ammonia–Water Power and Refrigeration Thermodynamic Cycle. *Sol. Energy* **2004**, *76* (1), 217–228.

Thorin, E. Thermophysical Properties of Ammonia–Water Mixtures for Prediction of Heat Transfer Areas in Power Cycles. *Int. J. Thermophys.* **2001**, *22* (1), 201–214.

Thorin, E.; Dejfors, C.; Svedberg, G. Thermodynamic Properties of Ammonia–Water Mixtures for Power Cycles. *Int. J. Thermophys.* **1998**, *19* (2), 501–510.

Tillner-Roth, R.; Friend, D. G. A Helmholz Free Energy Formulation of the Thermodynamic Properties of the Mixture {Water + Ammonia}. *J. Phys. Chem. Ref. Data* **1998**, *27* (1), 63–96.

Tsai, J. C.; Chen, Y. P. Application of a Volume-Translated Peng–Robinson Equation of State on Vapor–Liquid Equilibrium Calculations. *Fluid Phase Equilibr.* **1998**, *145* (2), 193–215.

Tsupari, E.; Arponen, T.; Hankalin, V.; Kärki, J.; Kouri, S. Feasibility Comparison of Bioenergy and CO_2 Capture and Storage in a Large Combined Heat, Power and Cooling System. *Energy* **2017**, *139*, 1040–1051.

Usvika, R.; Rifaldi, M.; Noor, A. Energy and Exergy Analysis of Kalina Cycle System (KCS) 34 with Mass Fraction Ammonia–Water Mixture Variation. *J. Mech. Sci. Technol.* **2009**, *23* (7), 1871–1876.

Valan Arasu, A.; Sornakumar, T. Design, Manufacture and Testing of Fibreglass Reinforced Parabola Trough for Parabolic Trough Solar Collectors. *Sol. Energy* **2007**, *81* (10), 1273–1279.

Valdimarsson, P.; Eliasson, L. In *Factors Influencing the Economics of the Kalina Power Cycle and Situations of Superior Performance*, Proceedings of International Geothermal Conference, Reykjavik, 2003; pp 31–39.

Victor, R. A.; Kim, J.-K.; Smith, R. Composition Optimisation of Working Fluids for Organic Rankine Cycles and Kalina Cycles. *Energy* 2013, *55*, 114–126.

Vidal, A.; Best, R.; Rivero, R.; Cervantes, J. Analysis of a Combined Power and Refrigeration Cycle by the Exergy Method. *Energy* 2006, *31* (15), 3401–3414.

Vijayaraghavan, S.; Goswami, D. Y. A Combined Power and Cooling Cycle Modified to Improve Resource Utilization Efficiency Using Distillation Stage. *Energy* 2006, *31* (8), 1177–1196.

Wagar, W. R.; Zamfirescu, C.; Dincer, I. Thermodynamic Performance Assessment of an Ammonia–Water Rankine Cycle for Power and Heat Production. *Energy Convers. Manag.* 2010, *51* (12), 2501–2509.

Wall, G.; Chuang, C, C.; Ishida, M. Exergy Study of the Kalina Cycle. *Anal. Des. Energy Syst.: Anal. Ind. Process.* 1989, *10* (3), 73–77.

Wang, E.; Yu, Z. A Numerical Analysis of a Composition-Adjustable Kalina Cycle Power Plant for Power Generation from Low-Temperature Geothermal Sources. *Appl. Energy* 2016, *180*, 834–848.

Wang, J.; Dai, Y.; Gao, L. Parametric Analysis and Optimization for a Combined Power and Refrigeration Cycle. *Appl. Energy* 2008, *85* (11), 1071–1085.

Wang, J.; Dai, Y.; Gao, L. Exergy Analyses and Parametric Optimizations for Different Cogeneration Power Plants in Cement Industry. *Appl. Energy* 2009, *86* (6), 941–948.

Wang, J.; Sun, Z.; Dai, Y.; Ma, S. Parametric Optimization Design for Supercritical CO_2 Power Cycle Using Generic Algorithm and Artificial Neural Network. *Appl. Energy* 2010, *87* (4), 1317–1324.

Wang, J.; Yan, Z.; Zhou, E.; Dai, Y. Parametric Analysis and Optimization of a Kalina Cycle Driven by Solar Energy. *Appl. Therm. Eng.* 2013, *20*, 408–415.

Wang, J.; Wang, J.; Zhao, P.; Dai, Y. Thermodynamic Analysis of a New Combined Cooling and Power System Using Ammonia–Water Mixture. *Energy Convers. Manag.* 2016, *117*, 335–342.

Weber, L. A. Estimating the Virial Coefficients of the Ammonia + Water Mixture. *Fluid Phase Equilibr.* 1999, *162* (1), 31–49.

Wormald, C. J.; Wurzberger, B. Second Virial Cross Coefficients for (Ammonia + Water) Derived from Gas Phase Excess Enthalpy Measurements. *J. Chem. Thermodyn.* 2001, *33* (9), 1193–1210.

Xu, F.; Goswami, D. Y. Thermodynamic Properties of Ammonia–Water Mixtures for Power-Cycle Applications. *Energy* 1999, *24* (6), 525–536.

Xu, F.; Goswami, D. Y.; Sunil Bhagwat, S. A Combined Power/Cooling Cycle. *Energy* 2000, *25* (3), 233–246.

Yari, M.; Mehr, A. S.; Zare, V.; Mahmoudi, S. M. S.; Rosen, M. A. Exergoeconomic Comparison of TLC (Trilateral Rankine Cycle), ORC (Organic Rankine Cycle), and Kalina Cycle Using Low Grade Heat Source. *Energy* 2015, *83*, 712–722.

Yue, C.; Han, D.; Pu, W.; He, W. Comparative Analysis of a Bottoming Transcritical ORC and a Kalina Cycle for Engine Exhaust Heat Recovery. *Energy Convers. Manag.* 2015, *89*, 764–774.

Zare, V.; Mahmoudi, S. M. S. A Thermodynamic Comparison between Organic Rankine and Kalina Cycles for Waste Heat Recovery from the Gas Turbine–Modular Helium Reactor. *Energy* **2015**, *79*, 398–406.

Zare, V.; Mahmoudi, S. M. S.; Yari, M.; Amidpour, M. Thermoeconomic Analysis and Optimization of an Ammonia–Water Power/Cooling Cogeneration Cycle. *Energy* **2012**, *47*, 271–283.

Zhang, C.; Yang, M.; Lu, M.; Shan, Y.; Zhu, J. Experimental Research on LiBr Refrigeration—Heat Pump System Applied in CCHP System. *Appl. Therm. Eng.* **2011**, *31* (17–18), 3706–3712.

Zhang, Z.; Guo, Z.; Chen, Y.; Wu, J, and Hua, J. Power Generation and Heating Performances of Integrated System of Ammonia–Water Kalina–Rankine Cycle. *Energy Convers. Manag.* **2015**, *92*, 517–522.

Zhao, P.; Wang, J.; Dai, Y. Thermodynamic Analysis of an Integrated Energy System Based on Compressed Air Energy Storage (CAES) System and Kalina Cycle. *Energy Convers. Manag.* **2015**, *98*, 161–172.

Zheng, D.; Chen, B.; Qi, Y.; Jin, H. Thermodynamic Analysis of a Novel Absorption Power/Cooling Combined-Cycle. *Appl. Energy* **2006**, *83* (4), 311–323.

Zhu, Z, Zhang Z, Chen, Y.; Wu, J. Parameter Optimization of Dual-Pressure Vaporization Kalina Cycle with Second Evaporator Parallel to Economizer. *Energy* **2016**, *112*, 420–429.

Ziegler, B.; Trepp, C. Equation of State for Ammonia–Water Mixtures. *Int. J. Refrig.* **1984**, *7* (2), 101–106.

INDEX

Milton Keynes UK
Ingram Content Group UK Ltd.
UKHW031140141024
449569UK00024B/1174

9 781774 634080